Lecture Notes in Mathematics

Edited by A. Dold and B. Eckmann

T0215389

420

Category Seminar

Proceedings Sydney Category Theory Seminar 1972/1973

Edited by Gregory M. Kelly

Springer-Verlag
Berlin · Heidelberg · New York 1974

Prof. Dr. Gregory M. Kelly
Department of Pure Mathematics
University of Sydney
New South Wales 2006
Australia

Library of Congress Cataloging in Publication Data

Category Seminar.
Proceedings Sydney Category Seminar 1972/1973.

(Lecture notes in mathematics ; 420)
Includes bibliography and index.
1. Categories (Mathematics)--Congresses.
2. Functor theory--Congresses. I. Kelly, G. M., ed.
II. Title. III. Series: Lecture notes in mathematics (Berlin) ; v. 420.
QA3.L28 no. 420 [QA169] 510'.8s [512'.55] 74-19483

AMS Subject Classifications (1970): 18A15, 18A35, 18A40, 18C15, 18D05, 18D15, 18D99, 18E35

ISBN 3-540-06966-6 Springer-Verlag Berlin · Heidelberg · New York
ISBN 0-387-06966-6 Springer-Verlag New York · Heidelberg · Berlin

Offsetdruck: Julius Beltz, Hemsbach/Bergstr.

PREFACE

The Sydney Category Theory Seminar was born in the middle of 1972, when the category-theorists at the three universities in Sydney conceived the idea of meeting for a whole day each week. The papers collected in this volume represent a large part of the mathematics that has emerged from its first eighteen months. The title-date "1972/1973" represents accurately enough the time at which the work was done, although the final details of it, along with the writing-up and the typing, have brought us past the middle of 1974.

As editor I was excessively sanguine about the time it would take to turn ideas into theorems; for the unforseen delays I owe apologies to Springer-Verlag and especially to Brian Day, whose first two papers were received more than a year ago.

The papers below are connected by more than our local proximity; the common theme is that of categories with structure. If categories are seen as analogous to sets, then pure category theory is the analogue of pure set theory, although a lot richer because of the "increase in dimension by 1". But then the analogue of sets with structure is categories with structure; and monoidal categories, monoidal closed categories, enriched categories, categories bearing a monad, and so on, become to the category theorist what his groups, rings, fields, and modules are to the algebraist.

If monoidal closed categories are in some sense the "fields" of category theory, then Brian Day is our field theorist. In fact his papers are concerned with two types of structure - monoidal and especially monoidal closed ones, and adjunctions - and with the relations between these two structures; all this moreover at the "enriched" level. His first paper gives a new adjoint-functor theorem, which has since been adapted by Mikkelsen to the elementary topos situation, and which may be seen as replacing by a two-step process the transfinite tower construction of Applegate and Tierney; along with applications to completions and to monoidal completions. His second combines his earlier work on monoidal closed structures for functor categories, and on the reflexion of monoidal closed structures, to study monoidal closed structures on reflective subcategories of functor categories; and gives a great many concrete applications. His third discusses the embedding of a non-monoidal closed category into a monoidal one, of great value for coherence problems, with extensions to bicategories; and his fourth uses his techniques to look deeply at certain cartesian closed categories of topological spaces. The reader will quickly perceive the definitive elegance of his theorems.

My own papers correspond in some sense to universal algebra. As equational algebras of any given species correspond to a monad on the category of sets - or on a more general category if we want things like topological groups - so categories with a given species of equational structure correspond to a 2-monad, also called a doctrine, on the 2-category of categories, or more generally on any 2-category. In this setting the "coherence problem" is that of finding the doctrine explicitly from a knowledge of its algebras. My first paper looks at a subclass of doctrines that admit a very concrete representation by what I call clubs, and a smaller subclass where the coherence problem can be formulated as a word-problem. My second paper concerns the interplay between equational structures and adjunctions; and my third gives some coherence results at the doctrine level, getting the fuller results available in the club case by specialization. The reader would do well to read the first paper last, except for §1 and §10, which alone are used in the last two papers.

The elementary 2-categorical background needed both for my papers and for Street's is contained in a joint expository paper. Street goes on in his papers to look more deeply at 2-categories. In one sense his papers are the most general; in another, the most fundamental. If the study of various structures borne by a set has led to category theory, that of structures borne by a category leads inexorably to 2-category theory. Much of category theory can be done inside any 2-category, and the arguments are then often more transparant. Street's first paper looks at some things that can be done in a <u>representable</u> 2-category - essentially the same thing as a finitely 2-complete one, except that it need not contain a terminal object, which is seldom needed. In particular he studies fibrations and bifibrations at this level, along with such things as pointwise Kan extensions. His second paper, which uses the first, investigates 2-categories with a structure so rich that we can imitate those arguments, including the Yoneda lemma, that depend upon the hom functor. Even applied to the 2-category of categories, it provides new proofs and thus contributes to the "elementary" theory of categories; applied to the 2-category of ordered objects, it throws new light on elementary topoi.

The investigations of Street and Kelly, at least, are to some degree tentative, and they mention many outstanding problems which may be of interest to others.

Because of the time the volume has been in preparation, I believe it appropriate to give dates of reception for the papers (although I don't quite know what it means to "receive" my own). The order is that of the table of contents. Day: Feb. 1973 revised May 1973; Apr. 1973; Feb. 1974; Feb. 1974. Kelly-Street: Oct. 1973. Street: July 1973; Feb. 1974. Kelly: Nov. 1973; Jan. 1974; May 1974.

G.M. Kelly
19 July 1974

TABLE OF CONTENTS

ON ADJOINT-FUNCTOR FACTORISATION

by

Brian Day

This note contains an alternative approach to a result of Applegate and Tierney ([2] and [3]) which states that an adjunction $S \dashv T: \mathcal{C} \to \mathcal{B}$ over a suitably complete category $\mathcal{B}$ can be factored through the full subcategory of $\mathcal{B}$ determined by the objects in $\mathcal{B}$ which are "orthogonal" to all the morphisms inverted by the functor $S: \mathcal{B} \to \mathcal{C}$. It is observed that a slight strengthening of the completeness hypothesis on $\mathcal{B}$ gives a simple proof of this result.

The factorisation of the given adjoint pair takes place in two stages, the first of which is a well-known epic-monic factorisation of the given adjunction unit. This produces a full reflective subcategory $\mathcal{B}'$ of $\mathcal{B}$ having the property that the class of objects which are orthogonal to any given class of morphisms inverted by the restriction to $\mathcal{B}'$ of S is reflective in $\mathcal{B}'$. The combined result contains a theorem of Fakir [11] which associates to each monad T on $\mathcal{B}$, the idempotent monad which inverts the same morphisms as T. For the relative $\mathcal{V}$-based case, where $\mathcal{V}$ is a complete symmetric monoidal closed category, the result is closely related to a theorem by Wolff [19] §5.6 using co-completeness hypotheses on $\mathcal{B}$.

Some of the observations made here are implicit in [2] and [3]. However, the relationship of category completion to epic-monic factorisation, and to relative categories of quasi-topological spaces, was not discussed in the relative $\mathcal{V}$-based version [9]. Thus, throughout this article, the concepts of category, functor, natural transformation, etc., are assumed to be relative to a suitable symmetric monoidal closed category $\mathcal{V}$; this category is assumed to be locally small with respect to a given cartesian closed category $\mathcal{S}$ of "small" sets and set maps.

The work relating to Example 2.5 and to the "minimal cartesian closure" of the category of topological spaces was done jointly with G.M. Kelly, but not published in view of [1] and [2] . The generalisations given here are useful in other contexts and the relationship of the factorisation of $S \dashv T\colon \mathcal{C} \to \mathcal{B}$ to a monoidal closed structure on $\mathcal{B}$ is discussed in §4.

The basic notations and the representation theorem are standard and are as given in the early parts of Eilenberg-Kelly [10] .

The section-headings are as follows:

§1 The preliminary factorisation.

§2 Factorisation of $\mathcal{M}$-adjunctions.

§3 Categories of relative cribles.

§4 Examples; completions and monoidal closure.

§5 The factorisation system for left adjoints.

---

§1 THE PRELIMINARY FACTORISATION

Recall that a morphism m in a $\mathcal{V}$-category $\mathcal{B}$ is called <u>monic in $\mathcal{B}$</u> if the morphism $\mathcal{B}(B,m)$ is monic in $\mathcal{V}$ for each $B \in \mathcal{B}$. A monic m is called a <u>strong monic in $\mathcal{B}$</u> if, for each epic e in $\mathcal{B}$ (that is, for each monic e in $\mathcal{B}^{op}$), the square

$$\begin{array}{ccc} \cdot & \xrightarrow{\mathcal{B}(e,1)} & \cdot \\ {\scriptstyle \mathcal{B}(1,m)}\downarrow & & \downarrow{\scriptstyle \mathcal{B}(1,m)} \\ \cdot & \xrightarrow{\mathcal{B}(e,1)} & \cdot \end{array}$$

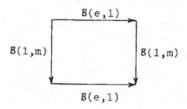

is a pullback diagram in $\mathcal{V}$. The usual properties (cf. [15] §3) are easily established. For example, any equaliser in $\mathcal{B}$ is a strong monic, any strong monic which is epic is an isomorphism, and if a composite fg of two morphisms is a strong monic then so is g.

The initial data are categories $\mathcal{C}$ and $\mathcal{B}$ and an adjunction $(\varepsilon,\eta)\colon S \dashv T\colon \mathcal{C} \to \mathcal{B}$. The category $\mathcal{B}$ is assumed to have canonical $(\mathcal{E},\mathcal{M})$ - factorisation for at least <u>one</u> of the following two cases:

(a) $\mathcal{E}$ = {all epics in $\mathcal{B}$} and $\mathcal{M}$ = {all strong monics in $\mathcal{B}$}

(b) $\mathcal{E}$ = {all strong epics in $\mathcal{B}$} and $\mathcal{M}$ = {all monics in $\mathcal{B}$}

Thus there will be two versions which may be compared but we shall <u>fix</u> $(\mathcal{E},\mathcal{M})$ throughout. In fact, $(\mathcal{E},\mathcal{M})$ could be taken to be any <u>proper</u> factorisation system between (a) and (b) in the sense of [13] §2.3.

Let $\mathcal{B}'$ be the full subcategory of $\mathcal{B}$ determined by the objects $B \in \mathcal{B}$ for which $\eta_B \in \mathcal{M}$.

<u>Proposition 1.1</u> An object $B \in \mathcal{B}$ is in $\mathcal{B}'$ if and only if there exists a morphism $B \to TC$ in $\mathcal{M}$.

The proof is clear.

<u>Proposition 1.2</u> The inclusion $\mathcal{B}' \subset \mathcal{B}$ has a left adjoint.

<u>Proof</u>. The reflection sends $B \in \mathcal{B}$ to the image of η_B; let

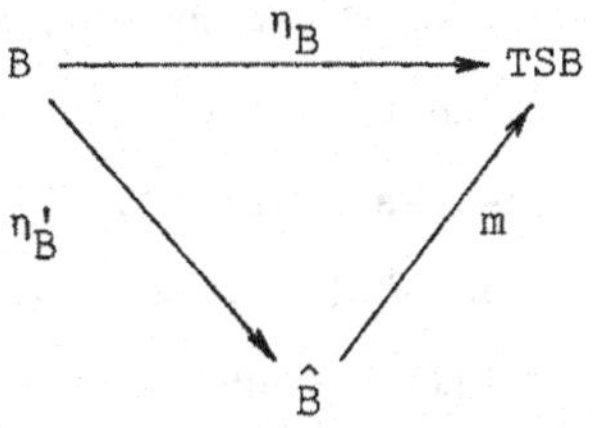

denote the factorisation of η_B with $\eta'_B \in \mathcal{E}$ and $m \in \mathcal{M}$. Then $S\eta'_B$ is an isomorphism because it is an epic in $\mathcal{C}$ with left inverse $\varepsilon_{SB} \cdot Sm$. Thus, for each $B' \in \mathcal{B}'$, the top arrow in the following pullback diagram is an isomorphism in $\mathcal{V}$, as required:

$$\begin{array}{ccc}
\mathcal{B}(\hat{B},B') & \xrightarrow{\mathcal{B}(\eta',1)} & \mathcal{B}(B,B') \\
\Big\downarrow{\scriptstyle \mathcal{B}(1,\eta)} & & \Big\downarrow{\scriptstyle \mathcal{B}(1,\eta)} \\
\mathcal{B}(\hat{B},TSB') & \xrightarrow{\mathcal{B}(\eta',1)} & \mathcal{B}(B,TSB') \\
\Big\downarrow{\scriptstyle \cong} & & \Big\downarrow{\scriptstyle \cong} \\
\mathcal{C}(S\hat{B},SB') & \xrightarrow[\mathcal{C}(S\eta',1)]{} & \mathcal{C}(SB,SB').
\end{array}$$

Moreover, because $S\colon \mathcal{B} \to \mathcal{C}$ inverts the unit η' of the reflection, there results an adjoint triangle:

$$\begin{array}{ccc}
\mathcal{B} & \xrightarrow{\ \hat{}\ } & \mathcal{B}' \\
& {\scriptstyle S}\searrow \quad \swarrow{\scriptstyle S'} & \\
& \mathcal{C} &
\end{array}$$

This process is clearly a closure operation for the given choice of M. In other words, $\mathcal{B}'$ has (E,M)-factorisations <u>as a subcategory</u> of $\mathcal{B}$, and is category equivalent to $\mathcal{B}''$. Following standard terminology, an adjunction such as $S' \dashv T$ with unit in M shall be called an <u>M-adjunction</u>.

<u>Example 1.3</u>. Let A be the category of finite sets, let $\mathcal{B}$ be the category $[A^{op},S]$ of all functors from A^{op} to S, and let $S\colon [A^{op},S] \to S$ be "evaluation-at-singleton". The category $\mathcal{B}'$ is the category of "simplicial complexes" (sets equipped with certain finite "spanning" subsets). If $M = \{$all monics in $\mathcal{B}'\}$ then $\mathcal{B}''$ remains equivalent to $\mathcal{B}'$, but if M is changed to $\{$all strong monics in $\mathcal{B}'\}$ then $\mathcal{B}''$ is equivalent to S.

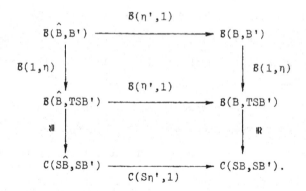

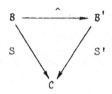

§2 FACTORISATION OF $\mathcal{M}$-ADJUNCTIONS

Under additional completeness hypotheses, the adjoint-functor factorisation of §1 reduces the given adjunction to one in which the following form of adjoint-functor theorem is applicable. The theorem is first established for ordinary set-based categories. We say that a category $\mathcal{C}$ is <u>$\mathcal{M}$-complete</u> if $\mathcal{M}$ is a subcategory of monics in $\mathcal{C}$ such that $\mathcal{C}$ has the following inverse limits and $\mathcal{M}$ contains each monic so formed:

(a) equalisers of pairs of morphisms.

(b) pullbacks of $\mathcal{M}$-monics (i.e. inverse $\mathcal{M}$-images).

(c) all intersections of $\mathcal{M}$-monics with a common codomain.

A functor $T\colon \mathcal{C} \to \mathcal{B}$ is <u>$\mathcal{M}$-continuous</u> if it preserves these inverse limits in $\mathcal{C}$.

<u>Theorem 2.1.</u> If $\mathcal{C}$ is an $\mathcal{M}$-complete category then a functor $T\colon \mathcal{C} \to \mathcal{B}$ has a left adjoint if and only if T is $\mathcal{M}$-continuous and there exists a "bounding" family $\{\beta_B\colon B \to TC_B;\ B \in \mathcal{B}\}$ of morphisms in $\mathcal{B}$ such that, for each $C \in \mathcal{C}$ and $f \in \mathcal{B}(B,TC)$, there exists a commuting square:

$$\begin{array}{ccc} B & \xrightarrow{\ \beta_B\ } & TC_B \\ {\scriptstyle f}\downarrow & & \downarrow{\scriptstyle Tg} \\ TC & \xrightarrow[\ Tm\ ]{} & TD \end{array}$$

with $m \in \mathcal{M}$.

<u>Proof.</u> If T has a left adjoint then T is clearly $\mathcal{M}$-continuous and the family $\{\eta_B;\ B \in \mathcal{B}\}$ of adjunction units has the required properties. Conversely, a left adjoint $S\colon \mathcal{B} \to \mathcal{C}$ is constructed by taking $h\colon SB \to C_B$ to be the intersection in $\mathcal{C}$ of all the $\mathcal{M}$-subobjects $n\colon M \to C_B$ such that β_B factors through Tn. Then β_B factors as $Th.\eta_B$ for some morphism $\eta_B\colon B \to TSB$. Moreover, each morphism $f \in \mathcal{B}(B,TC)$ factors uniquely through η_B. To see this, let $Tg.\beta_B = Tm.f\colon B \to TD$, with $m \in \mathcal{M}$ and let (p,q) be the pullback in $\mathcal{C}$ of (gh,m).

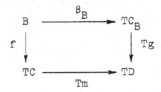

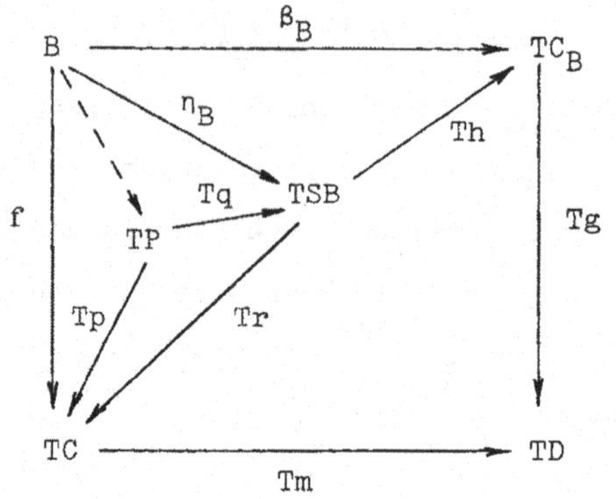

Then $q \in M$ and η_B factors through Tq so q is an isomorphism by the definition of SB. Thus f, which factors through Tp, factors through TSB as $T(pq^{-1})\eta_B$. This factorisation is unique. If $f = Tr.\eta_B$, let e be the equaliser in C of (p,rq). Then $e \in M$ and η_B factors through Te so e is an isomorphism by the definition of SB. This completès the proof.

Remark. Several of the standard statements (cf. [18] Chapter V, §6 and §8) of Freyd's adjoint functor theorems may be recovered from Theorem 2.1 under the additional hypothesis that enough products exist in C and are preserved by $T: C \to B$.

To obtain a V-based version of Theorem 2.1 we shall simply assume that C is cotensored as a V-category and that $T: C \to B$ preserves the cotensoring; this assumption provides a V-adjunction by [16] §4.1.

Our applications are based on the following:

Theorem 2.2 Let C be an M-complete full subcategory of B for which the inclusion $C \subset B$ is M-continuous. Then C is reflective in B if and only if there exists an endofunctor $S: B \to B$ and a natural transformation $\beta: 1 \to S$ such that β_B factors through an object of C for each $B \in B$ and $\beta_C \in M$ for each $C \in C$.

Proof. This follows from Theorem 2.1 and the fact that, for

each morphism $f: B \to C$ in $\mathcal{B}$, with codomain $C \in \mathcal{C}$, we have $Tf.\beta_B = \beta_C.f$ by the naturality of β, and $\beta_C \in M$ for all $C \in \mathcal{C}$ by hypothesis.

For a given class Z of morphisms in a V-category $\mathcal{B}$, let $\mathcal{B}_Z$ denote the full subcategory of all objects in $\mathcal{B}$ which are (following the terminology of [13]) orthogonal to Z. An object $B \in \mathcal{B}$ is called Z-orthogonal if $\mathcal{B}(s,B)$ is an isomorphism in V for each $s \in Z$ (that is, if B is Z-left-closed in the terminology of [14]).

Now suppose that $(\varepsilon,\eta): S \dashv T: \mathcal{C} \to \mathcal{B}$ is an M-adjunction for a proper factorisation system (E,M) on $\mathcal{B}$ and let Σ denote the class of morphisms in $\mathcal{B}$ inverted by S.

Corollary 2.3. If $\mathcal{B}$ is M-complete (and is cotensored as a V-category) and if $Z \subseteq \Sigma$ then $\mathcal{B}_Z \subset \mathcal{B}$ has a left M-adjoint.

Proof. Clearly $\mathcal{B}_Z$ is closed under limits (and cotensoring) in $\mathcal{B}$. Let $TS: \mathcal{B} \to \mathcal{B}$ be the desired endofunctor on $\mathcal{B}$. Because $TC \in \mathcal{B}_Z$ for all $C \in \mathcal{C}$, and $\eta_B \in M$, the conditions of Theorem 2.2 are satisfied and the reflection is an M-adjunction.

The class Σ is (orthogonally) closed in the sense that if $\mathcal{B}(f,B)$ is an isomorphism in V for all $B \in \mathcal{B}_\Sigma$ then $f \in \Sigma$. The operation of forming the closure of a subset of Σ reflects the class of subclasses of Σ onto a complete sublattice. Thus Corollary 2.3 implies:

Corollary 2.4. The class of all full reflective subcategories of $\mathcal{B}$ which contain $\mathcal{B}$ as a full reflective subcategory forms a complete lattice.

Example 2.5. Let *Top* denote the category of all topological spaces and continuous maps. Let $\mathcal{B}$ be the full subcategory of $[Top^{op},S]$ determined by all the subfunctors of representable functors (these functors are usually called "cribles"). Then $\mathcal{B}$ is locally S-small and contains *Top* as a full reflective subcategory. Furthermore, $\mathcal{B} = \mathcal{B}'$ if we take S: $[Top^{op}, S] \to S$ to be evaluation at the one-point space and M = {all monics in B}. Let T: $B \to B$ be the monad determined by the functor which evaluates each crible at the one-point space. Then Σ is the class of all bijections in B and, by inverting appropriate classes of bijections, one obtains reflective subcategories of quasi-topological spaces, limit spaces, and related structures, including the "minimal extension of *Top*" discussed in [1].

§3 CATEGORIES OF RELATIVE CRIBLES.

Suppose henceforth that the given symmetric monoidal closed category $V = (V, \otimes, [-,-], \dots)$ is S-complete and admits all intersections of M-subobjects, where M is fixed as either the class of monics in V or the class of strong monics in V. These completeness hypotheses imply that V has canonical (E,M)-factorisations for the corresponding class E of epics in V (cf. [15] Proposition 4.5).

Categories of relative cribles are a practical source of M-adjunctions. Given a category C, each functor M: $A \to C$ generates the ordinary category $\bar{A}_o$ of "M-cribles" or "M-preatlases" ([2] §2). An M-crible is a functor F: $A^{op} \to V$ for which there exists a natural transformation F $\to$ C(M-,C) each of whose components is in M. A morphism from F to G of M-cribles is a natural transformation from F to G.

We shall call the functor M: $A \to C$ extendable (with respect to M) if the limit $\int_A [FA, C(MA,C)]$ exists in V and has a representation $C(\bar{M}F,C)$ for each M-crible F.

Proposition 3.1. If M: $A \to C$ is extendable then $\bar{A}_o$ admits enrichment to a V-category $\bar{A}$ and $\bar{M}$: $\bar{A} \to C$ is a left V-adjoint functor.

Proof. For each pair of M-cribles F and G, define $\bar{A}(F,G) = \int_A [FA,GA]$ (cf. [5] §4). This limit exists in V by virtue of the $\mathcal{M}$-embedding

$$\int_A [FA,GA] \xrightarrow{\int_A [1,m]} \int_A [FA,\ \mathcal{C}(MA,C)] \cong \mathcal{C}(\bar{M}F,C)$$

with the following lemma.

<u>Lemma</u>. If m_{AB}: $S(AB) \to T(AB)$ is a natural family of $\mathcal{M}$-monics between two functors from $\mathcal{A}^{op} \otimes \mathcal{A}$ to $\mathcal{V}$ then the end of S exists in $\mathcal{V}$ if the end of T exists.

<u>Proof</u>. As in [9] Proposition III.2.2, the end $\int_A S(AA)$ is constructed directly as the intersection in $\int_A T(AA)$ of all the pullback diagrams:

$$\begin{array}{ccc} P_A & \longrightarrow & S(AA) \\ \big\downarrow & & \big\downarrow m_{AA} \\ \int_A T(AA) & \longrightarrow & T(AA). \end{array}$$

The components m_{AB} are <u>all</u> required to be monic in order that the induced family of morphisms $\int_A S(AA) \to S(AA)$ should be natural in A.

The functor category $\bar{\mathcal{A}}$ inherits equalisers, pullbacks of $\mathcal{M}$-monics, and intersections of $\mathcal{M}$-monics from $\mathcal{V}$. However, for $\bar{\mathcal{A}}$ to be cotensored we shall in general suppose that $\mathcal{C}$ is cotensored; this is because the pointwise cotensor $[X,F]$ of $X \in \mathcal{V}$ with $F \in \bar{\mathcal{A}}$ is then an $\mathcal{M}$-crible by virtue of the $\mathcal{M}$-embedding

$$[X,F] \xrightarrow{[1,m]} [X,\ \mathcal{C}(M-,C)] \cong \mathcal{C}(M-,\ [X,C]).$$

The category $\bar{\mathcal{A}}$ also inherits $(\mathcal{E},\mathcal{M})$-factorisation from $\mathcal{V}$ and, by Proposition 1.1, the relative Yoneda adjunction $\bar{M} \dashv T$: $\mathcal{C} \to \bar{\mathcal{A}}$ is an $\mathcal{M}$-adjunction. In other words, F: $\mathcal{A}^{op} \to \mathcal{V}$ is an $\mathcal{M}$-crible if and only if the associated natural transformation $F \to \mathcal{C}(\bar{M}-,\ MF)$ has components in $\mathcal{M}$.

If the functor M: $\mathcal{A} \to \mathcal{C}$ is $\mathcal{M}$-faithful in the sense that each component of the canonical transformation $A(-A) \to C(M-,MA)$ is in M, then every representable functor from A^{op} to V is an M-crible. Thus there is a dense Yoneda embedding $Y_A\colon A \to \bar{A}$ with respect to which $\bar{A}$ plays the role of the functor category $[A,V]$; however $\bar{A}$ is a well-defined V-category, even when A is large.

Proposition 3.2. If $M\colon A \to C$ is extendable and M-faithful then so is $\bar{M}\colon A \to C$.

Proof. To prove that $\bar{M}$ is extendable, let $K\colon \bar{A}^{op} \to V$ be an $\bar{M}$-crible with transformation $\alpha_F\colon KF \to C(\bar{M}F,C)$ in M. Because M is M-faithful, the category $\bar{A}$ contains all the representable functors from A^{op} to V. Thus we can substitute $F = A(-,A)$ in α and obtain an M-monic

$$K(A(-,A)) \to C(M(A(-,A)),C) \cong C(MA,C).$$

This makes $K(A(-,A))$, regarded as a functor in $A \in A^{op}$, into an M-crible so there exists a representation

$$C(\bar{M}(K(A(-,A))),-) \cong \int_A [K(A(-,A)), C(MA,-)]$$

Thus, on defining $\bar{\bar{M}}K = \bar{M}(K(A(-,-)))$, we obtain

$$C(\bar{\bar{M}}K,-) \cong \int_A [K(A(-,A)), C(MA,-)]$$

$$\cong \int_A [\int^F KF \otimes FA, C(MA,-)]$$

by the representation theorem,

$$\cong \int_F [KF, \int_A [FA, C(MA,-)]]$$

$$= \int_F [KF, C(\bar{M}F,-)],$$

as required for $\bar{M}$ to be extendable. To verify that $\bar{M}\colon \bar{A} \to C$ is M-faithful, consider the following commutative square:

$$\begin{array}{ccc} \int_A [FA,GA] & \xrightarrow{\bar{M}_{FG}} & C(\bar{M}F,\bar{M}G) \\ \downarrow {\scriptstyle \int_A [1,m]} & & \downarrow \wr\wr \\ \int_A [FA,C(MA,C)] & \xleftarrow[\int_A [1,C(1,\bar{m})]]{} & \int_A [FA,C(MA,\bar{M}G)] \end{array}$$

where $\bar{m}$: $\bar{M}G \to C$ corresponds to m: $G \to C(M-,C)$ under adjunction. Then $\bar{M}_{FG} \in \mathcal{M}$ because $\int_A [1,m] \in \mathcal{M}$. This completes the proof.

While the process of forming $\bar{M}$ from M is clearly not a closure operation, the preceding result leads us to ask whether it forms a monad. For a fixed category $\mathcal{C}$, the $\mathcal{M}$-faithful extendable functors M: $\mathcal{A} \to \mathcal{C}$ may be regarded as a category $\mathcal{M}(\mathcal{C})$ in which a morphism from M: $\mathcal{A} \to \mathcal{C}$ to N: $\mathcal{B} \to \mathcal{C}$ consists of a functor ϕ: $\mathcal{A} \to \mathcal{B}$ and a natural isomorphism $N\phi \cong M$. Each such morphism induces a restriction functor ϕ^*: $\bar{\mathcal{B}} \to \bar{\mathcal{A}}$ which maps $G \in \bar{\mathcal{B}}$ to $G\phi \in \bar{\mathcal{A}}$; this functor preserves limits but in general does not commute with the augmentations $\bar{M}$ and $\bar{N}$ into $\mathcal{C}$.

Example 3.3. The construction of $\bar{M}$ from M becomes a covariant "endofunctor" (composition being preserved only to within an isomorphism) on $\mathcal{M}(\mathcal{C})$ if we replace ϕ^* by its left adjoint $\bar{\phi}$. For $\mathcal{V} = \mathcal{S}$, the existence of $\bar{\phi}$ follows from Theorem 2.1. For a general $\mathcal{V}$, $\bar{\phi}$ exists as a left $\mathcal{V}$-adjoint if $\mathcal{C}$ is cotensored relative to $\mathcal{V}$. The resulting "precompletion monad" on $\mathcal{M}(\mathcal{C})$ has the Yoneda embedding Y_A: $\mathcal{A} \to \bar{\mathcal{A}}$ as its unit and Y_A^*: $\bar{\bar{\mathcal{A}}} \to \bar{\mathcal{A}}$ as its multiplication.

In the case where M: $\mathcal{A} \to \mathcal{C}$ has a right adjoint, $\bar{\mathcal{A}}$ is equivalent to the category of all cribles of $\mathcal{A}$. If, in addition, $\mathcal{A}$ is suitably complete (see Proposition 4.5) then the Yoneda embedding Y has a left adjoint which serves as an algebra structure for $\mathcal{A}$ with respect to this monad on $\mathcal{M}(\mathcal{C})$.

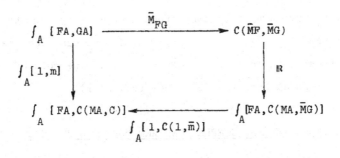

Conversely, if S: $\bar{\mathcal{A}} \to \mathcal{A}$ is the structure functor for any algebra of the precompletion monad then S is left adjoint to Y: $A \to \bar{A}$ with adjunction unit $\bar{S}\lambda$, where λ: $\bar{Y}_A \to Y_{\bar{A}}$ is the canonical natural transformation; this follows from Theorem 2.2. The role of λ was indicated to the author by Anders Kock (cf. [17]).

§4 EXAMPLES; COMPLETIONS AND MONOIDAL CLOSURE

Example 4.1. As in §3, the base category V is S-complete and admits all intersections of M-subobjects for the given choice of M. Let M: $A \to C$ be an M-faithful extendable functor and let K: $K \to A$ be a functor whose direct limit colim K exists in A. Then M preserves colim K if and only if $\bar{M}$ inverts the canonical transformation

$$s\colon \operatorname*{colim}_k A(-, Kk) \to A(-, \operatorname*{colim}_k Kk).$$

Because each representable functor from A^{op} to V is orthogonal to s there exists a largest (relative to M) full reflective subcategory $\bar{A}_s$ of $\bar{A}$ for which the Yoneda embedding $A \subset \bar{A}$ factors through $\bar{A}_s$ and preserves colim K.

This is the basis of many completion processes. In particular, if M: $A \to C$ is a strongly cogenerating and colimit-preserving extendable full embedding into a cotensored category C then the factorisation of §2 yields:

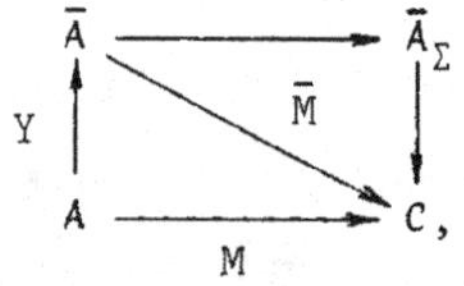

where Σ is the class of all morphisms inverted by $\bar{M}$. Thus one obtains a dense, strongly cogenerating, continuous and cocontinuous embedding $A \to \bar{A}_\Sigma$ and the functor $\bar{A}_\Sigma \to C$ reflects isomorphisms. This gives an alternative proof of [9] Theorem III.3.2.

Remark. The process of extending an M-faithful extendable functor M: $A \to C$ to $\bar{M}$: $\bar{A} \to C$ and then forming the category of fractions of $\bar{A}$ with respect to the class of all morphisms inverted by $\bar{M}$ defines a monad on the category $M(C)$ of M-faithful extendable functors over C. This monad is a quotient of the "precompletion monad" on $M(C)$ described in Example 3.3.

The following two examples concern the relationship of the factorisations of §1 and §2 to a given <u>monoidal</u> structure on the category B, and to the question of monoidal closure considered in monoidal localisation [8]. The monoidal structure on B is assumed to be symmetric for notational simplicity, however the results can be established in the more general setting of bicategories (in the sense of Bénabou [4]), biclosed bicategories, and their localisation.

We recall from [7] that a full reflective subcategory of B is called a <u>normal reflective subcategory</u> if the adjunction admits enrichment to a monoidal adjunction. The existence of such an enrichment implies that the reflecting functor preserves tensor products and, when B is monoidal closed, is equivalent to the subcategory being closed under exponentiation in B (by [7] Theorem 1.2).

<u>Example 4.2</u>. Let $S \dashv T\colon C \to B$ be an adjunction in which B has (E,M)-factorisations, as in §1, and let $P\colon B \to B'$ denote the reflection $B \mapsto \hat{B}$. If B has a monoidal structure then P is a monoidal localisation in the sense of [8] if $P(A \otimes \eta'_B)$ is an isomorphism for all $A,B \in B$. By Proposition 1.2, $P(A \otimes \eta'_B)$ is the unique morphism making both the following diagrams commute:

$$\begin{array}{ccccc}
A\otimes B & \xrightarrow{\eta'} & P(A\otimes B) & \xrightarrow{m} & TS(A\otimes B) \\
{\scriptstyle 1\otimes\eta'}\downarrow & & \downarrow{\scriptstyle P(1\otimes\eta')} & & \downarrow{\scriptstyle TS(1\otimes\eta')} \\
A\otimes PB & \xrightarrow[\eta']{} & P(A\otimes PB) & \xrightarrow[m]{} & TS(A\otimes PB)
\end{array}$$

where $\eta' \in E$ and $m \in M$. Thus an obvious sufficient condition for $P(1\otimes\eta')$ to be an isomorphism is that $A \otimes \eta'_B \in E$ and $TS(A \otimes \eta'_B) \in M$ for all $A,B \in B$.

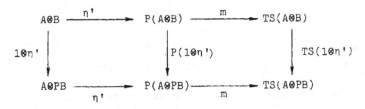

In the case where the monoidal structure on $\mathcal{B}$ is closed, the condition $A \otimes \eta'_B \in \mathcal{E}$ for all $A,B \in \mathcal{B}$ is automatically satisfied. Moreover, if $S \dashv T: \mathcal{C} \to \mathcal{B}$ is a monoidal adjunction then S necessarily preserves tensor products, so $TS(A \otimes \eta'_B)$ is always an isomorphism. Thus the category $\mathcal{B}'$ becomes a normal reflective subcategory of $\mathcal{B}$.

This example is related to the situation discussed in [8] where $\mathcal{B} = [A^{op},V]$ for a small monoidal category A over V, and $\mathcal{B}$ is assigned the convolution structure:

$$F \otimes G = \int^{AA'} FA \otimes GA' \otimes A(-, A \otimes A')$$

$$[F,G] = \int_A [FA, G(A\otimes -)] .$$

If $M: A \to \mathcal{C}$ is a functor into a cocomplete category $\mathcal{C}$ then the category $\bar{A}$ of M-cribles is a normal reflective subcategory of $[A^{op},V]$ if $\mathcal{C}(M(A\otimes -),C)$ is an M-crible for all $A \in A$ and $C \in \mathcal{C}$ (by [8] Proposition 1.1). This coincides with the preceding situation if $\mathcal{C}$ is monoidal closed and M preserves tensor products.

Example 4.3 Suppose $(\varepsilon,\eta): S \dashv T: \mathcal{C} \to \mathcal{B}$ is an M-adjunction in which $\mathcal{B}$ has equalisers, pullbacks of M-subobjects, and all intersections of M-subobjects. Then, by Corollary 2.3, $\mathcal{B}_Z \subset \mathcal{B}$ has a left adjoint if Z is any class of morphisms in $\mathcal{B}$ inverted by S. If Z is (orthogonally) closed then the reflection functor coincides with the projection of $\mathcal{B}$ onto the category of fractions of $\mathcal{B}$ with respect to Z.

Suppose that $\mathcal{B}$ has a monoidal closed structure and let $Z^o = \{s \in Z;\ B\otimes s \in Z \text{ for all } B \in \mathcal{B}\}$ denote the monoidal interior of Z with respect to this structure.

Proposition 4.4. If Z is a class of morphisms in $\mathcal{B}$ inverted by $S \dashv T: \mathcal{C} \to \mathcal{B}$ then $\mathcal{B}_{Z^o}$ is a normal reflective subcategory of $\mathcal{B}$.

Proof. The left adjoint of $\mathcal{B}_{Z^o} \subset \mathcal{B}$ exists by Corollary 2.3. To verify that $\mathcal{B}_{Z^o} = \{C \in \mathcal{B};\ \mathcal{B}(s,C) \text{ an isomorphism for all } s \in Z^o\}$ is closed under exponentiation in $\mathcal{B}$, choose objects $B \in \mathcal{B}$ and $C \in \mathcal{B}_{Z^o}$. Because $s\otimes B \in Z^o$ for each $s \in Z^o$ we have that $\mathcal{B}(s\otimes B,C) \cong \mathcal{B}(s,[B,C])$ is an isomorphism for all $s \in Z^o$. Thus $[B,C]$ is orthogonal to Z^o, as required.

It follows from this result that Z^o is orthogonally closed if Z is closed. Thus, the monoidal-interior operation on the class of subclasses of $\Sigma = \{f;\ Sf \text{ isomorphism}\}$ restricts to the lattice of closed subclasses of Σ, and each reflective subcategory of $\mathcal{B}$ which contains $\mathcal{B}_\Sigma$ as a reflective subcategory can be reflectively embedded in a "monoidal closure".

In particular, if $M: A \to C$ is M-faithful and extendable and if the Yoneda embedding $A \to \bar{A}$ has a left adjoint whose unit is inverted by $\bar{M}$ then each monoidal closed structure on $\bar{A}$ induces a monoidal closure of A itself (relative to M). A criterion for the existence of such an adjoint is the following:

<u>Proposition 4.5</u>. If $M: A \to C$ is an M-faithful extendable functor with a right adjoint R and if A is closed under the formation of equalisers, pullbacks of M-subobjects, and intersection of M-subobjects in $\bar{A}$, then A is reflective in $\bar{A}$.

<u>Proof</u>. This is a corollary of Theorem 2.2. The completeness hypotheses are satisfied by A and the functor $T: \bar{A} \to \bar{A}$ mapping F to $A(-, R\bar{M}F)$ admits a unit $\eta: 1 \to T$ with components $F \to C(M-, \bar{M}F) \cong A(-, R\bar{M}F)$ which all lie in M.

The motivating example is that where $A = Top$ and $M: Top \to S$ is the underlying-set functor. Here $\overline{Top}$ is cartesian closed and contains *Top* as a full reflective subcategory. The cartesian closure of *Top* with respect to M is the "minimal extension" discussed in [1] and it admits a normal reflective embedding into the cartesian closed category of limit spaces of Fischer [12].

<u>§5 THE FACTORISATION SYSTEM FOR LEFT ADJOINTS</u>.

The general process of factoring an adjoint pair of functors into a reflection followed by an isomorphism-reflecting embedding has a global interpretation. We consider the "category" $\mathcal{A}dj$ for which an object is a category which is $\mathcal{M}$-complete with respect to a suitable (see §§1 and 2) factorisation system $(\mathcal{E},\mathcal{M})$ on it and a morphism is a left-adjoint functor; composition of morphisms is just composition of the left adjoints. The class of reflection functors is denoted by $\mathcal{R}$ and the class of isomorphism-reflecting left adjoints is denoted by $\mathcal{N}$.

Proposition 5.1. To within natural isomorphism of functors, $(\mathcal{R},\mathcal{N})$ forms a factorisation system on $\mathcal{A}dj$.

Proof. We use several facts from [13] §2.3. Clearly every reflection is an "epimorphism" and every isomorphism-reflecting left adjoint, being faithful, is a "monomorphism" to within isomorphism of functors; thus the factorisation will be "proper". Moreover, every left-adjoint functor on an $\mathcal{M}$-complete category, where $(\mathcal{E},\mathcal{M})$ is a proper factorisation system, has a factorisation of the required form (by §1 and §2). Finally, we verify that if a diagram of left adjoints:

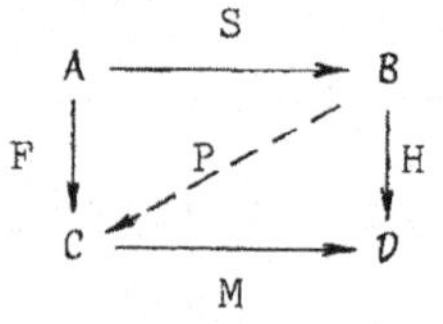

with $S \in \mathcal{R}$ and $M \in \mathcal{N}$, commutes to within an isomorphism then there exists a left adjoint P such that $PS \cong F$ and $MP \cong H$. Let $(\varepsilon,\eta)\colon S \dashv E$, $(\alpha,\beta)\colon F \dashv G$, $M \dashv N$, and $H \dashv K$ be the adjunctions. Because $HS \cong MF$ and M reflects isomorphisms, F factors through S as $P = FE$. Let $Q = SG$. To verify that $P \dashv Q$ it suffices to verify that G factors through E; that is, that the morphism

$$\eta_{GC}\colon GC \to ESGC$$

is an isomorphism for all $C \in \mathcal{C}$. But, for all $C \in \mathcal{C}$,

$$F\eta_{GC}\colon FGC \to FESGC$$

is an isomorphism because M reflects isomorphisms and $MF\eta_{GC} \cong HS\eta_{GC}$ which is an isomorphism because $S\eta$ is an isomorphism. Define ρ_C to be the composite:

$$\begin{array}{ccc} ESGC & \xrightarrow{\rho_C} & GC \\ {\scriptstyle\beta}\downarrow & & \uparrow{\scriptstyle\alpha} \\ GF(ESGC) & \xrightarrow[G(F\eta_{GC})^{-1}]{} & GFGC \end{array} .$$

Then $\rho_C \cdot \eta_{GC} = 1$ by naturality of β and the triangle axioms for the adjunction $(\alpha,\beta)\colon F \dashv G$. Because E is a full embedding, this implies $\eta_{GC} \cdot \rho_C = 1$. This completes the proof.

REFERENCES

[1] Antoine, P., Extension minimale de la catégorie des espaces topologiques, C.R. Acad. Sc. Paris, t, 262 (1966), 1389-1392.

[2] Applegate, H. and Tierney, M., Categories with models, Seminar on Triples and Categorical Homology Theory, Lecture Notes 80 (Springer 1969), 156-244.

[3] Applegate, H. and Tierney, M., Iterated cotriples, Reports of the Midwest Category Seminar IV, Lecture Notes 137 (Springer 1970), 56-99.

[4] Bénabou, J., Introduction to bicategories, Reports of the Midwest Category Seminar I, Lecture Notes 47 (Springer 1967), 1-77.

[5] Day, B.J. and Kelly, G.M., Enriched functor categories, Reports of the Midwest Category Seminar III, Lecture Notes 106 (Springer 1969), 178-191.

[6] Day, B.J., On closed categories of functors, Reports of the Midwest category Seminar IV, Lecture Notes 137 (Springer 1970), 1-38.

[7] Day, B.J., A reflection theorem for closed categories, J. Pure and Applied Algebra, Vol. 2, No. 1 (1972), 1-11.

[8] Day, B.J., Note on monoidal localisation, Bull. Austral. Math. Soc., Vol. 8 (1973), 1-16.

[9] Dubuc, E.J., Kan extensions in enriched category theory, Lecture Notes 145 (Springer 1970).

[10] Eilenberg, S. and Kelly G.M., Closed categories, in Proc. Conf. on Categorical Algebra, La Jolla, 1965 (Springer 1966), 421-562.

[11] Fakir, S., Monade idempotente associée à une monade, C.R. Acad. Sc. Paris, t. 270 (1970), 99-101.

[12] Fischer, H.R., Limesräume, Math. Annalen, Bd. 137 (1959) 269-303.

[13] Freyd, P. and Kelly, G.M., Categories of continuous functors I, J. Pure and Applied Algebra, Vol. 2, No. 3 (1972), 169-191.

[14] Gabriel, P. and Zisman, M., Calculus of fractions and homotopy theory, Springer-Verlag, Berlin, 1967.

[15] Kelly, G.M., Monomorphisms, epimorphisms, and pullbacks, J. Austral. Math. Soc., Vol. 9 (1969), 124-142.

[16] Kelly, G.M., Adjunction for enriched categories, Reports of the Midwest Category Seminar III, Lecture Notes 106 (Springer 1969), 166-177.

[17] Kock, A., Monads for which structures are adjoint to units (to appear).

[18] Maclane, S., Categories for the working mathematician, Springer-Verlag, New York, Heidelberg, Berlin, 1971.

[19] Wolff, H., V-localisations and V-monads, J. Algebra, Vol. 24, 1973, 405-438.

ON CLOSED CATEGORIES OF FUNCTORS II *

by

Brian Day

In many examples of monoidal biclosed categories the tensor product is constructed as an extension of a monoidal structure on a basic generating set of objects in the category. A simple example is the construction of the tensor product of two abelian groups from the free tensor product of two free abelian groups. We shall consider a formal generalisation of this construction by asking when a monoidal structure or, more generally, a promonoidal structure on a category A can be extended along a dense functor $N: A^{op} \to C$ to produce a monoidal biclosed structure on the category C.

We give an answer to this question by combining the two basic types of dense functor considered in [4] and [7]. The first is the Yoneda embedding of A^{op} into the category $[A,S]$ of all set-valued functors on A. To each promonoidal structure on a small category A there corresponds a monoidal biclosed structure on $[A,S]$. The second type is the reflection functor; that is, the left adjoint to a full embedding $C \subset B$. If B has a monoidal biclosed structure then the adjunction is monoidal if and only if C is closed under all exponentiation by the internal-hom functors of B.

The two results are combined in the following manner. If $N: A^{op} \to C$ is a dense functor then it can be decomposed into the Yoneda embedding of A^{op} in $[A,S]$ followed by a reflection from $[A,S]$ to

*The research here reported was supported by a grant from the Danish Research Council.

C provided C is sufficiently complete with respect to A. However, this over-all completeness hypothesis on C is generally unnecessary in order to produce a monoidal biclosed structure on C, as it is in the case when N itself is a reflection and A^{op} is monoidal biclosed. Thus we answer the original question by first considering a completion C* of C and finding conditions under which this completion is monoidal biclosed. The structure we obtain on C is simply the trace of the completed structure C*.

The procedure produces many known constructions of tensor-product functors and internal-hom functors. In particular, it produces the tensor product of algebras for a monoidal monad on a monoidal biclosed category, and gives a conceptual explanation of constructions by Linton [18] and Kock [16]. It also produces the canonical "convolution" structure on a functor category [A,B] when A is a promonoidal category and B is a suitably complete monoidal biclosed category.

One advantage of this approach is that the coherence of the structure produced on C follows from the coherence of C* which, in turn, follows from the coherence results already established in [4] and [7]. Completions can also be used to provide a concept of "change of V-universe" in the case where all categorical algebra is based on a fixed symmetric monoidal closed category V. This enables us to put a " V-structure" on any category [A,B] of V-functors from a (possibly large) V-category A to a suitably enriched category B. This, in turn, makes the relative completion process available for large V-categories (and ultimately leads to a reduction in algebraic computation).

The completion process is used in sections 5,6, and 7 to examine monoidal biclosed structures on categories of functors from a promonoidal category to a monoidal biclosed category. In section 6 we discuss biclosed categories of continuous functors and relate this to the work of Bastiani-Ehresmann [2]. Finally, section 8 contains a proof of the representation theorem for monads.

The unexplained notations and terminology used in this article are standard, and familiarity with the representation theorem is assumed (cf. [11] §1). This material is a development of results in [4], [7], [8], and [9] and is based on a doctoral thesis by the author ([5] and [6]). The thesis was supervised by Professor G.M. Kelly at the University of New South Wales. The author has also benefited from several discussions with A. Kock and R. Street.

The section-headings are as follows:

---

§1 REFLECTION IN CLOSED FUNCTOR CATEGORIES

The formulas needed for monoidal biclosed functor categories and reflection of monoidal biclosed structures are recalled from [4] and [7]. $\mathcal{S}$ denotes "the" cartesian closed category of small sets and $\mathcal{V} = (\mathcal{V}, \otimes, I, [-,-], \ldots)$ is a fixed symmetric monoidal closed base-category with small limits and colimits. All concepts of categorical algebra are henceforth assumed to be relative to this $\mathcal{V}$ unless otherwise stipulated.

An adjunction $(\varepsilon, \eta): S \dashv T: \mathcal{C} \to \mathcal{B}$ is called a monoidal reflection or normal reflection if T is a full embedding and the adjunction data admits monoidal enrichment. If $\mathcal{B} = (\mathcal{B},\otimes,I,/,\backslash,\ldots)$ is a monoidal biclosed category then $S \dashv T$ is monoidal if and only if one of the following pairs of conditions is satisfied (where T is omitted from the notation) for all $B,B' \in \mathcal{B}$ and $C \in \mathcal{C}$:

$$\left.\begin{array}{rl} \eta: & C/B \cong S(C/B) \\ \eta: & B\backslash C \cong S(B\backslash C) \end{array}\right\}$$

$$\left.\begin{array}{rl} \eta\backslash 1: & SB\backslash C \cong B\backslash C \\ 1/\eta: & C/SB \cong C/B \end{array}\right\}$$

$$\left.\begin{array}{rl} S(\eta\otimes 1): & S(B\otimes B') \cong S(SB\otimes B') \\ S(1\otimes\eta): & S(B'\otimes B) \cong S(B'\otimes SB) \end{array}\right\}$$

$$\left.\begin{array}{rl} S(\eta\otimes\eta): & S(B\otimes B') \cong S(SB\otimes SB'). \end{array}\right\}$$

A subcategory $\mathcal{A} \subset \mathcal{B}$ is called <u>strongly generating</u> in $\mathcal{B}$ if the class of functors $\{\mathcal{B}(A,-);\ A \in \mathcal{A}\}$ jointly reflects isomorphisms. The reflection theorem for monoidal biclosed categories (cf. [7] Theorem 1.2) states that the above conditions are equivalent to either of the pairs

$$\left.\begin{array}{rl} \eta: & D/A \cong S(D/A) \\ \eta: & A\backslash D \cong S(A\backslash D) \end{array}\right\} \qquad (1.1)$$

$$\left.\begin{array}{rl} S(\eta\otimes 1): & S(B\otimes A) \cong S(SB\otimes A) \\ S(1\otimes\eta): & S(A\otimes B) \cong S(A\otimes SB) \end{array}\right\}$$

for all $B \in \mathcal{B}$, $A \in \mathcal{A}$, and $D \in \mathcal{D}$, where $\mathcal{A}$ strongly generates $\mathcal{B}$ and $\mathcal{D}$ strongly cogenerates $\mathcal{C}$.

If $\mathcal{B}$ is a monoidal biclosed category and $\mathcal{C} \subset \mathcal{B}$ is a full subcategory we say that $\mathcal{C}$ is <u>closed under exponentiation</u> in $\mathcal{B}$ if $B\backslash C$ and C/B have isomorphs in $\mathcal{C}$ for <u>all</u> $B \in \mathcal{B}$ and $C \in \mathcal{C}$. Note that his is generally a <u>stronger</u> condition than requiring that the internal-hom functors $-\backslash-$ and $-/-$ have restrictions to $\mathcal{C}$.

Given a small category $\mathcal{A}$, each functor $F \in [\mathcal{A},\mathcal{V}]$ has an expansion:

$$F \cong \int^{A} FA \otimes \mathcal{A}(A,-) \ : \ \mathcal{A} \to \mathcal{V}$$

Thus each monoidal biclosed structure on the functor category is isomorphic to a structure of the following form:

$$F \otimes G = \int^{AB} FA \otimes GB \otimes P(AB\,-) \tag{1.2}$$

$$G/F = \int_{AB} [P(-AB),[FA,GB]]$$

$$F \backslash G = \int_{AB} [P(A-B),[FA,GB]] \tag{1.3}$$

where $P(AB-) = A(A,-) \otimes A(B,-)$ in $[A,V]$. The resulting functor $P\colon A^{op} \otimes A^{op} \otimes A \to V$ is the "structure functor" of $\otimes$ on $[A,V]$.

A monoidal biclosed structure on $[A,V]$ is determined by P and I and associativity and identity isomorphisms:

$$\alpha = \alpha_{ABC}\colon P(AX-)\circ P(BCX) \cong P(ABX)\circ P(XC-)$$

$$\lambda = \lambda_A \ \colon IX\circ P(XA-) \cong A(A,-)$$

$$\rho = \rho_A \ \colon IX\circ P(AX-) \cong A(A,-)$$

satisfying suitable coherence axioms ([4] §3), where "o" denotes profunctor composition. We call such a collection $(A,P,I,\alpha,\lambda,\rho)$ a promonoidal structure on A.

The extension of a promonoidal structure on A to $[A,V]$ is sometimes called the convolution of A with V and it is an internal hom of promonoidal categories, with:

$$P(FGH) = \int_{ABC} [P(ABC), [FA \otimes GB, HC]]$$

$$[A,V](F,G) = \int_{AB} [A(A,B), [FA,GB]]$$

$$I(F) = \int_{A} [IA, [I,FA]],$$

where P, Hom, and I are regarded as the 2-dimensional, 1-dimensional, and 0-dimensional components of promonoidal structure.

Two small promonoidal categories A and B have a promonoidal tensor product $A \otimes B$ whose (P,Hom,I) are the respective tensor products of those for A and B. The usual isomorphism of categories:

$$[A \otimes B, V] \simeq [A, [B, V]]$$

then becomes an isomorphism of monoidal biclosed categories.

To combine convolution with reflection, let $[A,V]$ have a monoidal biclosed structure and let $S \dashv T\colon C \to [A,V]$ be a full reflective embedding where C contains D as a strongly cogenerating subcategory. The Yoneda embedding $A^{op} \to [A,V]$ is strongly generating, thus we have:

Proposition 1.1. The adjunction $S \dashv T\colon C \to [A,V]$ is monoidal if and only if the functors

$$G/A(A,-) = \int_B [P(-AB), GB]$$

$$A(A,-)\backslash G = \int_B [P(A-B), GB]$$

have isomorphs in C for all $A \in A$ and $G \in D$.

Proof. This is a reformulation of Condition (1.1) with $B = [A,V]$. For example,

$$G/A(A,-) = \int_{B'B} [P(-B'B), [A(A,B'), GB]] \quad \text{by (1.3)}$$

$$\cong \int_B [P(-AB), GB] \text{ by the representation theorem}$$

applied to $B' \in A$.

Remark. The concept of convolution can be extended to include bicategories in the sense of Bénabou [3]. If

$$A = \{A_{xy};\ x,y \in \text{Obj } A\}$$

is a V-bicategory, with hom-categories A_{xy}, then

$$[A,V] = \{[A_{xy}, V];\ x,y \in \text{Obj } A\}$$

is a biclosed bicategory by formulas analogous to (1.2) and (1.3). This consideration, in turn, leads to the concept of a V-pro-bicategory. The extension of the results in this article to V-bicategories is a straightforward exercise once this conceptual framework is established. The reflection theorem for biclosed bicategories is an exact analogue of the "one-object" theorem above. A form of this theorem has been introduced by J. Meisen [19] in the study of relations in categories.

Remark on symmetry. The concept of a symmetric promonoidal category is defined in [4] §3. This produces an (obvious) analogous result-with-symmetry for each result verified in the sequel.

§2 THE COMPLETION PROCESS

For this section the base category V will be assumed to have all intersections of monomorphisms as well as all small limits and colimits.

To each small category C we will assign a completion C^* which is taken to be the largest full reflective subcategory of $[C^{op}, V]$ which contains the representable functors as a full strongly cogenerating subcategory. This coincides with the left-adjoint factorisation of the conjugation functor

$$[C^{op}, V] \to [C, V]^{op}; \; F \mapsto \int_C [FC, C(C,-)],$$

through a reflection followed by an isomorphism-reflecting embedding, denoted:

$$\begin{array}{ccc} [C^{op}, V] & \twoheadrightarrow & C^* \\ & \searrow & \downarrow \\ & & [C, V]^{op}, \end{array}$$

The resulting embedding of C into its completion will be denoted by:

$$E = E(C) : C \to C^*.$$

The basic properties of C^* can be verified by examining the cotriple-tower construction of C^* (Appelgate and Tierney [1] and Dubuc [10]) or by using a direct method of adjoint-functor factorisation ([9], Example 4.1). There is a closely related theorem in Wolff [22], §5.6.

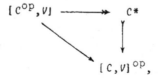

The completion C^* can be described explicitly as the full subcategory of $[C^{op}, V]$ of functors F such that $\int_C [s_C, FC]$ is an isomorphism whenever s is a morphism in $[C^{op},V]$ inverted by conjugation. As such, C^* is equivalent to the category of V-fractions of $[C^{op},V]$ with respect to the class of morphisms inverted by conjugation.

Remark. The assignment $C \mapsto C^*$ is canonically functorial (to within isomorphism) on functors $S: C \to B$ which have a right adjoint, and the image $S^*: C^* \to B^*$, $S^*E = ES$, has a right adjoint.

Proposition 2.1. If $N: A \to C$ is a dense functor then $EN: A \to C^*$ is dense.

This fact is established in [8] Corollary 3.2. More precisely, C^* is equivalent to the full subcategory of $[A^{op},V]$ of functors F such that $\int_A [s_A, FA]$ is an isomorphism whenever s is a morphism in $[A^{op},V]$ inverted by the restricted-conjugation functor:

$$[A^{op},V] \to [C,V]^{op};\ F \mapsto \int_A [FA, C(NA,-)].$$

Thus the reflection from $[A^{op},V]$ to C^* at $F \in [A^{op},V]$ has the value:

$$EN(F) = \int^A FA \cdot ENA,$$

and this is isomorphic to $E(\int^A FA \cdot NA)$ whenever the reflection $\int^A FA \cdot NA$ of F exists in C.

The second main observation of this section is that the completion process provides a "structural change of V-universe". Consider the case $V = S$ and $C = V$ and let S^* be a larger cartesian closed category of sets containing S and V, and any other categories that we want to regard as "small", as internal category objects. Let $W = V^*$ be the completion of V with respect to S^*. The basic properties of V^* (verified in [8]) are as follows:

Property 1. Because V^{op} is a symmetric monoidal category, the functor category $[V^{op},S^*]$ is symmetric monoidal closed and this structure extends the original structure of V. Because V is monoidal closed, the reflective embedding $W \subset [V^{op},S^*]$ is monoidal, hence the embedding $E: V \to W$ preserves tensor product and internal hom. This implies that the symmetric monoidal closed category V-Cat is fully embedded in W-Cat and the embedding preserves tensor product and internal hom.

Property 2. The embedding E: $V \to W$ is strongly cogenerating and dense, hence it preserves coends. Thus the biclosed bicategory V-$Prof$ (cf. [3], Introduction), consisting of V-categories and V-profunctors between them, is embedded (but not fully on morphisms) into the biclosed bicategory W-$Prof$ in such a way as to preserve profunctor composition.

Property 3. Because E: $V \to W$ is strongly cogenerating, the completion $[A,V]^*$ with respect to W of a functor category $[A,V]$ is equivalent to $[A,W]$ (as verified in [8] §4). This ensures that each V-promonoidal structure on a small V-category A corresponds to an essentially unique W-promonoidal structure on A when A is regarded as a W-category.

Because W has all S^*-small limits and colimits and all intersections of monomorphisms, the given base category V can be assumed to be an arbitrary symmetric monoidal closed category (with no completeness hypotheses on it).

Remark 2.2. If V is a cartesian closed category then W is cartesian closed and the reflection from $[V^{op},S^*]$ to W preserves finite products ([7] Corollary 2.1). Thus finite products commute with filtered colimits in any cartesian closed category, because this is so in $[V^{op},S^*]$.

§3 MONOIDAL CLOSED COMPLETION

Let N: $A^{op} \to C$ be a dense functor between two V-categories and let W be a completion of V with respect to a cartesian closed category S^* of sets which contains S, V, A, and C. Let $A = (A,P,I,\ldots)$ be a W-promonoidal category.

Proposition 3.1. The reflective embedding $\mathcal{C}^* \subset [\mathcal{A},\mathcal{W}]$ is monoidal if and only if the functors

$$\int_B [P(-AB),\ \mathcal{C}(NB,C)] \qquad (3.1)$$

$$\int_B [P(A-B),\ \mathcal{C}(NB,C)]$$

have isomorphs in $\mathcal{C}^*$ for all $A \in \mathcal{A}$ and $C \in \mathcal{C}$.

Proof. This is a restatement of Proposition 1.1, with the strongly cogenerating class $\mathcal{D} \subset \mathcal{C}^*$ being $\{\mathcal{C}(N-,C);\ C \in \mathcal{C}\}$.

Remark. The objects $C \in \mathcal{C}$ in (3.1) could be restricted further to lie in any subcategory $\mathcal{D} \subset \mathcal{C}$ for which $\{\mathcal{C}(N-,D);\ D \in \mathcal{D}\}$ strongly cogenerates $\mathcal{C}^*$. For example, if $\mathcal{D} \subset \mathcal{C}$ and each object $C \in \mathcal{C}$ is the limit of some functor (depending on C) with object values in $\mathcal{D}$, then the embedding $\mathcal{D} \subset \mathcal{C}$ followed by $E\colon \mathcal{C} \to \mathcal{C}^*$ strongly cogenerates $\mathcal{C}^*$.

We shall consider the special case where the functors (3.1) have isomorphs in $\mathcal{C}$, and let $\mathcal{C}(N-,H(AC))$ and $\mathcal{C}(N-,K(AC))$ be their representations. The internal-hom and tensor-product operations on $\mathcal{C}^*$ then provide functors $-/-\colon \mathcal{C} \otimes \mathcal{C}^{op} \to \mathcal{C}^*$, $-\backslash-\ \mathcal{C}^{op} \otimes \mathcal{C} \to \mathcal{C}^*$, and $-\otimes-\colon \mathcal{C} \otimes \mathcal{C} \to \mathcal{C}^*$ with values:

$$D/C = \int_A [\mathcal{C}(NA,C),\ \mathcal{C}(N-,H(AD))] \qquad (3.2)$$

$$C\backslash D = \int_A [\mathcal{C}(NA,C),\ \mathcal{C}(N-,K(AD))]$$

and

$$C \otimes D = \int^{AB} (\mathcal{C}(NA,C) \otimes \mathcal{C}(NB,D))\cdot Q(AB) \qquad (3.3)$$

where

$$Q(XY) = \int^A P(XYA)\cdot ENA.$$

The identity object of $\mathcal{C}^*$ is:

$$I = \int^A IA\cdot ENA. \qquad (3.4)$$

Remark. Note that, by the construction of the monoidal biclosed completion $\mathcal{C}^*$, the representing objects $H(AD)$ and $K(AD)$ in $\mathcal{C}$ are isomorphic in $\mathcal{C}^*$ to the exponentials ED/ENA and $ENA\backslash ED$ respectively.

§4 MONOIDAL MONADS

Let $\mathsf{T} = (T,\mu,\eta)$ be a <u>monoidal</u> monad on a monoidal category $\mathcal{B}$; that is, a monad where $T\colon \mathcal{B} \to \mathcal{B}$ has a monoidal functor structure $(T,\tilde{T},T^0)$ with respect to which $\mu\colon T^2 \to T$ and $\eta\colon 1 \to T$ are monoidal natural transformations.

Let $\mathcal{B}(\mathsf{T})$ denote the category of T-algebras over $\mathcal{B}$. This category exists over the base category $\mathcal{V}$ when $\mathcal{V}$ has the equalisers

$$\begin{array}{ccccc} \mathcal{B}(\mathsf{T})(CD) & \longrightarrow & \mathcal{B}(CD) & \xrightarrow{\mathcal{B}(\xi,1)} & \mathcal{B}(TC,D) \\ & & & {\scriptstyle T_{CD}} \searrow \quad \nearrow {\scriptstyle \mathcal{B}(1,\xi)} & \\ & & & \mathcal{B}(TC,TD) & \end{array}$$

for all T-algebras (C,ξ) and (D,ξ) (as constructed in [10] or [16]).

The monoidal axioms on (T,μ,η) imply that the Kleisli category $K(\mathsf{T})$ has a monoidal structure; the tensor-product functor

$$\tilde{\otimes}\colon K(\mathsf{T}) \otimes K(\mathsf{T}) \to K(\mathsf{T})$$

is defined on objects as it is in $\mathcal{B}$ and on morphism objects by the components:

$$\begin{array}{ccccc} K(\mathsf{T})(B,C) & & \xrightarrow{\quad A\,\tilde{\otimes}\,- \quad} & & K(\mathsf{T})(A\otimes B,A\otimes C) \\ \| & & & & \| \\ \mathcal{B}(B,TC) & \xrightarrow[A\,\otimes\,-]{} & \mathcal{B}(A\otimes B,A\otimes TC) & \xrightarrow[\mathcal{B}(1,\lambda)]{} & \mathcal{B}(A\otimes B,T(A\otimes C)) \end{array}$$

and

$$\begin{array}{ccccc} K(\mathsf{T})(A,C) & & \xrightarrow{\quad -\,\tilde{\otimes}\,B \quad} & & K(\mathsf{T})(A\otimes B,C\otimes B) \\ \| & & & & \| \\ \mathcal{B}(A,TC) & \xrightarrow[-\otimes B]{} & \mathcal{B}(A\otimes B,TC\otimes B) & \xrightarrow[\mathcal{B}(1,\rho)]{} & \mathcal{B}(A\otimes B,T(C\otimes B)), \end{array}$$

where λ and ρ are the left and right actions of $\mathcal{B}$ on T defined by commutativity of the triangles:

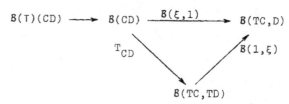

$$\begin{array}{ccccc} A \otimes TC & \xrightarrow{\eta \otimes 1} & TA \otimes TC & \xleftarrow{1 \otimes \eta} & TA \otimes C \\ & \searrow^{\lambda} & \downarrow & \swarrow_{\rho} & \\ & & T(A \otimes C). & & \end{array}$$

It is easily verified that the <u>interchange law</u>

$$\tilde{T} = \mu \cdot T\lambda \cdot \rho = \mu \cdot T\rho \cdot \lambda$$

holds and this is equivalent to $\tilde{\otimes}$ being a $\mathcal{V}$-bifunctor.

If T is a symmetric monoidal monad on a symmetric monoidal category $\mathcal{B}$ then the interchange law corresponds to the "commutativity law" of Kock [15]. The resulting monoidal structure on $K(T)$ is then symmetric.

<u>Lemma 4.1</u>. If a T-algebra is exponentiable in $\mathcal{B}$ then the exponents are evaluationwise T-algebras.

<u>Proof</u>. If (C,ξ) is a T-algebra and B is an object of $\mathcal{B}$ then the exponents C/B and $B\backslash C$ have algebra structures defined as the exponential transforms of the morphisms:

$$\begin{array}{ccccccc} B \otimes T(B\backslash C) & \xrightarrow{\lambda} & T(B \otimes (B\backslash C)) & \searrow^{Te} & & & \\ & & & & TC & \xrightarrow{\xi} & C \\ T(C/B) \otimes B & \xrightarrow[\rho]{} & T((C/B) \otimes B) & \nearrow_{Te} & & & \end{array}$$

where e denotes the respective evaluation transformations. The algebra axioms for these structures follow directly from those for ξ.

Thus, if all T-algebras are exponentiable in $\mathcal{B}$ then we obtain adjoint actions of $\mathcal{B}^{op}$ on $\mathcal{B}(T)$.

<u>Proposition 4.2</u>. The completion $\mathcal{B}(T)^*$ is monoidal biclosed if each T-algebra is exponentiable in $\mathcal{B}$.

<u>Proof</u>. Let (C,ξ) be a T-algebra. Then the adjunctions:

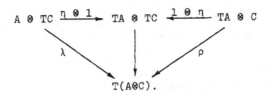

$$\mathcal{B}(-\otimes B, C) \cong \mathcal{B}(-, C/B)$$

$$\mathcal{B}(B\otimes -, C) \cong \mathcal{B}(-, B\backslash C)$$

provide adjunctions:

$$\mathcal{B}(\mathrm{T})(F(-\otimes B), C) \cong \mathcal{B}(\mathrm{T})(F-, C/B)$$

$$\mathcal{B}(\mathrm{T})(F(B\otimes -), C) \cong \mathcal{B}(\mathrm{T})(F-, B\backslash C) \quad ,$$

where $F\colon \mathcal{B} \to \mathcal{B}(\mathrm{T})$ is the free-algebra functor. If these isomorphisms remain natural when F is extended to the dense comparison functor $N\colon K(\mathrm{T}) \longrightarrow \mathcal{B}(\mathrm{T})$ (see §8) then $\mathcal{B}(\mathrm{T})^*$ is monoidal biclosed by Proposition 3.1. By Proposition 8.2, it suffices to show (working out one of the adjunctions) that the composites:

$$N(T(C/B)\ \tilde{\otimes}\ B) \underset{N(T\xi\tilde{\otimes}1)}{\overset{N(\mu\tilde{\otimes}1)}{\rightrightarrows}} N((C/B)\ \tilde{\otimes}\ B) \xrightarrow{\ Te\ } TC \xrightarrow{\ \xi\ } C$$

are equal. On composing both composites with η and filling in the definitions of $\tilde{\otimes}$ and $\xi_{C/B}$, we obtain a commuting diagram pair:

$$\begin{array}{ccc}
T(C/B)\otimes B & \xrightarrow{\ \eta\ } & T(T(C/B)\otimes A) \\
{\scriptstyle \eta\otimes 1}\downarrow & & \\
T^2(C/B)\otimes B & & {\scriptstyle N(\mu\tilde{\otimes}1)}\downarrow\downarrow{\scriptstyle N(T\xi\tilde{\otimes}1)} \\
{\scriptstyle \mu\otimes 1}\downarrow\downarrow{\scriptstyle T\xi\otimes 1} & & \\
T(C/B)\otimes B & \xrightarrow{\ \rho\ } & T((C/B)\otimes B) \\
{\scriptstyle \bar{\xi}_{C/B}}\downarrow & & \downarrow{\scriptstyle Te} \\
C & \xrightarrow{\ \xi\ } & TC
\end{array} \quad ,$$

where $\bar{\xi}_{C/B}$ is the adjoint-transform of $\xi_{C/B}$. Then the lefthand side is the adjoint-transform of

$$T(C/B) \xrightarrow{\ \eta\ } T^2(C/B) \underset{T\xi}{\overset{\mu}{\rightrightarrows}} T(C/B) \xrightarrow{\ \xi\ } C/B$$

and these composites are equal because $\xi_{C/B}$ is an algebra structure. This completes the proof.

Under the hypothesis of Proposition 4.2 we can determine when the trace of $\mathcal{B}(\mathrm{T})^*$ exists on $\mathcal{B}(\mathrm{T})$. First, for each T-algebra (C,ξ), there are natural transformations

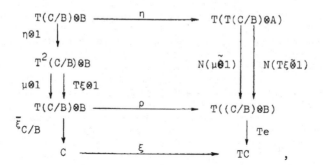

$$\overline{Te\cdot\lambda}:\quad C/B \longrightarrow TC/TB$$

$$\overline{Te\cdot\rho}:\quad B\backslash C \longrightarrow TB\backslash TC,$$

corresponding to the transformations:

$$(C/B) \otimes TB \xrightarrow{\lambda} T((C/B) \otimes B) \xrightarrow{Te} TC$$

$$TB \otimes (B\backslash C) \xrightarrow{\rho} T(B \otimes (B\backslash C)) \xrightarrow{Te} TC.$$

<u>Proposition 4.3</u>. The category $\mathcal{B}(T)$ is biclosed in $\mathcal{B}(T)^*$ if and only if the equalisers:

$$\begin{array}{ccccc} B\backslash\!\!\backslash C & \longrightarrow & B\backslash C & \xrightarrow{\xi\backslash C} & TB\backslash C \\ & & {\scriptstyle \overline{Te\cdot\rho}}\searrow & & \nearrow{\scriptstyle TB\backslash\xi} \\ & & & TB\backslash TC & \end{array}$$

and

$$\begin{array}{ccccc} C /\!\!/ B & \longrightarrow & C/B & \xrightarrow{C/\xi} & C/TB \\ & & {\scriptstyle \overline{Te\cdot\lambda}}\searrow & & \nearrow{\scriptstyle \xi/TB} \\ & & & TC/TB & \end{array}$$

exist in $\mathcal{B}$ for all T-algebras (B,ξ) and (C,ξ).

<u>Proof</u>. Because $K(T)$ is monoidal, the unit object NI lies in $\mathcal{B}(T)$. The equalisers are precisely the ends (3.2) by Proposition 8.2.

<u>Proposition 4.4</u>. If the reflective embedding $\mathcal{B}(T)^* \subset [K(T)^{op},\mathcal{W}]$ is monoidal then the tensor product in $\mathcal{B}(T)^*$ of two algebras (C,ξ) and (D,ξ) lies in $\mathcal{B}(T)$ if and only if the coequaliser of the reflective pair:

$$\begin{array}{ccc} T(TC\otimes TD) & & \\ {\scriptstyle T(\tilde{T})}\downarrow & \searrow{\scriptstyle T(\xi\otimes\xi)} & \\ T^2(C\otimes D) & \xrightarrow[\mu]{} & T(C\otimes D) \end{array}$$

exists in $\mathcal{B}(T)$.

<u>Proof</u>. The pair has a common right inverse $T(\eta\otimes\eta)$; that is, it

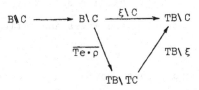

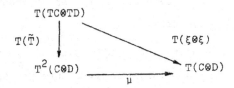

is reflective. Moreover, the coequaliser in $\mathcal{B}(T)$ is then the joint coequaliser of the pairs:

$$\begin{array}{ccccc} T(C\otimes TD) & \xrightarrow{T\lambda} & T^2(C\otimes D) & \xleftarrow{T\rho} & T(TC\otimes D) \\ & {\scriptstyle T(1\otimes\xi)}\searrow & \downarrow{\scriptstyle\mu} & \swarrow{\scriptstyle T(\xi\otimes 1)} & \\ & & T(C\otimes D) & & \end{array}$$

By Proposition 8.2, this coequaliser is the iterated coend (3.3);

$$C \otimes D = \int^{AB}(\mathcal{B}(T)(NA,C) \otimes \mathcal{B}(T)(NB,D))\cdot N(A\tilde{\otimes}B)$$

in $\mathcal{B}(T)$. Because $E\colon \mathcal{B}(T) \to \mathcal{B}(T)^*$ preserves and reflects colimits, this completes the proof.

The preceding propositions provide an alternative approach to earlier work by Kock [16] and Linton [18]. The use of completion leads to a significant reduction in coherence computations.

Another result of Kock [17], Theorem 2.6, may be established using the completion method. Namely, if the base category $\mathcal{V}$ is cartesian closed and $\mathcal{B}$ is cartesian closed over $\mathcal{V}$ then the category $\mathcal{B}(T)$ is cartesian closed if the functor $T\colon \mathcal{B} \to \mathcal{B}$ preserves finite products. This follows immediately from the fact that $K(T)$ has the cartesian monoidal structure, hence the completion $\mathcal{B}(T)^* \subset [K(T)^{op},\mathcal{W}]$ is cartesian closed.

As mentioned earlier, the Kleisli category $K(T)$ of a symmetric monoidal monad T has a symmetric monoidal structure, hence the completed structure on $\mathcal{B}(T)^*$ is symmetric. For a nonsymmetric monoidal monad on a symmetric monoidal closed category $\mathcal{B}$ the tensor product might exist on $\mathcal{B}(T)$ but have no symmetry.

<u>Example.</u> Let $\mathcal{V} = \mathcal{S}$ and let $\phi\colon \mathbb{N} \to \Delta$ be the monoidal inclusion of the discrete category $\mathbb{N}$ of finite integers into the simplical category Δ, both with the ordinal-sum monoidal structure. If $[\mathbb{N},\mathcal{S}]$ is given the convolution monoidal structure then the structure generated on $[\Delta^{op},\mathcal{S}]$ by the (monadic) adjoint pair

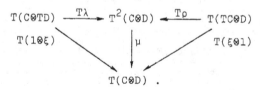

$$\phi \longmapsto [\phi^{op},1] : [\Delta^{op},S] \longrightarrow [\mathbb{N},S]$$

is isomorphic to the convolution monoidal structure on $[\Delta^{op},S]$. This structure extends the non-symmetric ordinal-sum structure on Δ.

§5 BICLOSED FUNCTOR CATEGORIES

When A and B are two V-categories the category $[A,B]$ of all V-functors from A to B exists as a V-category if V is suitably complete; that is, if the end:

$$[A,B](F,G) = \int_A B(FA,GA)$$

exists in V for all $F,G \in [A,B]$.

Lemma 5.1. If the category B admits the V-tensor-copowers $A(A,-)\cdot B$ for all $A \in A$ and $B \in B$ then the Yoneda functor

$$N: A^{op}\otimes B \to [A,B]; \; N(AB) = A(A,-)\cdot B,$$

is dense.

Proof.

$$\int_{A,B} [[A,B](N(AB),F), [A,B](N(AB),G)]$$

$$\cong \int_{A,B} [B(B,FA),B(B,GA)]$$

by definition of N,

$$\cong \int_A B(FA,GA)$$

by the representation theorem applied to $B \in B$,

$$= [A,B](F,G), \text{ as required.}$$

Remark. This result was established by F. Ulmer ([20], Theorem 1.33) for the case where V is the symmetric monoidal closed category Ab of abelian groups and group homomorphisms, with the usual tensor product of abelian groups. In fact, Ulmer considers a more general Yoneda functor $N: A^{op}\otimes\bar{B} \to [A,B]$, $N(AB) = A(A,-)\cdot MB$, where M: $\bar{B} \to B$ is an Ab-dense functor.

In order to investigate the existence of a monoidal biclosed structure on $[A,B]$, first take W to be the completion of V with respect to a cartesian closed category S^* of sets which contains A, B, V, and S, as "small" category objects. Note that the assumption that A,B exists as a V-category can be avoided by the use of such a completion W.

Let $A = (A,P,J,\ldots)$ be a W-promonoidal structure on A and let B be a monoidal biclosed category relative to V. Then the monoidal structure on B provides a promonoidal structure on B^{op} and consequently $A \otimes B^{op}$ has the tensor-product of promonoidal structures given by the expressions:

$$P((X,B),(Y,C),(-,-)) = P(XY-) \otimes B(-,B\otimes C)$$
$$I(X,B) = JX \otimes B(B,I).$$

<u>Proposition 5.2</u>. The completion $[A,B]^*$ of $[A,B]$ with respect to W is monoidal biclosed if the functor

$$\int_A [P(XYA), B(-,GA)] : B^{op} \to W$$

is representable for all $X,Y \in A$ and $G \in [A,B]$.

<u>Proof</u>. The first functor of (3.1), with the object $(a,b) \in A^{op} \otimes B$ marking the variable position, becomes:

$$\int_{XC} [P((a,b),(A,B),(X,C)),[A,B](N(XC),G)]$$

$$= \int_{XC} [P(aAX) \otimes B(C,b\otimes B), \int_Y B(A(X,Y)\cdot C, GY)]$$

by the definitions of N and $A \otimes B^{op}$,

$$\cong \int_X [P(aAX), B(b\otimes B,GX)]$$

by the representation theorem applied $Y \in A$ and $C \in B$,

$$\cong \int_X [P(aAX), B(b,GX/B)]$$

because B is biclosed,

$$\cong \mathcal{B}(b, H(A,B,G)(a))$$

by the representability hypothesis,

$$\cong \int_X \mathcal{B}(\mathcal{A}(a,X)\cdot b,\ H(A,B,G)(X))$$

by the representation theorem applied to $X \in \mathcal{A}$,

$$= [\mathcal{A},\mathcal{B}](N(ab),\ H(A,B,G)) \text{ by definition of } N,$$

as required. A similar computation reduces the second end in (3.1) to the required form.

Henceforth we assume that the hypothesis of Proposition 5.2 is satisfied, with natural isomorphisms:

$$\left.\begin{array}{l} \int_A [P(XYA), \mathcal{B}(-, GA/B)] \cong \mathcal{B}(-, H(Y,B,G)(X)) \\ \int_A [P(XYA), \mathcal{B}(-, B\backslash GA)] \cong \mathcal{B}(-, K(X,B,G)(Y)) \end{array}\right\} \quad (5.1)$$

for chosen representations H and K.

<u>Proposition 5.3</u>. The category $[\mathcal{A},\mathcal{B}]$ is biclosed in $[\mathcal{A},\mathcal{B}]^*$ if and only if the ends:

$$\left.\begin{array}{l} \int_A H(A,FA,G) \\ \int_A K(A,FA,G) \end{array}\right\}$$

exist in $[\mathcal{A},\mathcal{B}]$ for all $F,G \in [\mathcal{A},\mathcal{B}]$.

<u>Proof</u>. The internal-hom values (3.2) can be reduced as follows:

$$G/F = \int_{AC} [[\mathcal{A},\mathcal{B}](N(AC),F),\ [\mathcal{A},\mathcal{B}](N-,H(A,C,G))]$$

$$\int_{AC} [[\mathcal{A},\mathcal{B}](\mathcal{A}(A,-)\cdot C,F),\ [\mathcal{A},\mathcal{B}](N-,H(A,C,G))]$$

by the definition of N,

$$\cong \int_{AC} [\mathcal{B}(C,FA),\ [\mathcal{A},\mathcal{B}](N-,H(A,C,G))]$$

by the representation theorem applied to $A \in \mathcal{A}$,

$$\cong \int_A [\mathcal{A},\mathcal{B}](N-,H(A,FA,G))$$

by the representation theorem applied to $C \in \mathcal{B}$,

$$\cong [\mathcal{A},\mathcal{B}](N-, \textstyle\int_A H(A,FA,G)) \text{ if and only if the end}$$

$\int_A H(A,FA,G)$ exists in $[\mathcal{A},\mathcal{B}]$.

Similarly, we obtain:

$$F\backslash G \cong [\mathcal{A},\mathcal{B}](N-, \int_A K(A,FA,G)),$$

and this completes the proof.

Thus the resulting internal-hom functors on $[\mathcal{A},\mathcal{B}]$ have values:

$$\left.\begin{aligned} G/F &= \int_A H(A,FA,G) \\ F\backslash G &= \int_A K(A,FA,G) \end{aligned}\right\} \qquad (5.2)$$

and, by the definitions of H and K in Proposition 5.2, there exist natural isomorphisms:

$$G/N(AB) \cong H(A,B,G)$$

$$N(AB)\backslash G \cong K(A,B,G).$$

The internal-hom functors have an identity object if the identity of $[\mathcal{A},\mathcal{B}]^*$ lies in $[\mathcal{A},\mathcal{B}]$; this identity has the value:

$$I = \int^{AB} I(AB)\cdot EN(AB) \qquad \text{by } (3.4),$$

$$= \int^{AB} (JA \otimes \mathcal{B}(B,I))\cdot(\mathcal{A}(A,-)\cdot B)$$

$\cong J\cdot I$ in $[\mathcal{A},\mathcal{B}]^*$, by the representation theorem applied to $A \in \mathcal{A}$ and $B \in \mathcal{B}$.

The tensor product in $[\mathcal{A},\mathcal{B}]^*$ of two functors $F,G \in [\mathcal{A},\mathcal{B}]$ has the value:

$$F \otimes G = \int^{XBYC} (\mathcal{B}(B,FX) \otimes \mathcal{B}(C,GY))\cdot Q(X,B,Y,C) \text{ by } (3.3),$$

$$\cong \int^{XY} Q(X,FX,Y,GY)$$

by the representation theorem applied to $B,C \in \mathcal{B}$, where

$$Q(X,FX,Y,GY) = \int^{AB} (P(XYA) \otimes \mathcal{B}(B,FX\otimes GY))\cdot EN(AB)$$

by definition of the promonoidal category $\mathcal{A} \otimes \mathcal{B}^{op}$,

$$= \int^{AB} (P(XYA) \otimes \mathcal{B}(B,FX\otimes GY))\cdot E(\mathcal{A}(A,-)\cdot B)$$

by definition of N,

$$\cong \int^{AB} (P(XYA) \otimes \mathcal{B}(B,FX\otimes GY))\cdot(\mathcal{A}(A,-)\cdot EB)$$

because E preserves tensoring,

$$\cong P(XY-)\cdot(FX\otimes GY)$$

by the representation theorem applied to $A \in \mathcal{A}$ and $B \in \mathcal{B}$.

This establishes the following:

<u>Proposition 5.4</u>. The monoidal structure on $[\mathcal{A},\mathcal{B}]^*$ admits a restriction to $[\mathcal{A},\mathcal{B}]$ if the coend

$$F \otimes G = \int^{XY} P(XY-)\cdot(FX\otimes GY) \qquad (5.3)$$

exists in $[\mathcal{A},\mathcal{B}]$ for all $F,G \in [\mathcal{A},\mathcal{B}]$, and the identity object $J\cdot I$ of $[\mathcal{A},\mathcal{B}]^*$ lies in $[\mathcal{A},\mathcal{B}]$.

The coherence of the monoidal and biclosed structures induced on $[\mathcal{A},\mathcal{B}]$ is a consequence of the coherence of the monoidal biclosed structure on $[\mathcal{A},\mathcal{B}]^*$. In the case where $\mathcal{B} = \mathcal{V}$, the formulas (5.2) and (5.3) reproduce the original convolution structure (1.3) and (1.2) on $[\mathcal{A},\mathcal{V}]$.

<u>Remark.</u> It is straightforward to verify, using (5.3), that if $\mathcal{A}$ and $\mathcal{A}'$ are two small promonoidal categories and if $\mathcal{B}$ is a sufficiently complete and cocomplete monoidal biclosed category then the canonical isomorphism of categories:

$$[\mathcal{A} \otimes \mathcal{A}',\mathcal{B}] \cong [\mathcal{A},[\mathcal{A}',\mathcal{B}]]$$

admits enrichment to an isomorphism of monoidal biclosed categories, where $\mathcal{A} \otimes \mathcal{A}'$ has the tensor-product promonoidal structure and each functor category has the "convolution" monoidal biclosed structure defined by (5.2) and (5.3).

§6 BICLOSED CATEGORIES OF CONTINUOUS FUNCTORS

If A is a promonoidal category and B is a monoidal biclosed category for which the internal-hom functors defined by (5.2) exist on the functor category $[A,B]$ then we can ask whether a given full subcategory of $[A,B]$ is closed under all exponentiation by these internal-hom functors. The work of Bastiani-Ehresmann [2] on closed categories of sketched structures leads us to examine this question for those full subcategories of $[A,B]$ which consist of the Z-continuous functors from A to B for a given class Z of morphisms in $[A,V]$. Again, the base category is assumed to be V.

For a given category C let C^m denote the class of all morphisms in C. For each $Z \subset C^m$ let C_Z denote the full subcategory of C consisting of the objects orthogonal to Z (following terminology of [12]); that is,

$$C_Z = \{C \in C; C(s,C) \text{ is an isomorphism in } V \text{ for all } s \in Z\}.$$

Let $\bar{Z}$ denote the orthogonal closure of Z in C^m; that is,

$$\bar{Z} = \{s \in C^m; C(s,C) \text{ is an isomorphism in } V \text{ for all } C \in C_Z\}.$$

Then $C_Z = C_U = C_{\bar{Z}}$ for all $Z \subset U \subset \bar{Z}$.

For a given monoidal structure on the category C, a class $Z \subset C^m$ is called monoidal if $C \in C$ and $s \in Z$ imply that $C \otimes s \in Z$ and $s \otimes C \in Z$; that is, if Z is stable under the monoidal action of C on both sides of C^m.

Remark. It was established in [8] §1 that if Z is monoidal with respect to C then the category $C(Z^{-1})$ of V-fractions of C with respect to Z (as constructed in Wolff [21]) has a monoidal structure such that each monoidal functor on C which inverts the elements of Z has a unique monoidal-functor factorisation through the projection $C \to C(Z^{-1})$.

Proposition 6.1. If A is a strongly generating subcategory of a monoidal biclosed category C and $Z \subset C^m$ then the following are equivalent:

(a) $A \in \mathcal{A}$ and $s \in Z$ imply $A \otimes s \in \bar{Z}$ and $s \otimes A \in \bar{Z}$.

(b) $\mathcal{C}_Z$ is closed under exponentiation in $\mathcal{C}$.

(c) $\bar{Z}$ is monoidal.

Proof. (a) ⇒ (b). If $C \in \mathcal{C}_Z$ then $\mathcal{C}(A \otimes s, C) \cong \mathcal{C}(A, C/s)$ and $\mathcal{C}(s \otimes A, C) \cong \mathcal{C}(A, s\backslash C)$ are isomorphisms for all $A \in \mathcal{A}$ and $s \in Z$ by (a). Thus C/s and $s\backslash C$ are isomorphisms for all $s \in Z$ and $C \in \mathcal{C}_Z$ because $\mathcal{A}$ strongly generates $\mathcal{C}$. Hence $\mathcal{C}(s, B\backslash C) \cong \mathcal{C}(B, C/s)$ and $\mathcal{C}(s, C/B) \cong \mathcal{C}(B, s\backslash C)$ are isomorphisms for all $B \in \mathcal{C}$ and $s \in Z$, which implies that $B\backslash C$ and C/B both lie in $\mathcal{C}_Z$ for all $B \in \mathcal{C}$ and $C \in \mathcal{C}_Z$.

(b) ⇒ (c). If $B \in \mathcal{C}$ and $s \in \bar{Z}$ then $\mathcal{C}(B \otimes s, C) \cong \mathcal{C}(s, B\backslash C)$ and $\mathcal{C}(s \otimes B, C) \cong \mathcal{C}(s, C/B)$ are isomorphisms for all $C \in \mathcal{C}_Z$ by (b). Thus $B \otimes s$ and $s \otimes B$ are in $\bar{Z}$, as required.

(c) ⇒ (a) because $\mathcal{A} \subset \mathcal{C}$.

Let $\mathcal{C}$ be a functor category of the form $[\mathcal{A}, \mathcal{V}]$ and let $Z \subset [\mathcal{A}, \mathcal{V}]^m$. Then we can form (as is done in [13], §8.1) the full subcategory $[\mathcal{A}, \mathcal{B}]_Z \subset [\mathcal{A}, \mathcal{B}]$ consisting of the functors $F \in [\mathcal{A}, \mathcal{B}]$ such that $\mathcal{B}(B, F-) \in [\mathcal{A}, \mathcal{V}]_Z$ for all $B \in \mathcal{B}$. Such a functor F is interpreted as a "Z-continuous" model of $\mathcal{A}$ in $\mathcal{B}$. Clearly $[\mathcal{A}, \mathcal{B}]_Z = [\mathcal{A}, \mathcal{B}]_U = [\mathcal{A}, \mathcal{B}]_{\bar{Z}}$ for all $Z \subset U \subset \bar{Z}$.

We now suppose that $\mathcal{A} = (\mathcal{A}, P, J, \ldots)$ is a small promonoidal category over $\mathcal{V}$ and that $\mathcal{B}$ is a monoidal biclosed category over $\mathcal{V}$ for which the internal-hom functors defined by (5.2) exist on the functor category $[\mathcal{A}, \mathcal{B}]$.

Proposition 6.2. If Z is a monoidal class of morphisms in the convolution functor category $[\mathcal{A}, \mathcal{V}]$ then $[\mathcal{A}, \mathcal{B}]_Z$ is closed under exponentiation in $[\mathcal{A}, \mathcal{B}]$.

Proof. For all $s \in Z$, $B \in \mathcal{B}$, $F \in [\mathcal{A}, \mathcal{B}]$ and $G \in [\mathcal{A}, \mathcal{B}]_Z$, we have:

$$\int_X [s_X, \mathcal{B}(B,(G/F)(X))]$$

$$\cong \int_X [s_X, \mathcal{B}(B, \int_Y H(Y,FY,G)(X))] \qquad \text{by (5.2)},$$

$$\cong \int_Y \int_X [s_X, \mathcal{B}(B,H(Y,FY,G)(X))]$$

by interchanging limits,

$$\cong \int_Y \int_X [s_X, \int_A [P(XYA), \mathcal{B}(B,GA/FY)]] \text{ by (5.1)},$$

$$\cong \int_Y (\int_A [\int^X s_X \otimes P(XYA), \mathcal{B}(B \otimes FY, GA)])$$

by interchanging limits and using the tensor-hom adjunctions of $\mathcal{V}$ and $\mathcal{B}$,

$$\cong \int_Y (\int_A [(s \otimes \mathcal{A}(Y,-))(A), \mathcal{B}(B \otimes FY, GA)]) \quad (*)$$

because

$$(s \otimes \mathcal{A}(Y,-))(A) = \int^{XX'} s_X \otimes \mathcal{A}(Y,X') \otimes P(XX'A) \text{ by (1.2)},$$

$$\cong \int^X s_X \otimes P(XYA)$$

by the representation theorem applied to $X' \in \mathcal{A}$. The morphism (*) is an end over $Y \in \mathcal{A}$ of isomorphisms because $s \otimes \mathcal{A}(Y,-) \in Z$ since Z is monoidal and $G \in [\mathcal{A},\mathcal{B}]_Z$ by hypothesis. Thus G/F and, dually, $F \backslash G$ both lie in $[\mathcal{A},\mathcal{B}]_Z$ for all $F \in [\mathcal{A},\mathcal{B}]$ and $G \in [\mathcal{A},\mathcal{B}]_Z$.

Corollary 6.3. If Z is a class of morphisms in $[\mathcal{A},\mathcal{V}]$ for which $[\mathcal{A},\mathcal{V}]_Z$ is closed under exponentiation in the convolution functor category $[\mathcal{A},\mathcal{V}]$ then $[\mathcal{A},\mathcal{B}]_Z$ is closed under exponentiation in $[\mathcal{A},\mathcal{B}]$.

Proof. If $[\mathcal{A},\mathcal{V}]_Z$ is closed under exponentiation in $[\mathcal{A},\mathcal{V}]$ then $\bar{Z}$ is monoidal by Proposition 6.1. Thus $[\mathcal{A},\mathcal{B}]_Z$, which coincides with $[\mathcal{A},\mathcal{B}]_{\bar{Z}}$, is closed under exponentiation in $[\mathcal{A},\mathcal{B}]$ by Proposition 6.2.

Thus, if the Z-continuous models in $\mathcal{V}$ of a promonoidal category $\mathcal{A}$ form an exponentially biclosed subcategory of the convolution $[\mathcal{A},\mathcal{V}]$ then so do the Z-continuous models of $\mathcal{A}$ in any suitably complete monoidal biclosed category $\mathcal{B}$ which is based on $\mathcal{V}$.

This result contains an alternative approach to constructions in [2], Chapter III. For example (cf. [2] §12), let $V = S$ and let Δ be the simplicial category. Then the category *Cat* of small categories and functors is fully embedded in $[\Delta^{op}, S]$ as a cartesian closed reflective subcategory. This implies that *Cat* is defined by a (cartesian) monoidal class of morphisms in $[\Delta^{op}, S]$; namely, the class of <u>all</u> morphisms inverted by the reflection. It now follows from Corollary 6.3 that the category $[\Delta^{op}, B]_Z$ of all category objects in B is closed under exponentiation in $[\Delta^{op}, B]$ whenever $[\Delta^{op}, B]$ admits the internal-hom functors (5.2). Moreover, if the monoidal biclosed structure considered in §5 exists on $[\Delta^{op}, B]$ and the embedding

$$[\Delta^{op}, B]_Z \subset [\Delta^{op}, B]$$

has a left adjoint then this adjunction admits monoidal enrichment by the reflection theorem of §1. Note that, by Proposition 6.1, Z may be replaced by <u>any</u> class of morphisms in $[\Delta^{op}, S]$ which defines *Cat* as its class of orthogonal objects.

<u>Remark</u>. The general question of the existence of a left-adjoint functor to an inclusion of the form $[A, B]_Z \subset [A, B]$ has been studied in some detail by Freyd-Kelly [12] and by Gabriel-Ulmer [13] §8. Results bearing on special cases of this problem have appeared in [1], [9], [10], and [22].

§7 COMPLETION OF FUNCTOR CATEGORIES

This section is supplementary to §5 and the same V and W will be used. We will verify that the completion $[A, B]^*$ of a functor category $[A, B]$ is equivalent to the functor category $[A, B^*]$ when the Yoneda functor

$$N: A^{op} \otimes B \to [A, B];\ N(AB) = A(A,-)\cdot B,$$

exists.

First recall from §2 that $[A, B]^*$ is characterised as the largest full reflective subcategory of $[[\mathcal{A},\mathcal{B}]^{op},\mathcal{W}]$ containing the representable functors as a strongly cogenerating subcategory and that $\mathcal{B}^*$ is the largest full reflective subcategory of $[\mathcal{B}^{op},\mathcal{W}]$ containing the representable functors as a strongly cogenerating subcategory.

Proposition 7.1. The full embedding

$$[1,E] : [\mathcal{A},\mathcal{B}] \to [\mathcal{A},\mathcal{B}^*]$$

is dense and strongly cogenerating.

Proof. Because E is a full embedding, so is $[1,E]$. The functor $[1,E]$ is strongly cogenerating because the class $\{EB;\ B \in \mathcal{B}\}$ strongly cogenerates $\mathcal{B}^*$, hence the class $\{ \mathcal{A}(-,A),EB] ;\ A \in \mathcal{A},\ B \in \mathcal{B}\}$ of $\mathcal{V}$-cotensor powers in $[\mathcal{A},\mathcal{B}^*]$ strongly cogenerates $[\mathcal{A},\mathcal{B}^*]$. Moreover, because $\mathcal{B}^*$ is $\mathcal{W}$-cocomplete, so is $[\mathcal{A},\mathcal{B}^*]$. This yields the adjunction:

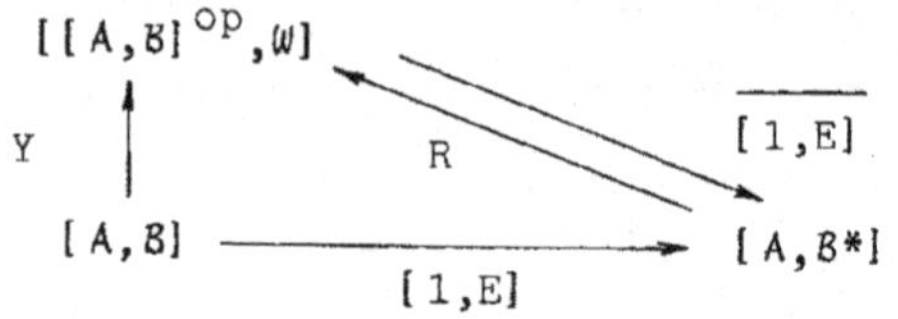

where $\overline{[1,E]}$ is the left Kan extension of $[1,E]$ along the Yoneda embedding Y, and R is its right adjoint. The counit of this adjunction has the canonical components:

$$\int^F [\mathcal{A},\mathcal{B}^*](EF,G)\cdot EFA \to GA.$$

It follows from the representation theorem (on taking $G = EF$ and evaluating at the identity morphism of EF) that this transformation coincides with the following composite of isomorphisms:

$$\int^F [\mathcal{A},\mathcal{B}^*](EF,G)\cdot EFA \cong (\int^F [\mathcal{A},\mathcal{B}^*](EF,G))\cdot(\int^B \mathcal{B}(B,FA)\cdot EB)$$

by the representation theorem,

$$\cong \int^B (\int^F [\mathcal{A},\mathcal{B}^*](EF,G) \otimes \mathcal{B}(B,FA)))\cdot EB$$

on interchanging colimits,

$$\cong \int^B (\int^F ([\mathcal{A},\mathcal{B}^*](EF,G) \otimes [\mathcal{A},\mathcal{B}]\ (N(AB),F)))\cdot EB$$

by definition of N,

$$\cong \int^B ([\mathcal{A},\mathcal{B}^*](E(N(AB)),G)\cdot EB)$$

by the representation theorem,

$$\cong \int^B ([\mathcal{A},\mathcal{B}^*](\mathcal{A}(A,-)\cdot EB,G)\cdot EB$$

because E preserves tensoring,

$$\cong \int^B \mathcal{B}^*(EB,GA)\cdot EB$$

by the representation theorem,

$$\cong GA \text{ because } E \text{ is dense.}$$

Thus $\overline{[1,E]} \dashv R$ is a full reflective embedding hence the embedding $[1,E]$ is dense.

<u>Proposition 7.2</u>. $[\mathcal{A},\mathcal{B}^*]$ is a full reflective subcategory of $[\mathcal{A},\mathcal{B}]^*$.

<u>Proof</u>. This follows from Proposition 7.1 and the fact that $[\mathcal{A},\mathcal{B}]^*$ contains every dense and strongly cogenerated completion of $[\mathcal{A},\mathcal{B}]$ as a full reflective subcategory.

<u>Proposition 7.3</u>. $[\mathcal{A},\mathcal{B}^*]$ is equivalent to $[\mathcal{A},\mathcal{B}]^*$.

<u>Proof</u>. Because N is dense (Lemma 5.1), both categories are fully embedded in $[\mathcal{A}\otimes\mathcal{B}^{op},\mathcal{W}]$ by the restriction functor

$$[N,1] : [[\mathcal{A},\mathcal{B}]^{op},\mathcal{W}] \to [\mathcal{A}\otimes\mathcal{B}^{op},\mathcal{W}] .$$

Consider the following diagram:

$$\begin{array}{ccccc}
[\mathcal{A},[\mathcal{B}^{op},\mathcal{W}]] \cong [\mathcal{A}\otimes\mathcal{B}^{op},\mathcal{W}] & \underset{[N,1]}{\overset{\bar{N}}{\rightleftarrows}} & [[\mathcal{A},\mathcal{B}]^{op},\mathcal{W}] & & \\
\uparrow Y & & \uparrow Y & \overset{R}{\nwarrow}\ \searrow \overline{[1,E]} & \\
\mathcal{A}^{op}\otimes\mathcal{B} & \xrightarrow[N]{} & [\mathcal{A},\mathcal{B}] & \xrightarrow[{[1,E]}]{} & [\mathcal{A},\mathcal{B}^*] ,
\end{array}$$

where $\bar{N} \dashv [N,1]$ and $[N,1]R$ is a full embedding whose value at $G \in [\mathcal{A},\mathcal{B}^*]$ is given by:

$$R(G)N(AB) = [\mathcal{A},\mathcal{B}^*]([1,E]N(AB),G)$$

by the definition of R,

$$= \int_X \mathcal{B}^*(\mathcal{A}(A,X)\cdot EB, GX)$$

by the definition of N,

$$\cong \mathcal{B}^*(EB, GA)$$

by the representation theorem applied to $X \in \mathcal{A}$.

Thus $[N,1]R$ is isomorphic to the embedding

$$[\mathcal{A},\mathcal{B}^*] \to [\mathcal{A},[\mathcal{B}^{op},\mathcal{W}]]$$

induced by the inclusion $\mathcal{B}^* \subset [\mathcal{B}^{op},\mathcal{W}]$. Recall from §2 that this inclusion is $[\mathcal{B}^{op},\mathcal{W}]_\Sigma \subset [\mathcal{B}^{op},\mathcal{W}]$ where

$$\Sigma = \{\beta \in [\mathcal{B}^{op},\mathcal{W}]^m;\ \int_B [\beta_B, \mathcal{B}(B,C)] \text{ is an isomorphism for all } C \in \mathcal{B}\}.$$

Also $[\mathcal{A},\mathcal{B}]^* = [\mathcal{A}\otimes\mathcal{B}^{op},\mathcal{W}]_Z$ where

$$Z = \{\alpha \in [\mathcal{A}\otimes\mathcal{B}^{op},\mathcal{W}]^m;\ \int_{XB} [\alpha_{XB}, \mathcal{B}(B,FX)] \text{ is an isomorphism for all } F \in [\mathcal{A},\mathcal{B}]\}.$$

Thus is remains to verify that HA is orthogonal to Σ for all $A \in \mathcal{A}$ if H is orthogonal to Z. But $\int_B [\beta_B, H(AB)]$ is an isomorphism for all $A \in \mathcal{A}$ if and only if $\int_{XB} [\mathcal{A}(A,X)\otimes\beta_B, H(XB)]$ is an isomorphism for all $A \in \mathcal{A}$ by the representation theorem applied to $X \in \mathcal{A}$. Thus $\beta \in \Sigma$ implies $\mathcal{A}(A,-)\otimes\beta \in Z$ for all $A \in \mathcal{A}$. This completes the proof.

<u>Corollary 7.4</u>. If $\mathcal{A}$ has a $\mathcal{W}$-promonoidal structure and $\mathcal{B}$ is monoidal biclosed relative to $\mathcal{V}$ then $[\mathcal{A},\mathcal{B}]^*$ is monoidal biclosed.

<u>Proof</u>. If $\mathcal{B}$ is monoidal biclosed then so is $\mathcal{B}^*$ by the application of Proposition 3.1 with $N = 1: \mathcal{B} \to \mathcal{B}$. Thus $[\mathcal{A},\mathcal{B}^*]$ has the convolution monoidal biclosed structure (of §5). The result now follows from the equivalence of $[\mathcal{A},\mathcal{B}]^*$ to $[\mathcal{A},\mathcal{B}^*]$.

<u>Remark</u>. It follows from the above construction that the monoidal biclosed structure produced on $[\mathcal{A},\mathcal{B}]^*$ by the equivalence of $[\mathcal{A},\mathcal{B}]^*$ to $[\mathcal{A},\mathcal{B}^*]$ makes the inclusion $[\mathcal{A},\mathcal{B}]^* \subset [\mathcal{A}\otimes\mathcal{B}^{op},\mathcal{W}]$ into a monoidal reflection. Thus the representability hypothesis of Proposition 5.2 is

unnecessary in order that $[A,B]^*$ alone should be monoidal biclosed.

§8. DENSENESS PRESENTATIONS

For a given dense functor $N: A^{op} \to C$, with A promonoidal, the monoidal (biclosed) structure:

$$C \otimes D = \int^{AB} (C(NA,C) \otimes C(NB,D)) \cdot Q(AB) \qquad (3.3)$$

obtained in §3 is derived from the coend presentation:

$$C \cong \int^{A} C(NA,C) \cdot NA.$$

In applications, such as in §4 on monoidal monads, an alternative presentation of denseness may be more useful and results in a different colimit expansion of the tensor product (3.3). For this purpose we shall formulate the general concept of a <u>denseness presentation</u> and describe some basic examples.

Given a functor $N: A \to C$, an N-presentation of C will consist of:

(a) an <u>index</u> functor $J: K \to A$

(b) a <u>coefficient</u> functor $K: K^{op} \otimes |C| \to V$

(c) a <u>structure</u> coend:

$$\xi_C: \int^{k} K(k,C) \cdot NJk \cong C$$

for each $C \in |C|$, such that the induced morphism:

$$\int^{k} K(k,C) \otimes A(A,Jk) \to C(NA,C) \qquad (8.1)$$

is an isomorphism for all $A \in A$ and $C \in C$. Here $|C|$ denotes the discrete V-category on C.

<u>Remark</u>. If A is small with respect to V and N is a full embedding then (8.1) is an isomorphism if and only if the representation functor:

$$C \to [A^{op}, V], C \mapsto C(N-,C),$$

preserves the structure coend for each $C \in C$.

<u>Proposition 8.1</u> If C has an N-presentation then N is dense and the structure coend induces an isomorphism:

$$\int_A [C(NA,C), GA] \cong \int_k [K(k,C), GJk] \qquad (8.2)$$

for each $C \in \mathcal{C}$ and $G\colon A^{op} \to \mathcal{V}$.

Proof. For each functor $G\colon A^{op} \to \mathcal{V}$ and object $C \in \mathcal{C}$ we have:

$$\int_A [C(NA,C),GA] \cong \int_A [\int^k K(k,C) \otimes A(A,Jk),GA]$$

by hypothesis on (8.1),

$$\cong \int_k [K(k,C), \int_A [A(A,Jk),GA]]$$

on interchanging limits,

$$\cong \int_k [K(k,C),GJk]$$

by the representation theorem.

This establishes the isomorphism (8.2). On taking $G = C(N-,D)$ in this isomorphism, we obtain:

$$\int_A [C(NA,C),C(NA,D)] \cong \int_k [K(k,C), C(NJk,D)]$$

$$\cong C(\int^k K(k,C) \cdot NJk,D)$$

because $C(-,D)$ preserves limits,

$$\cong C(C,D) \text{ by hypothesis on } \xi,$$

thus N is dense.

Remarks. (a) Neither (8.1) nor (8.2) is required to be natural in $C \in \mathcal{C}$. However, a degree of naturality may occur in practice.

(b) For N to be dense (without Condition (8.2)) it suffices that the morphism (8.1) should be an epimorphism in $\mathcal{V}$ for all $A \in \mathcal{A}$ and $C \in \mathcal{C}$.

Example (a); If $N\colon \mathcal{A} \to \mathcal{C}$ is dense then the coend presentation:

$$\int^A C(NA,C) \cdot NA \cong C$$

with the "evaluation" structure is natural in $C \in \mathcal{C}$.

Example (b); If $N\colon \mathcal{A} \to \mathcal{C}$ is dense then $\mathcal{C}$ has a comma-category presentation:

$$\int^k C(Qk,C) \cdot NPk \cong C$$

which is natural in $C \in \mathcal{C}$. Here P and Q denote the projections:

$$A \xleftarrow{P} N/C \xrightarrow{Q} C .$$

The isomorphisms (8.1) and (8.2) follow from the isomorphism:

$$\int_A B(FNA,GA) \cong \int_k B(FQk,GPk)$$

which holds for all functors F: $C^{op} \to B$ and G: $A^{op} \to B$.

Example (c); Suppose T = (T,μ,η) is a monad on C with Kleisli category A and algebra category C(T). Let N: A → C(T) be the comparison functor. Let D be the free V-category on the category:

$$\{2 \overset{m}{\underset{s}{\rightrightarrows}} 1\},$$

and let K = D⊗C(T) with index functor J: K → C(T) defined by:

$$J(m,(C,\xi)) = \mu_C : T^2C \to TC$$

$$J(s,(C,\xi)) = T\xi : T^2C \to TC$$

for each T-algebra (C,ξ). The coefficient functor $K: K^{op} \otimes C(T) \to V$ is then defined by:

$$K(d,(C,\xi),(D,\xi)) = I \otimes C(T)(C,D) \cong C(T)(C,D).$$

Then we have:

$$\int^{dD} K((d,D),C) \cdot NJ(d,D) \cong \int^{dD} C(T)(C,D) \cdot NJ(d,D)$$

$$\cong \int^{d} NJ(d,C)$$

by the representation theorem applied to D ∈ C(T),

$$= \operatorname{colim} NJ(d,C)$$

$$\cong C$$

because

$$T^2C \overset{\mu}{\underset{T\xi}{\rightrightarrows}} TC \xrightarrow{\xi} C$$

is a coequaliser diagram in C(T). This isomorphism is natural in C ∈ C(T). The coequaliser is preserved by each functor C(T)(NA,-): C(T) → V hence (8.1) is an isomorphism.

On applying the underlying-set functor V: V → S, Proposition 8.1 becomes:

Proposition 8.2. (the representation theorem for monads). The comparison functor $N\colon \mathcal{A} \to \mathcal{C}(T)$ is dense and, for each algebra $(C,\xi) \in \mathcal{C}(T)$, the natural transformations from $\mathcal{C}(T)(N-,C)$ to a prealgebra $G\colon \mathcal{A}^{op} \to \mathcal{V}$ correspond to the elements in the equaliser of:

$$VGC \underset{VGT\xi}{\overset{VG\mu}{\rightrightarrows}} VGTC,$$

where μ and $T\xi$ are regarded as morphisms in $\mathcal{A}$.

In conclusion we note two examples of presentations which are not fully natural in $C \in \mathcal{C}$. But first we note the following fact. Suppose the data (J,K,ξ) are given such that (8.1) is an isomorphism but ξ is not known to be an isomorphism.

Proposition 8.3. If (J,K,ξ) are such that (8.1) is an isomorphism then ξ is an isomorphism if and only if N strongly generates $\mathcal{C}$ and

$$\mathcal{C}(NA, \int^k K(k,C)\cdot NJk) \cong \int^k K(k,C) \otimes \mathcal{A}(A,Jk)$$

for all $A \in \mathcal{A}$ and $C \in \mathcal{C}$.

Example (d); Suppose $\mathcal{V} = \mathcal{S}$ and $\mathcal{C}$ is a category with canonical $\mathcal{E}$-$\mathcal{M}$ factorisations for a proper $\mathcal{E}$-$\mathcal{M}$ system on $\mathcal{C}$ (in the sense of [12]). Suppose that $\mathcal{A}$ is a full subcategory of $\mathcal{C}$ which is closed under $\mathcal{E}$-images. Let $N\colon \mathcal{A} \subset \mathcal{C}$ denote the inclusion and let $\mathcal{K} = \mathcal{A} \cap \mathcal{M}$ with

$$K\colon \mathcal{K}^{op} \times \mathcal{M} \to \mathcal{S}$$

defined by $K(A,C) = \mathcal{M}(A,C)$. Then

$$\int^A \mathcal{M}(A,C) \times \mathcal{C}(B,A) \cong \mathcal{C}(B,C) \tag{8.3}$$

naturally in $C \in \mathcal{M}$ for each $B \in \mathcal{C}$. Thus, if each object $C \in \mathcal{C}$ is the "union" of its subobjects in $\mathcal{K}$ in the sense that

$$\int^A \mathcal{M}(A,C)\cdot A \cong C, \tag{8.4}$$

then $N\colon \mathcal{A} \subset \mathcal{C}$ is dense and

$$\int_{A\in\mathcal{A}} [\,\mathcal{C}(A,C), GA] \cong \int_{A\in\mathcal{M}} [\,\mathcal{M}(A,C), GA]$$

for each functor $G: A^{op} \to S$. It follows from Proposition 8.3 that the morphism (8.4) is an isomorphism if A strongly generates C and each functor $C(A,-): C \to S$ preserves the coend (8.4). For the case $C = S$, any category $A \subset S$ which is closed under surjective images in S satisfies these requirements.

Example (e); discrete presentations. We suppose that V is sufficiently cocomplete. It is always the case that if $N: A \to C$ is a functor and, for each $C \in |C|$, there exists a free V-category K_C and a functor

$$P = P_C: K_C \to A,$$

together with a colimit cone

$$\xi_C: \operatorname{colim}_k NP(k) \cong C,$$

such that:

$$\operatorname{colim}_k A(A,Pk) \cong C(NA,C)$$

for all $A \in A$, then (P,K,ξ) provides a presentation of C. This is achieved by putting:

$$K = \sum_{|C|} K_C \text{ and } J = (P_C): K \to A$$

and defining $K: K^{op} \otimes |C| \to V$ by:

$$\begin{cases} K(k,C) = I \text{ if } k \in K_C, \\ K(k,C) = 0 \text{ if } k \notin K_C. \end{cases}$$

For example, if $V = S$, then the "discrete" comma-category presentation of a dense functor is given by $K_C = N/C$ for each $C \in C$. Furthermore, given any functor $N: A \to C$, the condition

$$\operatorname{colim}_k A(A,Pk) \cong C(NA,C)$$

is satisfied for all $A \in A$ and $C \in C$ by the representation theorem. Thus, by Proposition 8.3, N is dense if and only if N is strongly generating and

$$C(NA, \operatorname{colim}_k NPk) \cong \operatorname{colim}_k A(A,Pk)$$

for all $A \in A$ and $C \in C$.

In particular, let C be an S-based category and let α be a regular cardinal number. Following the terminology of Gabriel-Ulmer [13], an object $A \in C$ is called α-presentable if $C(A,-): C \to S$ preserves α-filtered colimits. Let C_α denote the full subcategory of C determined by the α-presentable objects. Then, by the preceding observation, a composite inclusion $A \subset C_\alpha \subset C$ is dense if A strongly generates C and A is α-filtered. Thus the observation reproduces a result of Gabriel-Ulmer [13] Theorem 7.4; in this connection, recall that a cocomplete category C is said to be locally α-presentable if it contains a small strongly generating set of α-presentable objects.

REFERENCES

[1] Appelgate, H. and Tierney, M., Categories with models, Seminar on Triples and Categorical Homology Theory (Springer Lecture Notes, Vol. 80, 1969), 156-244.

[2] Bastiani, A. and Ehresmann, C., Categories of sketched structures Cahiers de topologie et géométrie différentielle, Vol. XIII, 1972, 105-214.

[3] Benabou, J., Introduction to bicategories, Reports of the Midwest Category Seminar I (Springer Lecture Notes, Vol. 47, 1967), 1-77.

[4] Day, B., On closed categories of functors, Reports of the Midwest Category Seminar IV (Springer Lecture Notes, Vol. 137, 1970) 1-38.

[5] Day, B., Construction of biclosed categories, Ph.D. thesis, University of New South Wales, 1970.

[6] Day, B., Construction of biclosed categories, Abstracts of Australn. Ph.D. theses, Bull. Austral. Math. Soc., Vol. 5, 1971, 139-140.

[7] Day, B., A reflection theorem for closed categories, J. of Pure and Applied Algebra, Vol. 2, 1972, 1-11

[8] Day, B., Note on monoidal localisation, Bull. Austral. Math. Soc., Vol. 8, 1973, 1-16.

[9] Day, B., On adjoint-functor factorisation, (submitted).

[10] Dubuc, E.J., Kan extensions in enriched category theory, Springer Lecture Notes, Vol. 145, 1970.

[11] Eilenberg, S. and Kelly, G.M., Closed categories, Proc. Conference on Categorical Algebra (La Jolla, 1965), Springer-Verlag 1966, 421-562.

[12] Freyd, P. and Kelly, G.M., Categories of continuous functors I, J. Pure and Applied Algebra, Vol. 2, 1972, 169-191.

[13] Gabriel, P. and Ulmer, F., Lokal präsentierbare Kategorien, Springer Lecture Notes, Vol. 221, 1971.

[14] Gabriel, P. and Zisman, M., Calculus of fractions and homotopy theory, Ergebnisse der Mathematik und ihrer Grenzgebiete, Band 35, Springer-Verlag, 1967.

[15] Kock, A., Monads on symmetric monoidal closed categories, Archiv der Mathematik, Vol. XXI, 1970, 1-10.

[16] Kock, A., Closed categories generated by commutative monads, J. Austral. Math. Soc., Vol. XII, 1971, 405-424.

[17] Kock, A., Bilinearity and cartesian closed monads, Math.Scand., 29, 1971, 161-174.

[18] Linton, F.E.J., Coequalisers in categories of algebras, Seminar on Triples and Categorical Homology Theory (Springer Lecture Notes, Vol. 80, 1969), 75-90.

[19] Meisen, J., Relations in categories, A.M.S. Notices, No. 695-A3, Vol. 19, 1972, A549-A550.

[20] Ulmer, F., Representable functors with values in arbitrary categories, J. Algebra, Vol. 8, 1968, 96-129.

[21] Wolff, H., V-localisations and V-fractional categories, to appear.

[22] Wolff, H., V-localisations and V-monads, J. Algebra, Vol. 24, 1973, 405-438.

AN EMBEDDING THEOREM FOR CLOSED CATEGORIES

by

Brian Day

0. Introduction.

The concept of a monoidal closed category V has been used in categorical algebra as a generalisation of the category S of small sets and set maps. For algebra relative to V the monoidal structure serves to extend the concept of colimit and the closed structure to extend that of limit.

The less complete structures of monoidal category and closed category occur by themselves and have lead to promonoidal categories as a common generalisation. In this note we discuss the possibility of embedding a promonoidal category into a monoidal closed category so as to preserve as much of the original limit structure as possible.

The "several objects" form of the embedding is discussed in § 4 and provides an embedding for bicategories.

For the basic notation and terminology we refer to [2], [4], and [6].

The sections are:

1. Promonoidal categories; review.
2. Embedding of promonoidal categories.
3. Special cases.
4. The embedding for bicategories.

1. Promonoidal categories; review.

Throughout the article we suppose that all categorical algebra is relative to a fixed symmetric monoidal closed "ground" category V. Furthermore, V is assumed to have all small limits and colimits.

A promonoidal category (over V) consists of a category A together with functors:

promultiplication P: $A^{op} \otimes A^{op} \otimes A \to V$

V-valued-hom Hom : $A^{op} \otimes A \to V$

proidentity J : $A \to V$,

and natural identity and associativity isomorphisms:

λ : $JX \circ P(XA-) \cong A(A-)$

ρ : $JX \circ P(AX-) \cong A(A-)$

α : $P(AX-) \circ P(BCX) \cong P(ABX) \circ P(XC-)$

satisfying coherence axioms (PC1,2 of [4]) where "$\circ$" denotes profunctor composition.

A symmetry of the promonoidal structure consists of a natural isomorphism:

σ : $P(AB-) \cong P(BA-)$

satisfying the additional axioms PC3($\sigma^2 = 1$) and PC4 of [4].

Examples of promonoidal categories are monoidal categories (where $P(ABC) = A(A \otimes B, C)$) and biclosed categories (where $P(ABC) = A(B, A\backslash C) \cong A(A, C/B)$).

By weakening the above data so that the associativity α is not required to be an isomorphism the (not necessarily monoidal) closed categories of Eilenberg-Kelly [6] appear as "promonoidal" structures. In general, a promonoidal category for which α is not necessarily an isomorphism will be called non-associative.

The concept of a monoidal category with "several objects" is due to Benabou ([2] for $V = S$); such a structure is called a bicategory and will be denoted by $A = \{A_{xy};\ x,y \in \mathrm{Obj}\ A\}$. In a natural way this leads to "probicategories". From the conceptual point of view a probicategory structure on $\mathcal{A}$ is essentially a closed bicategory structure on the collection $\{[\mathcal{A}_{xy}, \mathcal{V}];\ x, y \in \mathrm{Obj}\ \mathcal{A}\}$ of all $\mathcal{V}$-valued functors on $\mathcal{A}$.

2. Embedding of promonoidal categories.

First consider a given <u>small</u> promonoidal category $\mathcal{A} = (\mathcal{A}, P, J, \cdots)$ over $\mathcal{V}$. Let $F = [\mathcal{A}, \mathcal{V}]$ be the convolution of $\mathcal{A}$ with $\mathcal{V}$. That is, F is monoidal biclosed with respect to the following operations :

$$F * G = \int^{AA'} FA \otimes GA' \otimes P(AA'-)$$

$$F \setminus G = \int_{AA'} [P(A-A') \otimes FA, GA']$$

$$G / F = \int_{AA'} [P(-AA') \otimes FA, GA']$$

The detailed description of this structure is given in [4] §3 .

Let $\mathcal{B} \subset F$ be a small monoidal subcategory of F containing $\mathcal{A}^{op}$ and let $E : \mathcal{A} \to [\mathcal{B}, \mathcal{V}]$ denote the evaluation embedding. If $[\mathcal{B}, \mathcal{V}]$ is given the convolution structure then E is a promonoidal embedding; that is, E preserves the promonoidal structure of A to within isomorphism.

<u>Proposition 2.1.</u> <u>The embedding E preserves $\lim Kk$ in $\mathcal{A}$ if and only if $B(\lim Kk) \cong \lim BKk$ for all $B \in \mathcal{B}$.</u>

The next step is to reduce the codomain of E so that suitable colimits are preserved by E. For this purpose we consider the monoidal closure of a completion of $\mathcal{B}^{op} \subset [\mathcal{B}, \mathcal{V}]$ (in the sense of Lambek - see [1]) .

Let $\mathcal{C}$ be a small-cocomplete category and let $M : \mathcal{B}^{op} \to \mathcal{C}$ be a full colimit-preserving embedding which strongly cogenerates $\mathcal{C}$.

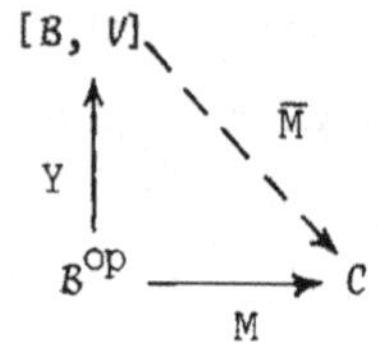

The left Kan extension of M along the Yoneda embedding Y is given by the coend formula :

$$\bar{M}F = \int^{B} FB \cdot MB$$

Let $Z = \{s \in [B,V]^{2};\ \bar{M}s$ is an isomorphism$\}$. Then the full subcategory $[B,V]_Z$ of $[B,V]$ determined by the functors orthogonal to Z is a completion of B^{op} ; $[B,V]_Z = \{F \in [B,V]\ ;\ \int_B [s_B, FB]$ is an isomorphism for all $s \in Z\}$.

An important (in fact, the extremal) case is obtained from taking $M = Y^{op} : B^{op} \to [B^{op}, V]^{op}$. Here $\bar{M}$ is the Isbell conjugation functor : $\bar{M}F = \int_B [FB, B(-B)]$ and $[B,V]_Z$ is the largest reflective subcategory of $[B,V]$ to contain the representable functors as a strongly cogenerating set.

Let $Z^{o} = \{s \in Z;\ F * s \in Z$ and $s * F \in Z$ for all $F \in [B,V]\}$ be the monoidal interior of Z (as in [5]). Then the reflective embedding

$$[B,V]_{Z^o} \subset [B,V]$$

is monoidal or, equivalently, $[B,V]_{Z^o}$ is closed under exponentiation in $[B,V]$. This results in a promonoidal embedding:

$$E : A \to B^{op} \to [B,V]_{Z^o}$$

Footnote: Given a functor $K : K \to D$, we use $\lim Kk$ to denote $\lim_k Kk$, the inverse limit in D of K over $k \in K$. Dually, $\operatorname{colim} Kk = \operatorname{colim}_k Kk$.

<u>Proposition 2.2.</u> <u>The embedding $\mathcal{A} \subset [\mathcal{B},\mathcal{V}]_{Z^o}$ preserves colim Kk in $\mathcal{A}$ if and only if</u>

$$\underline{B * \lim \mathcal{A}(Kk,-) \cong \lim(B * \mathcal{A}(Kk,-))}$$

$$\underline{\lim \mathcal{A}(Kk,-) * B \cong \lim(\mathcal{A}(Kk,-) * B)}$$

<u>in $\mathcal{F} = [\mathcal{A},\mathcal{V}]$ for all $B \in \mathcal{B}$</u>

<u>Proof</u>. The embedding E preserves colim Kk if and only if the comparison morphism

$$s : \operatorname{colim} \mathcal{B}(\mathcal{A}(Kk,-),-) \to \mathcal{B}(\mathcal{A}(\operatorname{colim} Kk,-),-)$$

lies in Z^o. By the representation theorem it suffices to check if $\mathcal{B}(B-) * s \in Z$ and $s * \mathcal{B}(B-) \in Z$ for all $B \in \mathcal{B}$. Equivalence of these conditions to the stated ones follows from the strongly cogenerating property of M. For example, define $\bar{s}$ by means of the diagram :

$$\begin{array}{ccc}
\mathcal{B}(B-) * \operatorname{colim} \mathcal{B}(\mathcal{A}(Kk,-),-) & \xrightarrow{\mathcal{B}(B-) * s} & \mathcal{B}(B-) * \mathcal{B}(\mathcal{A}(\operatorname{colim} Kk,-),-) \\
\wr\| & & \wr\| \\
\operatorname{colim} \mathcal{B}(B * \mathcal{A}(Kk,-),-) & \xrightarrow{\bar{s}} & \mathcal{B}(B * \mathcal{A}(\operatorname{colim} Kk,-),-) .
\end{array}$$

Then $M\bar{s}$ is an isomorphism iff

$$\operatorname{colim} M(B * \mathcal{A}(Kk,-)) \cong M(B * \mathcal{A}(\operatorname{colim} Kk,-)) .$$

Because M strongly cogenerates $\mathcal{C}$, this is equivalent to:

$\lim(B * \mathcal{A}(Kk,-)) \cong B * \mathcal{A}(\operatorname{colim} Kk,-)$ and the result follows .

<u>3. Special cases.</u>

In this section we suppose that $\mathcal{B}$ is the smallest monoidal subcategory of $\mathcal{F} = [\mathcal{A},\mathcal{V}]$ containing the representable functors. Explicitly, the objects of $\mathcal{B}$ are those functors in $\mathcal{F}$ which are isomorphic either to J or to a finite "path" $\mathcal{A}(A_1-) * \cdots * \mathcal{A}(A_n-)$ of representable functors.

Example 3.1. Let $A = (A,\otimes,I,\ldots)$ be a monoidal category. Then the monoidal structure on $[A,V]$ extends that on A^{op} hence $B = A^{op}$. By Proposition 2.1, the embedding $E : A \subset [B,V]_{Z^o}$ preserves all limits. By Proposition 2.2, E preserves those colimits which are preserved by all endofunctors $-\otimes A$ and $A\otimes-$ for $A \in A$. In particular, E preserves all colimits if the monoidal structure on A is biclosed.

Example 3.2. Suppose $A = (A,[-,-],I,\ldots)$ is a closed category (Eilenberg-Kelly [6]) with internal-hom functor $[-,-] : A^{op}\otimes A \to A$. We say that A is associative if the promonoidal structure $P(A_1A_2A_3) = A(A_1[A_2A_3])$ is associative; that is, if the canonical morphism from

$$A(A_1-)*(A(A_2-)*A(A_3-)) \cong \int^D A(A_1[D-]) \otimes A(A_2[A_3D])$$

to

$$(A(A_1-)*A(A_2-)) * A(A_3-) \cong A(A_1[A_2[A_3-]])$$

is an isomorphism. In particular, if the promonoidal structure of A is symmetric then it is associative.

If A is an associative closed category then each object of B is isomorphic to one of the form $A(A_1[A_2[A_3\ldots[A_n-]\ldots])$. Thus, by Proposition 2.1, E preserves the limits which are preserved by each of the functors $[A_1[A_2[A_3\ldots[A_n-]\ldots] : A \to A$. By Proposition 2.2, E preserves those colimits preserved by each of the functors $[A_1[A_2[A_3\ldots[-A]\ldots] : A^{op} \to A$.

Finally, E maps the internal-hom functor of A to one of the internal-hom functors of $[B,V]$, even if A is non-associative. For all $A,A' \in A$ the convolution structure on $[B,V]$ gives:

$$\begin{aligned}(EA'/EA)B &\cong \int_{B'} [B'(A)\,,\,(B*B')A'] \\ &\cong \int_{B'} [B(A(A-),B')\,,\,(B*B')A']\end{aligned}$$

by the representation theorem,

$$\cong (B* A(A-))A'$$

by the representation theorem applied to $B' \in B$,

$$= \int^{A''} B[A'',A'] \otimes A(AA'')$$

by definition of $*$,

$$\cong B[AA']$$

by the representation theorem applied to $A'' \in A$,

$$= E[AA'](B)$$

for all $B \in B$, as required.

Example 3.3. (V cartesian closed). For each small V- category A the functor category $[A,V]$ is cartesian closed. Thus A is a symmetric promonoidal category with $P(A_1 A_2 -) = A(A_1 -) \times A(A_2 -)$. In this case $E:A \to [B,V]_Z{}^o$ preserves all limits. By Proposition 2.2, E preserves colim Kk if and only if:

$$A(A_1-)\times\cdots\times A(A_n-) \times (\lim A(Kk,-))$$
$$\cong \lim (A(A_1-) \times\cdots\times A(A_n-) \times A(Kk,-))$$

for all $A_1,\ldots,A_n \in A$. It is easily seen that this condition is not always satisfied by considering the case where A has finite coproducts.

Note (3.4): The computations of §2 apply equally to the more general case where the promonoidal structure on A is assumed to be non-associative. It is necessary to consider such non-associative structures if we wish to study all closed categories (in the sense of Eilenberg-Kelly [6]). Note that the monoidal-interior operation is not available in the above form unless the monoidal structure is associative.

Note (3.5). The embedding $E:A\to[A,S],\hat{S}]$ (where $S \in \hat{S}$) is used in[7] to study coherence in closed categories. Here the authors describe the closed-functor structure of the embedding E in detail.

Note (3.6): If the category $\mathcal{B}$ is required to be biclosed in addition to being monoidal then we can choose $\mathcal{B}$ to consist of all the isomorphs in $\mathcal{F} = [A,\mathcal{V}]$ of:

$$\cap\{\mathcal{D}\ ;\ A^{op} \subset \mathcal{D} \subset \mathcal{F} \text{ with } \mathcal{D} \text{ closed under the monoidal biclosed structure of } \mathcal{F}\}.$$

That is, $\mathcal{B}$ is equivalent to the full subcategory of $\mathcal{F}$ consisting of all finite paths constructed from the operations $*$, $/$, and $\backslash$ acting on the representable functors in $\mathcal{F}$. Thus $\mathcal{B}$ is determined, up to category equivalence, by:

$$\bigcup_{n \in N} \{\mathcal{D}_n\ ;\ \mathcal{D}_{n+1} = \mathcal{D}'_n \text{ where } \mathcal{D}'_n \text{ consists of all finite } *\text{-products of internal-homs of pairs of objects from } \mathcal{D}_n\}\ .$$

Note that, as a consequence, the set of isomorphism classes of $\mathcal{B}$ is countable if obj$A$ is countable.

A more efficient embedding of this type is obtained by first forming $\mathcal{F}_\Sigma o$ where :

$$\Sigma = \{s \in \mathcal{F}^{2};\ \int_A [s_A\ ,\ A(-A)] \text{ is an isomorphism}\},$$

and then taking $\mathcal{B}$ to be the full subcategory of $\mathcal{F}$ determined by :

$$\cap\{\mathcal{D}\ ;\ A^{op} \subset \mathcal{D} \subset \mathcal{F}_\Sigma o \text{ with } \mathcal{D} \text{ closed under the monoidal biclosed structure of } \mathcal{F}_\Sigma o\}\ .$$

For this choice of $\mathcal{B}$ the embedding E preserves limKk if and only if it is preserved by each of the functors $A(A_1-)/A(A_2-)$ and $A(A_2-)\backslash A(A_1-)$ where $A_1\ ,\ A_2 \in A$.

4. The embedding for bicategories

The preceding results can be readily extended to the case where the structure of A has "several objects".

First we consider closed bicategory structures (over V) of the form $F_{xy} = [A_{xy},V]$ with $x,y,\dots \in \mathrm{Obj}F$. Such a structure consists of a bicategory $F = [A,V]$ whose composition V-functors

$$F_{yz} \otimes F_{xy} \xrightarrow{\ *\ } F_{xz}$$

have right adjoints in each variable separately. Because the Yoneda embedding $A^{op} \subset [A,V]$ is locally dense each such composition functor is determined (up to an isomorphism) as the extension of its value on the representable 1-cells of F. Thus the structure of $A = \{A_{xy}\}$ determines the extension. In particular, each bicategory A is such a probicategory and $\{[A_{xy},V]\}$ is its convolution with the V regarded as a one-object bicategory.

Given such a closed bicategory $F = \{[A_{xy},V];\ x,y \in \mathrm{Obj}\ F\}$ we construct a subbicategory $B = \{B_{xy}\}$ by taking the 1-cells of B_{xy} to be the finite paths (under $*$) from x to y of the representable and identity 1-cells of F. That is, B is the smallest bicategory containing A^{op}

On forming the convolution $\{[B_{xy},V];\ x,y \in \mathrm{Obj}\ F\}$ we locally (i.e. for each (x,y)) form the category of fractions with respect to the class $Z^o = \{Z^o_{xy}\}$, where Z_{xy} is the class of morphisms inverted by the conjugation functor $[B_{xy},V] \to [B^{op}_{xy},V]^{op}$ and $Z^o_{xy} = \{s \in Z_{xy};\ F{\bullet}s \in Z$ and $s{\bullet}G \in Z$ for all s-composable 1-cells $F,G \in B^{op}\}$. The class Z^o is monoidally closed by associativity of the bicategory composition on B. The resulting closed bicategory :

$$\{[B_{xy},V]_{Z^o_{xy}}\ ;\ x,y \in \mathrm{Obj}\ F\}$$

then locally has the properties analogous to those given in §2 for the one-object case.

References

[1] Appelgate, H. and Tierney, M., Iterated cotriples, Reports of the Midwest Category Seminar IV, Lecture Notes 137 (Springer 1970), 56-99.

[2] Bénabou, J., Introduction to bicategories, Reports of the Midwest Category Seminar I, Lecture Notes 47 (Springer 1967), 1-77 .

[3] Bénabou, J., Bicategories and distributors, Seminar Reports, Mathematics Institute, University of Louvain.

[4] Day, B.J., On closed categories of functors, Reports of the Midwest Category Seminar IV, Lecture Notes 137 (Springer 1970), 1-38.

[5] Day, B.J., Note on monoidal localisation, Bull. Austral. Math. Soc., Vol 8, 1973, 1-16.

[6] Eilenberg, S. and Kelly, G.M., Closed categories, Proc. Conf. on Cat. Alg. (La Jolla, 1965), Springer-Verlag 1966, 421-562.

[7] Laplaza, M., and Day, B.J., The coherence problem for closed categories.

[8] Lawvere, F.W., Metric spaces, generalised logic and closed categories, Lecture Notes, Instituto di Matematica, Università di Perugia .

LIMIT SPACES AND CLOSED SPAN CATEGORIES

by

Brian Day

0. Introduction.

The main aim of this note is to characterise the category of those quasi-topological spaces which occur as sheaves for the canonical Grothendieck topology on the category *Top* of all topological spaces and continuous maps. The general process is applicable to similar situations.

In §1 we recall some standard properties of closed span categories. These are finitely complete categories C such that C/C is cartesian closed for all $C \in C$. The relationship of such a structure to an adjoint pair of functors is discussed further in §2.

In §3 we describe the smallest closure of this type to contain *Top*. This particular closure is contained in the category of all limit spaces (as described, for instance, by Binz-Keller [3]) and it is necessarily larger than the minimal extension of *Top* defined by Antoine [1].

The notation and unexplained terminology used are standard (cf. Benabou [2], Lawvere [11]). Throughout the article, S denotes "the" cartesian closed category of small sets and set maps.

The author wishes to thank Professor G.M. Kelly for helpful discussions.

The sections are:

1. Closed span categories.

A finitely complete category C is called a closed span category (a cs-category) if C/C is cartesian closed for each object $C \in C$. A cs-functor is a functor $T: C \to B$ together with a cartesian monoidal structure on each of the induced functors $T_C: C/C \to B/TC$; $x \in C(XC) \mapsto Tx \in B(TX,TC)$.

To each finitely complete category C there is an associated bicategory SpanC ([2]§2.6). The objects are those of C while SpanC (AB) = $C/A{\times}B$, spans being composed by pullback. The bicategory is closed if the composition functors have right adjoints in each variable. It is well-known (see Proposition 4.1) that a finitely complete category C is a closed span category if and only if SpanC is closed as a bicategory.

A cs-functor induces a map of the associated span bicategories and if C is a cs-category then so is C/C because $(C/C)/f \cong C/B$ for all $f \in C(BC)$.

Proposition 1.1. If B is a cs-category and $R \dashv T: C \to B$ is a full reflective embedding for which R preserves pullbacks over all objects of the form TC then C is a cs-category.

This follows from the fact that each reflective embedding $R_C \dashv T_C : C/C \to B/TC$ is cartesian (by [6] Theorem 1.2).

The hypothesis is satisfied if C is a (left) exact retract of B ; that is, if R preserves finite limits. Suppose B is a cs-category and Z is an orthogonally closed class of morphisms in B (in the sense of [10]) for which the inclusion $B_Z \subset B$ of Z-orthogonal objects has a left adjoint R. Then, if Z is stable under pullback we can apply [6] Theorem 1.2 to each of the reflective embeddings $B_Z/RB \subset B/B$ and obtain:

<u>Proposition 1.2.</u> R is left exact if and only if Z is stable.

Common conditions for stability are given in Proposition 4.2.

<u>Examples.</u> An elementary topos of Lawvere-Tierney (a finitely complete cartesian closed category with a subobject classifier Ω) is a cs-category (see Lawvere [11]).

Similarly, let C be a cartesian closed category of topological spaces for which the inclusion $C \subset Top$ has a right adjoint W . Call $B \in C$ a <u>W-hausdorff</u> object if the diagonal $B \subset W(B \times B)$ is closed.

For each W-hausdorff object B , C/B is cartesian closed. To verify this, let Ω (indifferently) denote the coreflection in C of the two-point space $\{0,1\}$ <u>either</u> with $\{0\}$ open and $\{1\}$ not open <u>or</u> with the trivial topology. Then the resulting projections $p : B \times \Omega \to B$ strongly cogenerate C/B . For each $f \in C(CB)$ define $[f,p]$ as the equaliser of the pair:

$$\begin{array}{ccc}
[C\Omega] \times B & \xrightarrow{\text{projn.}} & [C\Omega] \\
{\scriptstyle 1\times s}\downarrow & & \uparrow{\scriptstyle \wedge} \\
[C\Omega] \times [B\Omega] & \xrightarrow[1 \times [f,1]]{} & [C\Omega] \times [C\Omega]
\end{array}$$

where $s : B \to [B\Omega]$ characterises the closed diagonal $B \subset W(B \times B)$ and $\wedge$ denotes intersection of closed subsets - that is;

$\wedge : [C\Omega] \times [C\Omega] \xrightarrow{\times} [C \times C, \Omega \times \Omega] \xrightarrow{[\Delta,\wedge]} [C\Omega]$. The exponentiation of p then extends to all of C/B (see [6] Theorem 3.4).

Thus the bicategory of those spans in C which have a W-hausdorff codomain is closed. In particular :

<u>Proposition 1.3.</u> The W-hausdorff objects of $C \subset Top$ form a cs-category.

2. Adjoint-functor factorisation.

If $\mathcal{B}$ is a suitably complete closed span category and $(\varepsilon,\eta) : S \dashv T : \mathcal{C} \to \mathcal{B}$ is an adjunction over $\mathcal{B}$ with η a monomorphism then the monad $(TS,\eta,T\varepsilon S)$ factors through an associated <u>exact</u> idempotent monad on $\mathcal{B}$ (analogous to the factorisation of Fakir [9]). We suppose that $\mathcal{B}$ is a cs-category with a proper factorisation system E-M (in the sense of Freyd-Kelly [10]). Further, E is assumed to be stable and $\mathcal{B}$ to be M-complete (in the sense of Day [8]).

Let $(\varepsilon,\eta) : S \dashv T : \mathcal{C} \to \mathcal{B}$ be an adjunction and let $\mathcal{B}' \subset \mathcal{B}$ consist of those objects $B \in \mathcal{B}$ such that $\eta_B \in M$. Then $\mathcal{B}'$ is reflective in $\mathcal{B}$.

<u>Proposition 2.1.</u> If $\mathcal{B}$ is a cs-category then $\mathcal{B}'$ is a cs-category if the canonical morphism:

$$\begin{array}{ccc} TS(x \wedge y) & \longrightarrow & TSx \wedge TSy \\ \downarrow & \searrow & \downarrow \\ TS(X \times Y) & \longrightarrow & TSX \times TSY \end{array}$$

lies in M for all $B \in \mathcal{B}'$ and $x \in \mathcal{B}(XB)$, $y \in \mathcal{B}(YB)$.

The proof is immediate from Proposition 1.1. Note that the hypothesis is satisfied if TS is left exact but this does <u>not</u> imply that the reflection $\mathcal{B} \to \mathcal{B}'$ is left exact*.

If $(\varepsilon,\eta) : S \dashv T : \mathcal{C} \to \mathcal{B}$ is now assumed to have $\eta \in M$ and $\Sigma = \{f ;\ Sf$ is an isomorphism$\}$ then, for each $Z \subset \Sigma$, the inclusion $\mathcal{B}_Z \subset \mathcal{B}$ has a left adjoint ([8] Corollary 2.3). Moreover, if $\overline{Z}$ denotes the orthogonal closure of Z then $\overline{Z}$ is stable if and only if $\overline{Z}_C$ is closed under cartesian product in $\mathcal{B}/TC$ for all $C \in \mathcal{C}$. Thus, from Proposition 1.2, we obtain:

<u>Proposition 2.2.</u> $\mathcal{B}_Z$ is a left exact retract of $\mathcal{B}$ if and only if $\overline{Z}_C$ is closed under cartesian product in $\mathcal{B}/TC$ for all $C \in \mathcal{C}$.

---

* In the particular case where S is an associated-sheaf functor between topoi $\mathcal{B}$ and $\mathcal{C}$ the category $\mathcal{B}'$ is just the category of separated presheaves in $\mathcal{B}$.

Finally, the class of stable subclasses of Σ is closed under union, thus each $Z \subset \Sigma$ has a <u>stable interior</u> Z^* :

$$Z^* = \bigcup\{V \subset Z \; ; \; V \text{ stable}\} .$$

Thus each left retract of $\mathcal{B}$ through which S factors has a left exact "closure".

<u>Example 2.3</u>. Let A be a category with a generator I and let $M = A(I-) : A \to S$. Then the left Kan extension of M along the Yoneda embedding $Y : A \to [A^{op}, S]$ is evaluation-at-I with right adjoint T ; $TX = S(M-,X) : A^{op} \to S$.

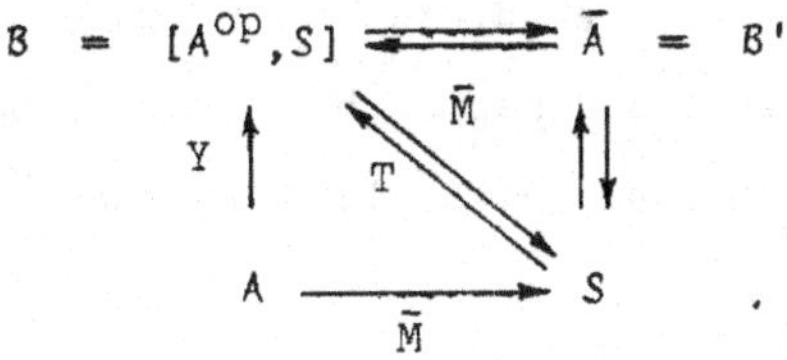

In this example, $\mathcal{B}'$ is the full subcategory $\bar{A}$ of M-cribles. $\mathcal{B}'$ is locally small, even if A is large, and is a closed span category because $\bar{M}$ preserves all limits.

Because M is faithful the Yoneda embedding lands in $\bar{A}$. If $Y : A \subset \bar{A}$ has a left adjoint whose unit is inverted by $\bar{M}$ then A is embedded in a minimum left exact retract of $\bar{A}$. In particular, this is so if M has a right adjoint and A is sufficiently complete ([8] Proposition 4.5).

An <u>M-topology</u> on A is a Grothendieck topology with $\bar{M}i$ an isomorphism for each covering crible $i : G \to A(-A) : A^{op} \to S$. It is easy to verify (Proposition 4.2) that the M-topologies on A are in one-one correspondence with the left exact retracts of $\bar{A}$ through which $\bar{M}$ factors. Thus each family of M-topologies has a union.

3. Limit spaces.

In the preceding example take $A = \mathit{Top}$, the category of all topological spaces and continuous maps, and let $M : \mathit{Top} \to S$ be the underlying-set functor (usually omitted from the notation). While Top has a proper class of objects, the category $\overline{\mathit{Top}}$ of M-cribles is locally small and is a closed span category.

The category $\overline{\mathit{Top}}$ is alternatively described as follows. An object of $\overline{\mathit{Top}}$ is a set X together with, for each $A \in \mathit{Top}$, a set $\mathrm{Ad}(AX)$ of admissible set maps from A to X. These sets satisfy the axioms:

Q1. Each constant set map is admissible.

Q2. If $f \in \mathit{Top}(AB)$ and $g \in \mathrm{Ad}(BX)$ then $gf \in \mathrm{Ad}(AX)$.

A morphism $h \in \overline{\mathit{Top}}(X,Y)$ is a set map such that $hf \in \mathrm{Ad}(AY)$ if $f \in \mathrm{Ad}(AX)$. The functor $M : \overline{\mathit{Top}} \to S$ is simply the underlying-set functor.

The Yoneda embedding $\mathit{Top} \subset \overline{\mathit{Top}}$ has a left adjoint which assigns to $X \in \overline{\mathit{Top}}$ the same set as X with the final topology with respect to $\{\mathrm{Ad}(AX);\ A \in \mathit{Top}\}$.

The M-topologies on Top are the topologies for which covers are bijective at the underlying-set level. Each stable class of identification maps in Top provides such a topology, hence describes an exact left retract of $\overline{\mathit{Top}}$.

Example 3.1. Take Z to be the class of morphisms in $\overline{\mathit{Top}}$ generated by the class of identification maps $f : A \to B$ in Top with the property: for each point $b \in B$ there exists a non-empty finite set $\{a_1, \cdots, a_n\} \subset f^{-1}b$ such that f maps each neighbourhood of $\{a_1,\cdots,a_n\}\subset A$ to a neighbourhood of $b\in B$. This class is stable under pullback in *Top* (cf. [5] Theorem 1). The category $\overline{Top}_Z$ is equivalent to the category of limit spaces and continuous maps. We recall (from Binz-Keller [3]) that a limit space is a set X together with, for each $x\in X$, a set $C(x)$ of "convergent" filters on X ; the sets $C(x)$ satisfy the axioms :

L1. $\langle x\rangle \in C(x)$.

L2. If $\theta\geqslant\phi$ and $\phi\in C(x)$ then $\theta\in C(x)$.

L3. If $\theta\in C(x)$ and $\phi\in C(x)$ then $\theta\wedge\phi\in C(x)$.

<u>Example 3.2</u>. Let Z be the (larger) class of morphisms generated by the identification maps $f:A \to B$ with the following property:
for each point $b\in B$ and open covering $\{G_\lambda ; \lambda\in\Lambda\}$ of $f^{-1}b$ there exists a finite set $\{\lambda_1,\cdots,\lambda_n\}\subset\Lambda$ such that $fG_{\lambda_1}\cup\cdots\cup fG_{\lambda_n}$ is a neighbourhood of $b\in B$ (see [5] Theorem 1). The resulting category $\overline{Top}_Z$ is equivalent to the category of those limit spaces which satisfy the following axioms :

L1. $\langle x\rangle \in C(x)$.

L2. If $\theta\geqslant\phi$ and $\phi \in C(x)$ then $\theta \in C(x)$.

L3'. $\theta \in C(x)$ if there exists a set of filters $\{ \theta_\lambda ; \lambda\in\Lambda$ and $\theta_\lambda\in C(x) \}$ such that each set $\{V_\lambda ; \lambda\in\Lambda$ and $V_\lambda \in \theta_\lambda\}$ contains a finite subset $\{V_{\lambda_i} ; i = 1,\cdots,n\}$ such that $\cup V_{\lambda_i} \in \theta$.

The class $\bar{Z}$ of morphisms obtained here is the <u>stable</u> interior of the reflection from $\overline{Top}$ to *Top* and it is strictly smaller than the <u>monoidal</u> interior (see [5] §3 for a counterexample).

Finally, we note that in order to completely describe the canonical Grothendieck topology on *Top* it is necessary to consider all covering classes $\{f_\alpha : A_\alpha \to B ; \alpha \in \Gamma\}$ in *Top*. Such a class of morphisms into a given space B covers in the canonical topology if and only if it has the property: for each point $b \in B$ and open covering $\{G_\lambda ; \lambda \in \Lambda$ and $\Lambda \in S\}$ of $\cup_\alpha f_\alpha^{-1} b$ there exists a finite subset $\{\lambda_1, \ldots, \lambda_n\} \subset \Lambda$ such that $\cup_{\alpha, i} f_\alpha(G_{\lambda_i})$ is a neighbourhood of $b \in B$. However, it is easy to verify that the introduction of these covers imposes no new restriction on the limit spaces under consideration. In other words it suffices to take Γ small so that $\bar{Z}$ is generated by the identification maps already described above.

4. Stability conditions and relations.

In this section we note several "known" facts about general closed span categories.

Proposition 4.1. A finitely complete category C is a closed span category if and only if Span C is closed as a bicategory.

Proof. For each morphism $f \in C(BC)$ let $f^* : C/C \to C/B$ denote pullback-along-f and let $f_* : C/B \to C/C$ denote the left adjoint of f^*. Span C is closed if and only if f^* has a right adjoint Πf for all f. The result then follows from the fact that if $[f,-] \vdash - \wedge f$ exists then it has a unique limit-preserving extension along the comonadic functor f_* :

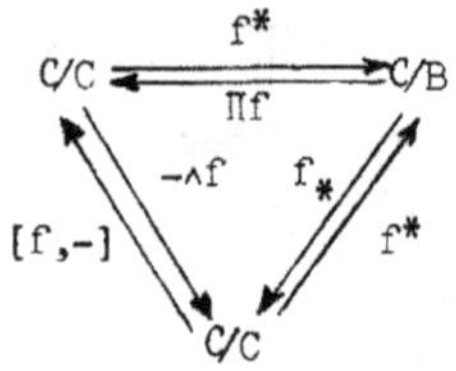

Suppose $\mathcal{B}$ is a cs-category containing a strongly generating class $\mathcal{A}$ such that each $B \in \mathcal{B}$ has a presentation as a coequaliser: $\Sigma A_x \rightrightarrows \Sigma A_y \to B$. A class Z of morphisms in $\mathcal{B}$ is <u>A-stable</u> if $f^* s \in Z$ for all $s \in Z$ and $f : A \to B$ with $A \in \mathcal{A}$. Let $Z(\mathcal{A}) = \{s: B \to A: s \in Z \text{ and } A \in \mathcal{A}\}$

<u>Proposition 4.2</u>. (a) $\bar{Z} = \overline{Z(\mathcal{A})}$ if Z is $\mathcal{A}$-stable.

(b) If $Z = \bar{Z}$ then Z is stable if Z is $\mathcal{A}$-stable.

(c) If Z consists of epimorphisms then $\bar{Z}$ is stable if Z is stable.

The verifications are straightforward.

If Z is a stable class of morphisms in a finitely complete category $\mathcal{B}$ then Z is closed with respect to span composition. Thus the class of categories of fractions of $\{\mathcal{B}/A \times B \;;\; A, B \in \mathcal{B}\}$ with respect to Z forms a new bicategory (cf. [7]) with the evident universal property. In particular, suppose $\mathcal{B}$ is a closed span category with a proper factorisation system $\mathcal{E}$-$\mathcal{M}$ where $\mathcal{E}$ is stable. Then, on taking $Z = \mathcal{E}$, one obtains the bicategory of <u>$\mathcal{M}$-relations</u> in $\mathcal{B}$. Because Span$\mathcal{B}$ is closed the subbicategory of relations is closed under exponentiation in Span$\mathcal{B}$ (by the "several objects" form or the reflection theorem for closed categories [6]) and the exponentiation provides a form of universal quantification (cf. [11]).

Bicategories (not necessarily closed) of relations and spans have also been considered in [12] by J. Meisen.

References

[1] Antoine, P., Extension minimale de la catégorie des espaces topologiques, C.R. Acad. Sc. Paris, t.262 (1966), 1389-1392.

[2] Bénabou, J., Introduction to bicategories, Reports of the Midwest Category Seminar I, Lecture Notes 47 (Springer 1967), 1-77.

[3] Binz, E. and Keller, H.H., Funktionenräume in der Kategorie der Limesräume, Annales Acad. Sc. Fen., A.I.383 (1966), 4-21.

[4] Day, B.J., Relationship of Spanier's quasi-topological spaces to k-spaces, M.Sc. Thesis, Univ. Sydney, 1968.

[5] Day, B.J. and Kelly, G.M., On topological quotient maps preserved by pullbacks or products, Proc. Cambridge Phil. Soc. 67 (1970), 553-558.

[6] Day, B.J., A reflection theorem for closed categories, J. Pure and Appl. Alg., Vol.2, No.1 (1972), 1-11.

[7] Day, B.J., Note on monoidal localisation, Bull. Austral. Math. Soc., Vol. 8 (1973), 1-16.

[8] Day, B.J., On adjoint-functor factorisation, Proc. Sydney Cat. Conf., to appear.

[9] Fakir, S., Monade idempotente associée à une monade, C.R. Acad. Sc. Paris, t.270 (1970), 99-101.

[10] Freyd, P. and Kelly, G.M., Categories of continuous functors I, J. Pure and Appl. Alg., Vol.2, No.3 (1972), 169-191.

[11] Lawvere, F.W., ed., Toposes, algebraic geometry and logic, Lecture Notes 274 (Springer 1972), Introduction and references.

[12] Meisen, J., On bicategories of relations and pullback spans, preprint, University of B.C., (1973).

REVIEW OF THE ELEMENTS OF 2-CATEGORIES

by

G.M. Kelly and Ross Street

The purpose of this review is to serve as a common introduction to the authors' papers in this volume, by collecting together some basic notions needed by each of us, and establishing some notation, to avoid later duplication. Nothing of substance here is original, but we could find no connected account in the literature that exactly satisfied our needs. In §1 we rehearse the most elementary facts about 2-categories, partly in the hope of making our papers below self-contained for such beginners as may care to read them, but chiefly to introduce our notation and especially the operation of pasting that we use constantly. In §2 we use the pasting operation to give a treatment, which seems to us simpler and more complete than any we have seen, of the isomorphism (bu,u'a) = (f'b,af) arising from adjunctions $f \dashv u$ and $f' \dashv u'$ in any 2-category, and of its naturality. In §3 we recall the basic properties of monads in a 2-category, and then mention some enrichments of these that become available in the 2-category of 2-categories (because it is really a 3-category).

§1 DOUBLE CATEGORIES AND 2-CATEGORIES

1.1 Both notions are originally due to Ehresmann; see [6] and [7]. We recall first the notion of double category. We denote by *CAT*, *SET*, *Cat*, *Set*, respectively the categories of all categories, of all sets, of small categories, and of small sets. Conceptually, a double category is a category object in *CAT*; but it admits the following elementary description.

It has objects A etc.; horizontal arrows a etc.; vertical arrows x etc.; and squares α etc.; there are various domain and codomain functions sufficiently indicated by the diagrams

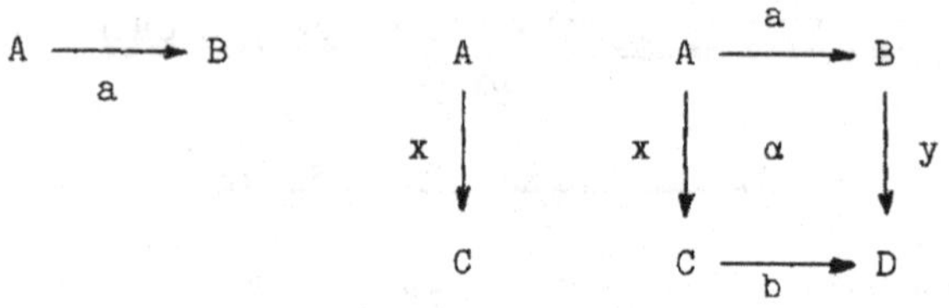

The objects and the horizontal arrows form a category, with identities $h_A: A \to A$; the objects and the vertical arrows form a category, with identities $\begin{smallmatrix} A \\ \downarrow v_A \\ A \end{smallmatrix}$. The squares have horizontal and vertical laws of composition, represented by

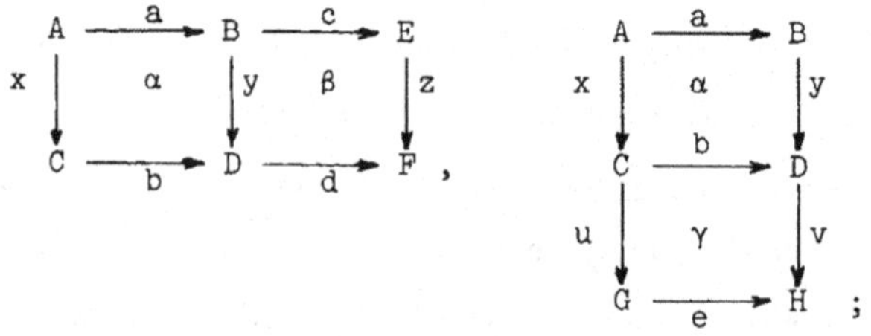

under each of these laws they form a category, with respective identities

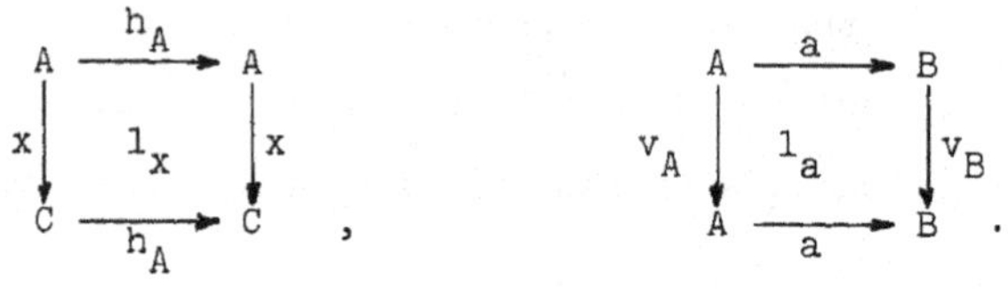

In the situation

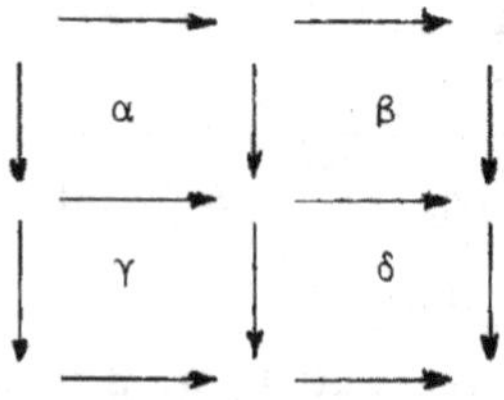

the result of composing first horizontally and then vertically is to be the same as the result of composing first vertically and then horizontally. The composite

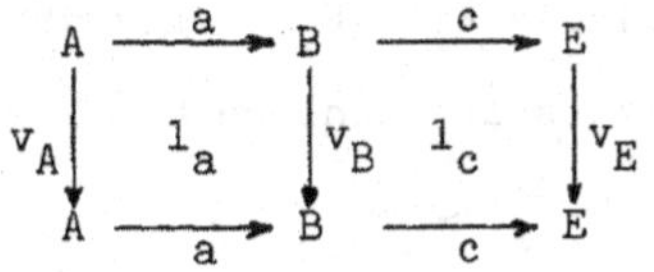

is to be 1_{ca}, and similarly for vertical composition of horizontal identities. Finally the horizontal and vertical identities

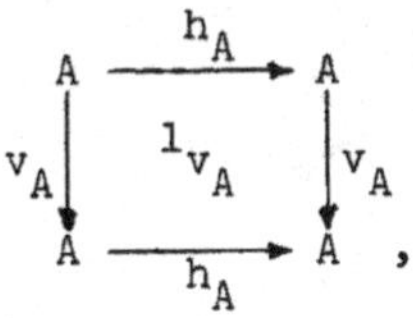 , 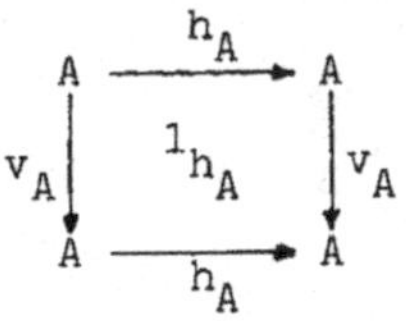 ,

are to coincide.

Examples of double categories may be found in Palmquist [21], and in §2.2 below.

1.2. A 2-category K may be thought of as a double category in which all the vertical arrows are identities. A more direct description is as follows.

K has objects or 0-cells A etc., arrows or morphisms or 1-cells $f: A \to B$ etc., and 2-cells

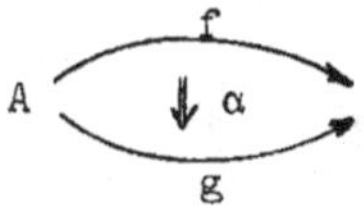 or 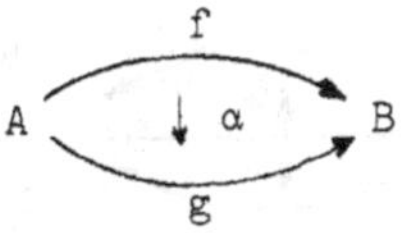 .

(Kelly tends to use double arrows for 2-cells, and Street single ones - it is purely a matter of taste.)

The objects and the arrows form a category K_0, called the underlying category of K, with identities $1_A: A \to A$; when the context shows what is meant, we sometimes write K for K_0.

For fixed A and B, the arrows $A \to B$ and the 2-cells between them form a category $K(A,B)$ under the operation known as vertical composition:

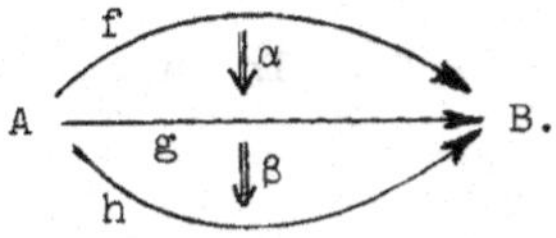

The vertical composite $f \Rightarrow h$ above is denoted by $\beta\cdot\alpha$ or $\beta.\alpha$, or rarely

by $\beta\alpha$ when no confusion is likely with the horizontal composite to be introduced below; its identities are denoted by

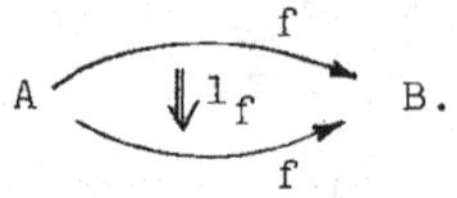

There is also a law of horizontal composition of 2-cells, whereby from 2-cells

we get a 2-cell

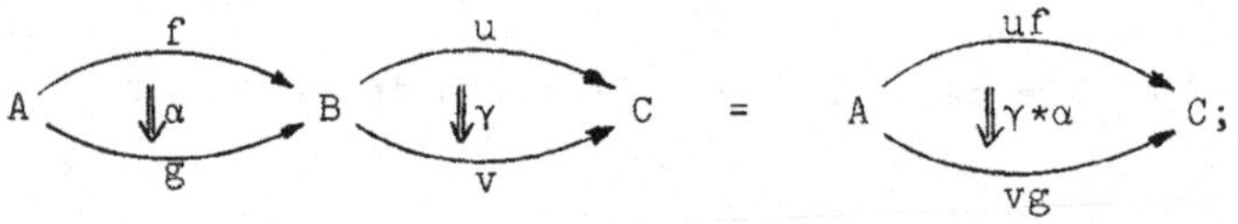

this composite $\gamma*\alpha$ is also denoted by $\gamma\alpha\colon uf \Rightarrow vg$. Under this law the 2-cells are to form a category, with identities

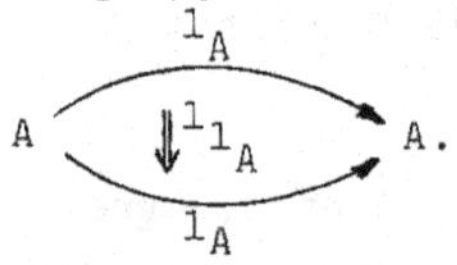

We require finally that, in the situation

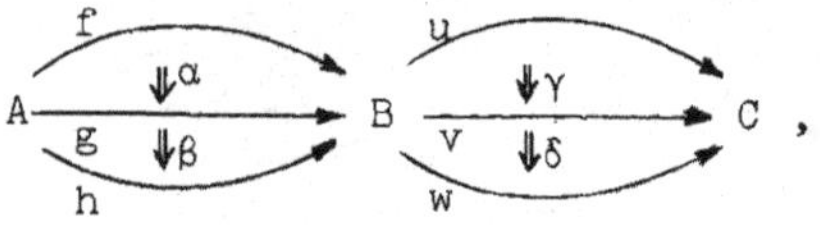

the composites $(\delta*\beta)\cdot(\gamma*\alpha)$ and $(\delta\cdot\gamma)*(\beta\cdot\alpha)$ coincide; and that in the situation

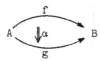

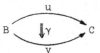

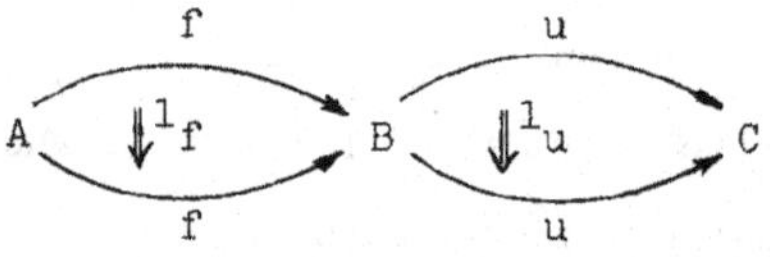

we have $1_u*1_f = 1_{uf}$.

We also freely use the convention whereby the name of an object A or of an arrow f is also used as the name of its identity arrow 1_A or its identity 2-cell 1_f. In particular the horizontal composite

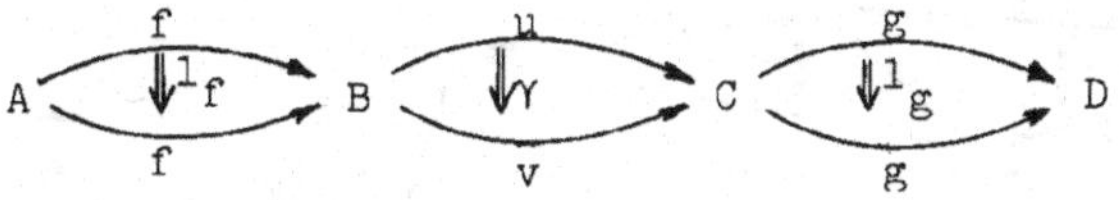

is also written as

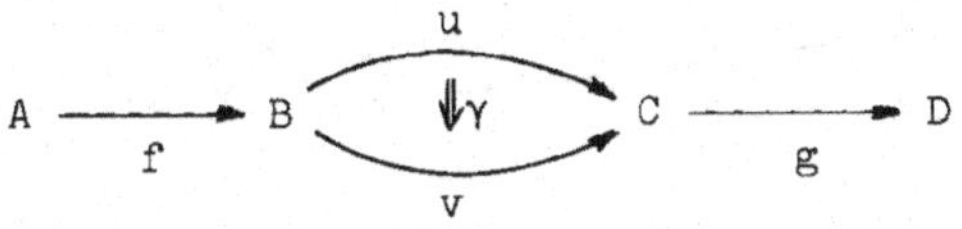

and denoted by $g\gamma f$.

The above operations on 2-cells can be combined to give the more general operation of <u>pasting</u>, introduced by Benabou [1]. The two basic situations are

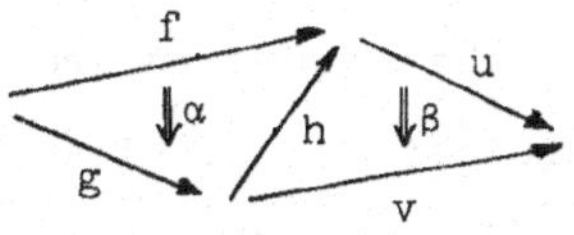

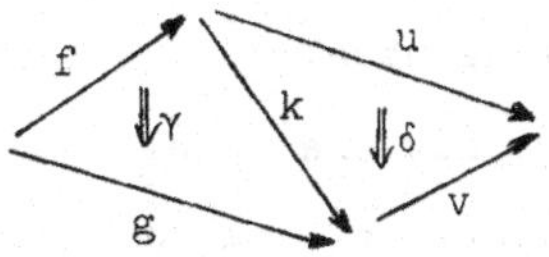

The first of these is meant to indicate the 2-cell $\beta g \cdot u\alpha : uf \Rightarrow uhg \Rightarrow vg$, and the second to indicate the 2-cell $v\gamma \cdot \delta f : uf \Rightarrow vkf \Rightarrow vg$. Thus we give meaning to such composites as

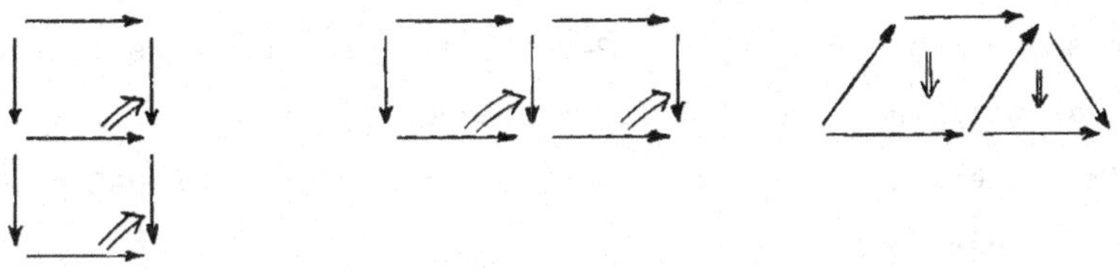

If in a diagram such as

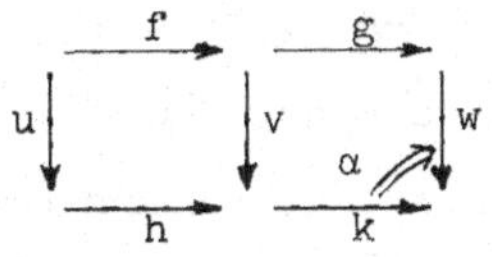

one region has no 2-cell marked in it, it is to be understood that the identity 2-cell is meant, which implies that $vf = hu$.

One can generalize the pasting operation further still, so as to give meaning to such multiple composites as

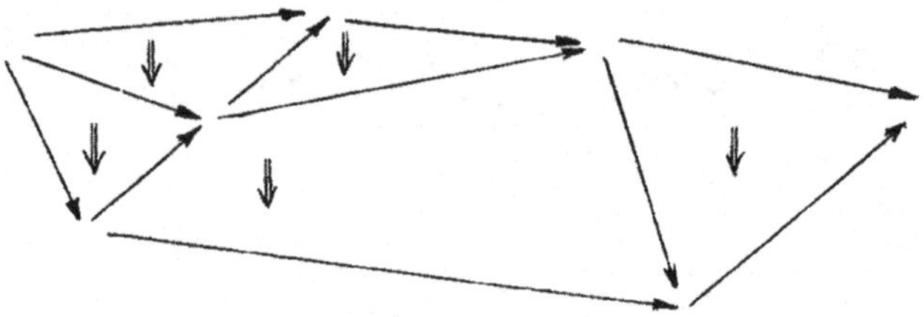

This is meant to indicate a vertical composite of horizontal composites of the form

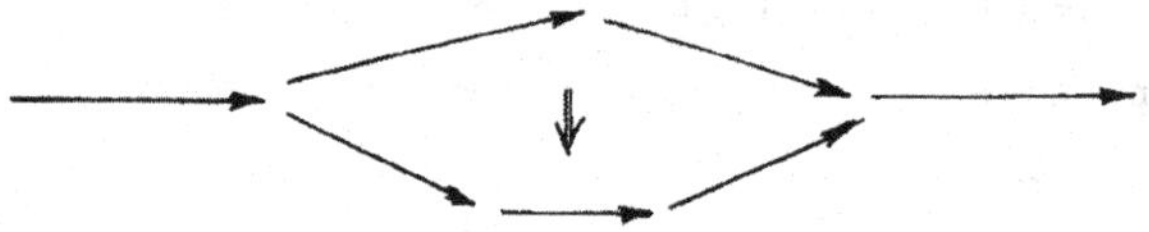

;

there is usually a choice of the order in which the composites are taken, but the result is independent of this choice; this is clear in simple cases, and can be proved inductively in the general case after an appropriate formalization in terms of polygonal decompositions of the disk.

1.3. As for examples of 2-categories, the paradigmatic one is *CAT*, just as the paradigmatic category is *SET*. The objects are categories, the arrows are functors, and the 2-cells are natural transformations. The context will show when *CAT* is considered as a 2-category and when merely the underlying category is meant. There is the sub-2-category *Cat* of small categories.

For a monoidal category *V*, we have the 2-category *V-CAT* of *V*-categories, *V*-functors, and *V*-natural transformations, in the sense of [8]; again with the sub-2-category *V-Cat* of small ones. Monoidal categories themselves, with monoidal functors and monoidal natural transformations in the sense of [8], form a 2-category *Mon CAT*.

The category *K* of ordered objects in any category A becomes a 2-category when we observe that the hom-set *K*(A,B) has a natural order, and can therefore be regarded as a category.

For a category $\mathcal{A}$, the comma category $\mathcal{CAT}/\mathcal{A}$, an object of which is a category $\mathcal{B}$ together with a functor $B: \mathcal{B} \to \mathcal{A}$, and an arrow of which from $(\mathcal{B},B)$ to $(\mathcal{C},C)$ is a functor $T: \mathcal{B} \to \mathcal{C}$ with $CT = B$, becomes a 2-category when we take the 2-cells $\alpha: T \Rightarrow S: \mathcal{B} \to \mathcal{C}$ to be the natural transformations $\alpha: T \Rightarrow S$ for which $C\alpha = \mathrm{id}$. Many other examples will arise in the papers below.

1.4. Besides the elementary definition of 2-category in §1.2 above, there is an equivalent but non-elementary one: the category $\mathcal{CAT}$ is cartesian closed, and a 2-category $\mathcal{K}$ is just a $\mathcal{CAT}$-category, its $\mathcal{CAT}$-valued hom being $\mathcal{K}(A,B)$. This definition determines what we shall mean by 2-functor and by 2-natural transformation: namely $\mathcal{CAT}$-functor and $\mathcal{CAT}$-natural transformation. Similarly for 2-adjoint, = $\mathcal{CAT}$-adjoint; see [14] for a general treatment of $\mathcal{V}$-adjunction. Note that we do not use "2-natural" or "2-adjoint" in the more general senses given to them by Gray in [10].

In elementary terms, a 2-functor $D: \mathcal{K} \to \mathcal{L}$ sends objects of $\mathcal{K}$ to objects of $\mathcal{L}$, arrows of $\mathcal{K}$ to arrows of $\mathcal{L}$, and 2-cells of $\mathcal{K}$ to 2-cells of $\mathcal{L}$, preserving domains and codomains and all types of identity and composition. A 2-natural transformation $\eta: D \Rightarrow E: \mathcal{K} \to \mathcal{L}$ assigns to each object A of $\mathcal{K}$ an arrow η_A or $\eta A: DA \to EA$ in $\mathcal{L}$, which is not only natural in the ordinary sense that, for $f: A \to B$, we have $\eta B.Df = Ef.\eta A$, but also 2-natural in the sense that, for each 2-cell $\alpha: f \Rightarrow g$ in $\mathcal{K}$, we have

$$(1.1)\qquad DA \underset{Dg}{\overset{Df}{\rightrightarrows}} DB \;(\Downarrow D\alpha)\; \xrightarrow{\eta B} EB \;=\; DA \xrightarrow{\eta A} EA \underset{Eg}{\overset{Ef}{\rightrightarrows}} EB \;(\Downarrow E\alpha).$$

As in general the $\mathcal{V}$-natural transformations from the $\mathcal{V}$-functor D to the $\mathcal{V}$-functor E form not only a set but an object of $\mathcal{V}$, so here where $\mathcal{V} = \mathcal{CAT}$ the 2-natural transformations $D \Rightarrow E$ form a category; in other words 2-$\mathcal{CAT}$ is really a 3-category, i.e. a 2-$\mathcal{CAT}$-

category. We follow Benabou [1] in calling morphisms of 2-natural transformations <u>modifications</u>. A modification

$$\rho\colon \eta \to \zeta\colon D \Rightarrow E\colon K \to L,$$

also written

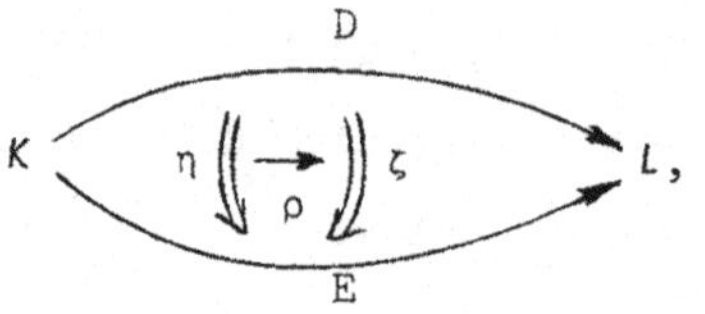

assigns to each object A of K a 2-cell $\rho A\colon \eta A \Rightarrow \zeta A$ such that, for $f\colon A \to B$, we have

(1.2)

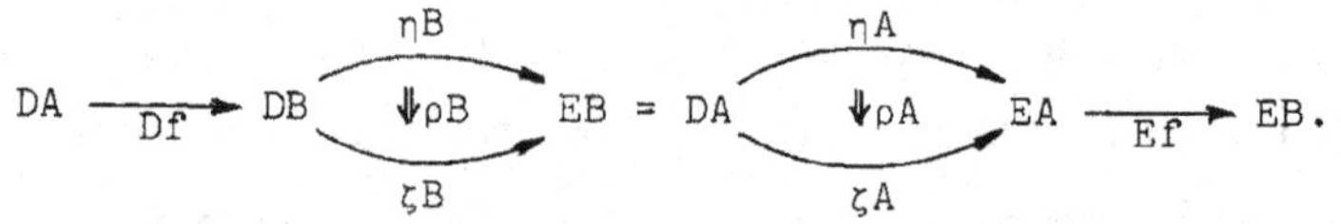

In particular, taking a fixed 2-category K, we get a 2-category End K of endomorphisms of K; its objects are endo-2-functors $D\colon K \to K$ of K; its arrows are 2-natural transformations $\eta\colon D \Rightarrow E$; and its 2-cells are modifications $\rho\colon \eta \to \zeta$. When working totally within this 2-category Kelly reserves the right to write $\eta\colon D \to E$ and $\rho\colon \eta \Rightarrow \zeta$ if he feels like it.

The non-elementary definition of a 2-category K also determines what K^{op} shall mean: we have $K^{op}(A,B) = K(B,A)$, so that we reverse the 1-cells but not the 2-cells. We write K^{co} for the other dual, induced by the ordinary duality on CAT, so that $K^{co}(A,B) = K(A,B)^{op}$; thus we reverse the 2-cells but not the 1-cells. Observe that $K^{coop} = K^{opco}$.

To say that 2-functors $D\colon K \to L$ and $E\colon L \to K$ are 2-<u>adjoint</u> is to say that they are CAT-adjoint; that is, that there is an isomorphism of categories $K(EB,A) \cong L(B,DA)$ which is 2-natural in A and B. Specializing the result of [14] on V-adjunction to the case $V = CAT$, we see that it comes to the same thing to have 2-natural transformations $\eta\colon 1 \Rightarrow DE$ and $\varepsilon\colon ED \Rightarrow 1$ satisfying the usual conditions; so that 2-adjunction is just adjunction <u>in</u> the 2-category 2-CAT, in the sense of §2.1 below.

1.5. A great many notions definable in V-CAT for any V admit in 2-CAT generalizations that would make no sense for an arbitrary V; and this because 2-CAT is a 3-category. For where a diagram was to commute in the basic notion, we can now allow it to commute only to within a given isomorphism, or even go further and replace this isomorphism by a given morphism (in a given sense); the given isomorphisms or morphisms will then usually be required to satisfy some "coherence axioms". Moreover experience shows that these "relaxed notions" do indeed occur in nature; in some cases they seem in fact to be the more normal thing, that we must willy-nilly consider; in others we can prove a "coherence theorem" allowing them to be replaced by the simpler "strict" notion.

It is not our intention to go into this systematically here; such things are treated in detail in the forthcoming book (doubtless to appear before this one) of Gray [12]. We should however like to indicate a systematic nomenclature that we shall use.

Where the original "strict" notion N is taken as the norm, we say "pseudo-N" for the relaxed notion with equality replaced by an isomorphism, and "lax N" for the still more relaxed notion with equality replaced by a mere morphism: we have no general principle to offer for our various choices of the sense of this morphism, but having chosen a sense by some criterion of usefulness, we call the same notion with the sense reversed an "op-lax N".

An example is the concept of lax functor $D: K \to L$ between 2-categories, which has 2-cells $Dg.Df \Rightarrow D(gf)$ and $1_{DA} \Rightarrow D1_A$ instead of equalities; these have been considered by Street [24] and Roberts [22], and are a special case of Benabou's morphisms of bicategories [1]. Indeed a bicategory in Benabou's sense is itself a pseudo-2-category, and this illustrates a general point: when one relaxes some "morphism-like" notion, the definition continues to make sense when the domain and the codomain are themselves "lax". Lax functors as such do not occur in the present book, but pseudo-functors occur in the paper [25] of Street below, and go back to Grothendieck [13].

A second example is the notion of lax natural transformation in place of 2-natural transformation; these do occur below in the paper of Kelly [15], and are the things called "2-natural" by Gray in [10] and "quasi-natural" (now adopted by Gray) in Bunge [2]. Another is the notion of lax monad (relaxed 2-monad) of Bunge [3]; the corresponding pseudo-monads are considered by Zöberlein [27]. Then there are the lax algebras or the pseudo-algebras over these, and their lax morphisms...

This brings us to the second aspect of our nomenclature. Even with strict 2-monads, and strict algebras for them, the lax morphisms of algebras are the normal thing, as is shown by the special case of monoidal functors, where strict ones (preserving the tensor product and identity on the nose) are quite rare in nature. In these cases we take the lax notion as our basic notion N; we then call the pseudo-notion "strong N" and the strict notion "strict N"; thus "monoidal functor" (with a comparison $\phi A \otimes \phi B \to \phi(A \otimes B)$), "strong monoidal functor" (with an isomorphism $\phi A \otimes \phi B \cong \phi(A \otimes B)$), and "strict monoidal functor" (with equality $\phi A \otimes \phi B = \phi(A \otimes B)$).

We say no more here; such lax or pseudo notions as we need are introduced in the individual papers below.

§2. ADJUNCTION IN A 2-CATEGORY

2.1. Other accounts of the matter below, of varying degrees of completeness, can be found in [19], [14], and [21]. The utility of pasting for the neat expression of the adjunction equations (2.1) and (2.2) below we learnt from R.F.C. Walters.

An adjunction $\eta, \varepsilon \colon f \dashv u \colon A \to B$ in a 2-category K consists of arrows $u \colon A \to B$ and $f \colon B \to A$ together with 2-cells $\eta \colon 1 \Rightarrow uf$ and ε: fu ⇒ 1 satisfying the axioms

(2.1) 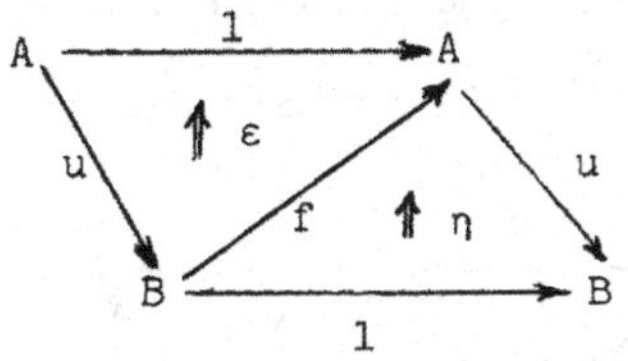 equals identity,

(2.2) 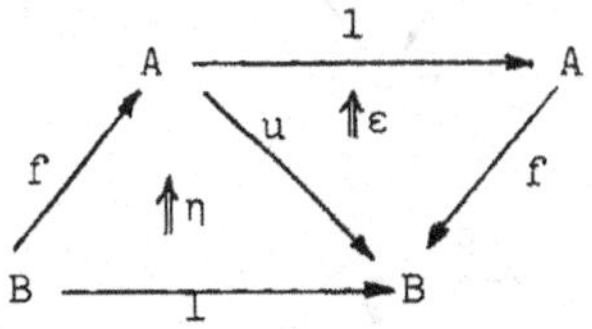 equals identity.

We say that f is <u>left adjoint</u> to u, and that u is <u>right adjoint</u> to f; we call η the <u>unit</u>, and ε the <u>counit</u>, of the adjunction.

When *K* is *V-CAT* for a symmetric monoidal closed *V*, it has been shown by Kelly in [14] that adjunctions f ⊣ u are in bijection with *V*-natural isomorphisms A(fb,a) ≅ B(b,ua); in particular we get the usual notion of adjunction when *K* = *CAT*.

If η_1, ε_1: $f_1 \dashv u_1$: B → C is a second adjunction, we clearly get a composite adjunction η_2, ε_2: $ff_1 \dashv u_1 u$: A → C if we define η_2, ε_2 as the composites

(2.3) 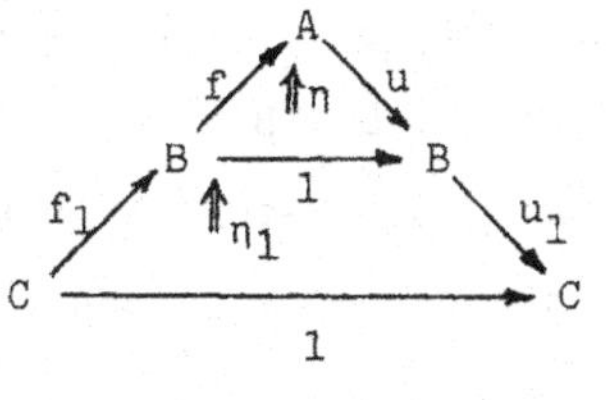 , 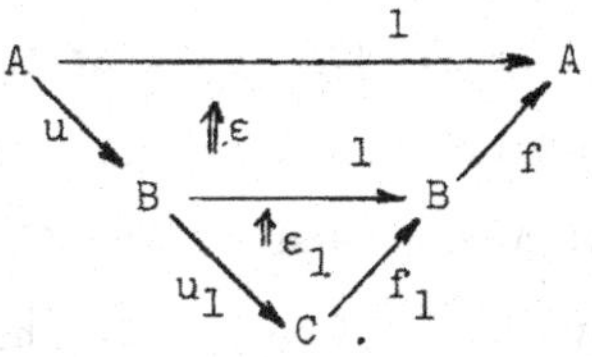 .

Thus adjunctions in *K* form a category, with 1,1: 1 ⊣ 1: A → A as identities.

2.2. If we look upon (2.1) and (2.2) as asserting that the 2-cells η and ε are mutually inverse under the indicated pasting operations, the following proposition becomes evident:

<u>Proposition 2.1</u>. <u>Let</u> $\eta,\varepsilon\colon f \dashv u\colon A \to B$ <u>and</u> $\eta',\varepsilon'\colon f' \dashv u'\colon A' \to B'$. <u>Let</u> $a\colon A \to A'$ <u>and</u> $b\colon B \to B'$. <u>Then there is a bijection between</u> 2-<u>cells</u> $\lambda\colon bu \Rightarrow u'a$ <u>and</u> 2-<u>cells</u> $\mu\colon f'b \Rightarrow af$, <u>where</u>

(2.4) μ <u>is the composite</u>

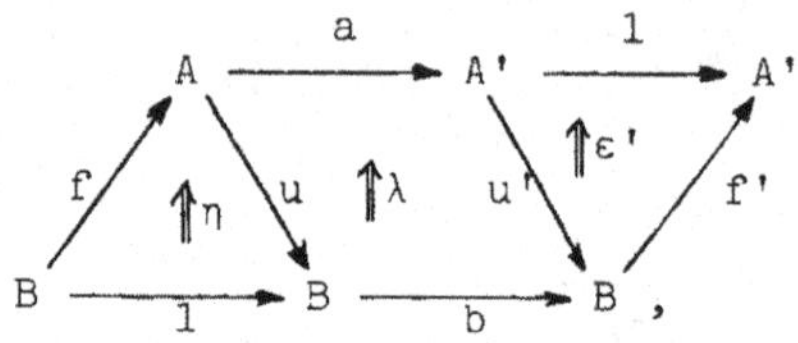

(2.5) λ <u>is the composite</u>

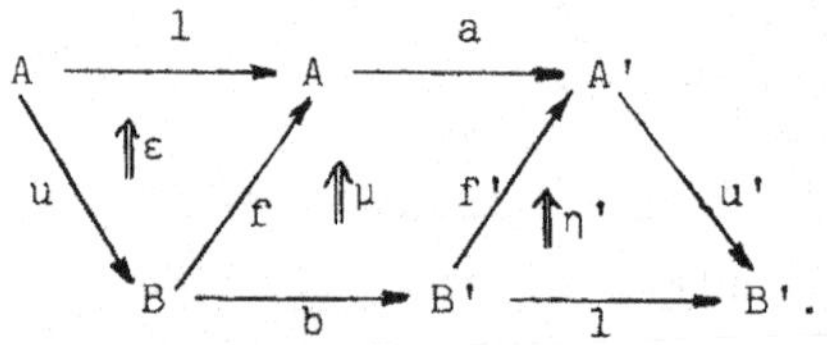

□

The naturality of this bijection between λ and μ may be expressed as follows. Consider two double categories. In each the objects are those of K, the horizontal arrows are the arrows of K, and the vertical arrows are the adjunctions in K. Composition of horizontal arrows is just composition in K, while composition of vertical arrows is the composition of adjunctions of §2.1 above. (In our diagrams "vertical" is conveniently shown as "oblique".) In the first double category, a square with sides $a\colon A \to A'$, $b\colon B \to B'$, $\eta,\varepsilon\colon f \dashv u\colon A \to B$, and $\eta',\varepsilon'\colon f' \dashv u'\colon A' \to B'$, is a 2-cell λ in K such as appears in (2.4). In the second double category, a square with the same sides is a 2-cell μ in K such as appears in (2.5). In both double categories, horizontal and vertical composition of squares are given by pasting, horizontally or vertically, the corresponding 2-cells of K. The "naturality" in question is now expressed by:

<u>Proposition 2.2</u>. <u>The above bijection between</u> λ <u>and</u> μ <u>is an isomorphism between the two double categories we have just described. That is to say, the bijection respects composition and identities, vertical and horizontal.</u>

Proof. As regards vertical identities: if $f \dashv u$ and $f' \dashv u'$ are respectively $1,1\colon 1 \dashv 1\colon A \to A$ and $1,1\colon 1 \dashv 1\colon A' \to A'$, we have by (2.4) that $\mu = \lambda$ and in particular that, when $a = b$, $\mu = 1$ if and only if $\lambda = 1$. For horizontal identities, we have a and b both identities, and $f \dashv u$ coinciding with $f' \dashv u'$; then if $\lambda = 1$ we have $\mu = 1$ by (2.4) and (2.2), and similarly if $\mu = 1$ then $\lambda = 1$.

For vertical composition we have only to write

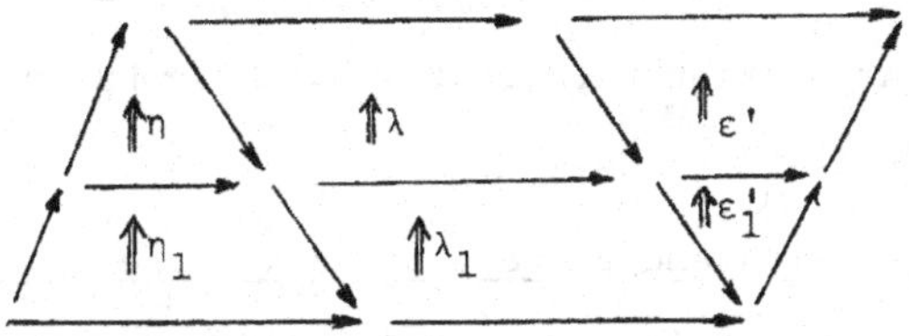

and to look at (2.3) and (2.4). For horizontal composition we paste together two diagrams like (2.4) to get

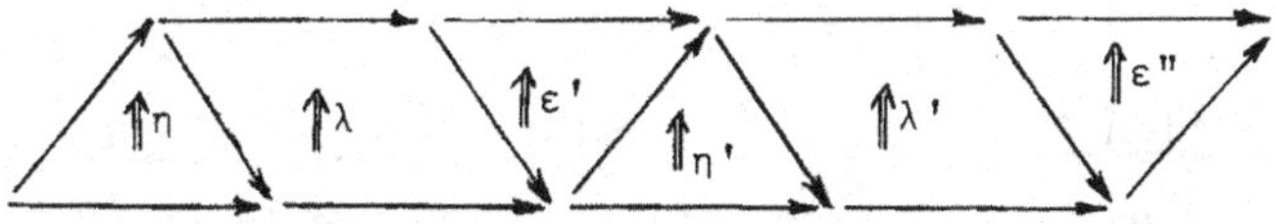

,

and observe that the central triangles ε' and η' cancel out by (2.1).□

It will usually, in context, be unambiguous if we call λ and μ mates under the adjunctions $f \dashv u$ and $f' \dashv u'$, without explicit mention of a and b.

Proposition 2.3. If $f \dashv u$ and $f' \dashv u$ then f and f' are canonically isomorphic.

Proof. Let the mates of

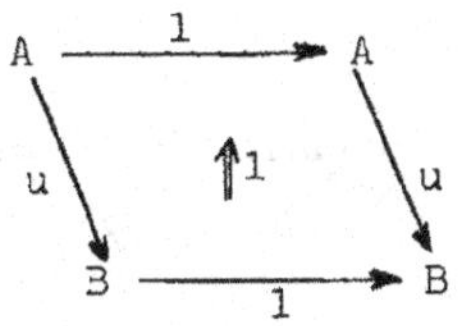

under the adjunctions $f \dashv u$ and $f' \dashv u$, and under the adjunctions $f' \dashv u$ and $f \dashv u$, be respectively

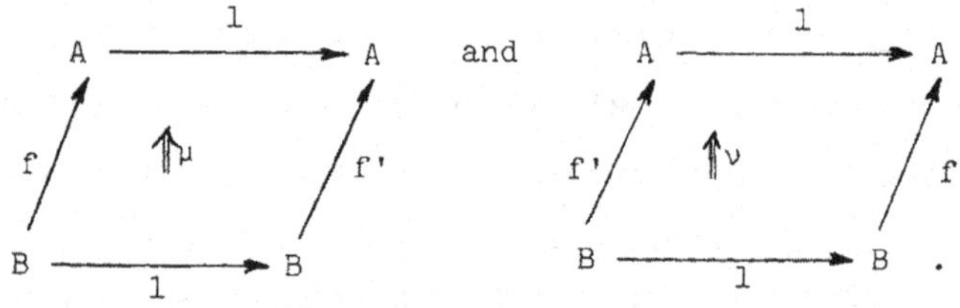

Then by the "horizontal" part of Proposition 2.2, μ and ν are mutually inverse. □

2.3. If $D: K \to L$ is a 2-functor, an adjunction $\eta,\varepsilon: f \dashv u: A \to B$ in K clearly gives an adjunction $D\eta, D\varepsilon: Df \dashv Du: DA \to DB$ in L. By applying D to (2.4) and (2.5) we get:

Proposition 2.4. *If λ and μ are mates under the adjunctions $f \dashv u$ and $f' \dashv u'$ in K, and if $D: K \to L$ is a 2-functor, then $D\lambda$ and $D\mu$ are mates under the adjunctions $Df \dashv Du$ and $Df' \dashv Du'$ in L.* □

Of course it is not in general true that μ is the identity 2-cell when λ is; but we have:

Proposition 2.5. *Let $D,E: K \to L$ be 2-functors and let $\alpha: D \Rightarrow E$ be a 2-natural transformation. Then the identity 2-cells*

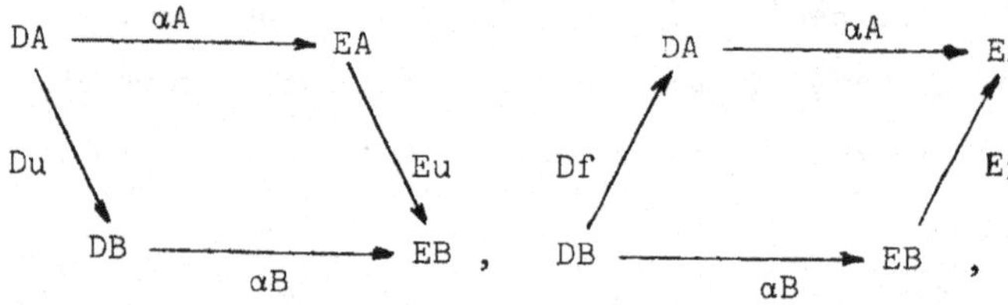

are mates under the adjunctions $Df \dashv Du$ and $Ef \dashv Eu$ in L.

Proof. The 2-naturality of α expressed in the form (1.1) above gives

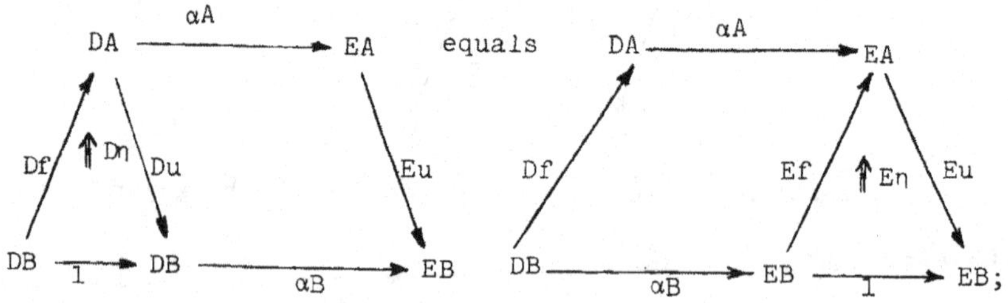

pasting $E\varepsilon$ on the right of each and using (2.2) gives what we want in the form (2.4). □

§3 MONADS IN A 2-CATEGORY

3.1 The notion of monad (= triple) makes sense in any 2-category K; the classical case is that where $K = CAT$. A monad in K, on the object B of K, is an endomorphism $t: B \to B$ together with 2-cells $\eta: 1 \Rightarrow t$, $\mu: t^2 \Rightarrow t$ called respectively the unit and the multiplication of the monad t; these are to satisfy the usual equations

$$(3.1) \qquad \mu \cdot t\eta = 1, \qquad \mu \cdot \eta t = 1, \qquad \mu \cdot t\mu = \mu \cdot \mu t.$$

A detailed treatment of monads in this generality has been given by Street in [23]; here we recall some of the simpler aspects and then pass to the special case where K = 2-CAT.

An action of the monad t above on an arrow $s: A \to B$ is a 2-cell $\nu: ts \Rightarrow s$ satisfying

$$(3.2) \qquad \nu \cdot \eta s = 1, \qquad \nu \cdot t\nu = \nu \cdot \mu s.$$

An arrow s together with such an action ν is called a t-algebra (with domain A). A morphism of t-algebras (with common domain A) is a $\sigma: s \Rightarrow s'$ such that

$$(3.3) \qquad \nu' \cdot t\sigma = \sigma \cdot \nu.$$

Thus the t-algebras with domain A form a category $Alg(A,t)$, with a forgetful functor U_A to $K(A,B)$ sending (s,ν) to s and σ to itself.

For any $r: A \to B$ it is clear from (3.1) that tr, with action $\mu r: t^2 r \Rightarrow tr$, is a t-algebra; and that for any $\rho: r \Rightarrow r'$, $t\rho: tr \Rightarrow tr'$ is a morphism of t-algebras. This gives a functor $F_A: K(A,B) \to Alg(A,t)$. It is further clear from (3.2) that, for a t-algebra s, the action $\nu: ts \Rightarrow s$ is also a morphism of t-algebras, when ts is taken with the action μs. From these remarks we easily verify the following proposition, in the light of which we call the t-algebras tr free t-algebras:

Proposition 3.1 F_A is left adjoint to U_A. In detail, let $s: A \to B$ be a t-algebra and $r: A \to B$ any arrow. Then there is a bijection between 2-cells $\alpha: r \Rightarrow s$ and t-algebra morphisms $\sigma: tr \Rightarrow s$, given by

(3.4) $$\sigma = \nu \cdot t\alpha, \qquad \alpha = \sigma \cdot \eta r.$$

In the classical case $K = CAT$ the phrase "t-algebra" is more commonly restricted to those with domain the unit category $\underline{1}$. The t-algebra $s: \underline{1} \to B$ is then identified with the corresponding object s of B, and $\nu: ts \Rightarrow s$, $\sigma: s \Rightarrow s'$ are morphisms in B. The category $\mathcal{Alg}(\underline{1},t)$ of t-algebras in this primary sense is classically denoted by B^t. Identifying $K(\underline{1},B)$ with B, we write the adjoint functors $U_{\underline{1}}$ and $F_{\underline{1}}$ in this classical case as $u^t: B^t \to B$ and $f^t: B \to B^t$. Of course in the general case the monad t on B induces a classical monad $K(A,t)$ on the category $K(A,B)$, and $\mathcal{Alg}(A,t)$ is just the category $K(A,B)^{K(A,t)}$ of $K(A,t)$-algebras.

3.2 If t' is another monad on the same B, a map of monads is a 2-cell $\tau: t \Rightarrow t'$ such that

(3.5) $$\mu' \cdot \tau^2 = \tau \cdot \mu, \qquad \eta' = \tau \cdot \eta;$$

here τ^2 denotes $\tau\tau: tt \Rightarrow t't'$. If (s,ν') is a t'-algebra with domain A, then clearly (s,ν) is a t-algebra where

(3.6) $$\nu = \nu' \cdot \tau s;$$

and a morphism $\sigma: s \Rightarrow s'$ of t'-algebras is also a morphism of t-algebras. This gives a functor $\mathcal{Alg}(A,\tau)$ rendering commutative

(3.7)

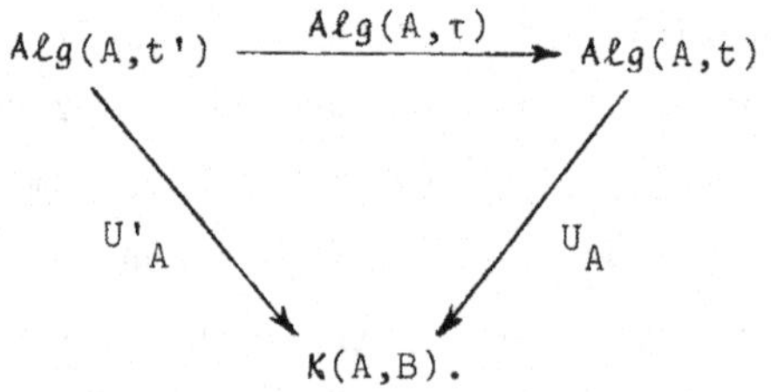

In particular the t'-algebra t' itself is a t-algebra under the action

(3.8) $$\theta\colon tt' \underset{\tau t'}{\Rightarrow} t't' \underset{\mu'}{\Rightarrow} t';$$

and τ can be recovered from θ in the form

(3.9) $$\tau\colon t \underset{t\eta'}{\Rightarrow} tt' \underset{\theta}{\Rightarrow} t',$$

since $\mu'\cdot t'\eta' = 1$. If an <u>arbitrary</u> action $\theta\colon tt' \Rightarrow t'$ of t on t' is given, the necessary and sufficient condition for it to arise as in (3.8) from a map (3.9) of monads is that $\mu'\colon t't' \Rightarrow t'$ be a t-algebra morphism, when t't' is taken with the t-action $\theta t'$; that is, that

(3.10) $$\theta\cdot t\mu' = \mu'\cdot\theta t'.$$

If $\eta',\varepsilon'\colon f' \dashv u'\colon C \to B$ is any adjunction in K, it is immediate that (t',η',μ') is a monad on B where $t' = u'f'$ and $\mu' = f'\varepsilon'u'$. We call this the monad <u>generated</u> by the adjunction $f' \dashv u'$. Observe that u' is a t'-algebra with action

(3.11) $$u'\varepsilon'\colon t'u' = u'f'u' \Rightarrow u'.$$

If t' is this monad and if t is any monad on B, there is by Proposition 2.1 a bijection between 2-cells $\tau\colon t \Rightarrow t' = u'f'$ and 2-cells $\nu\colon tu' \Rightarrow u'$, given by

(3.12) $$\nu = u'\varepsilon'\cdot\tau u', \qquad \tau = \nu f'\cdot t\eta'.$$

<u>Proposition 3.2</u>. <u>τ is a map of monads if and only if ν is an action of</u> t <u>on</u> u'.

<u>Proof</u>. If τ is a map of monads, $u'\varepsilon'$ is an action of t' on u' by (3.11) and thus ν is an action of t on u' by (3.6). If ν is an action of t on u', then $\theta = \nu f'$ is an action of t on $u'f' = t'$, which trivially satisfies (3.10); whence by (3.9) τ is a map of monads. □

<u>3.3</u> We say that K <u>admits the construction of algebras</u> if, for every monad t in K, the notion of t-algebra can be "internalized" in

the sense that the 2-functor $A \mapsto \mathcal{A}lg(A,t)$ from K^{op} to CAT is representable, so that

$$\mathcal{A}lg(A,t) \cong K(A,B^t) \tag{3.13}$$

(2-naturally in A) for some B^t in K, called the <u>object of</u> t-<u>algebras</u>. When this is so, the adjunction of Proposition 3.1 becomes an adjunction $K(A,B^t) \to K(A,B)$ which, because of its 2-naturality in A, arises from an adjunction $\eta^t,\varepsilon^t\colon f^t \dashv u^t\colon B^t \to B$. As the notation suggests, CAT does admit the construction of algebras, and in that case the B^t,f^t,u^t are those at the end of §3.1.

When $K = V\text{-}CAT$ for a symmetric monoidal closed category V, K admits the construction of algebras provided that V has equalizers (at least of pairs with a common left inverse); this case is treated in detail by Dubuc [5]. Here again the <u>primary</u> meaning of t-algebra is one with domain $\underline{1}$, which now denotes not the unit category but the unit V-category; so in this sense a t-algebra is an object s of the V-category B with an action of t on it. The category $\mathcal{A}lg(\underline{1},t)$ of these t-algebras admits a canonical enrichment to a V-category, which is B^t. In particular, taking $V = CAT$, this applies to the case $K = 2\text{-}CAT$.

The best general result we know of — it is easy to prove and a more general result is contained in Gray [12] — is that K admits the construction of algebras if it is <u>finitely complete as a 2-category</u>. In accordance with the general definition of completeness for V-categories in [4], this means that K has all finite limits, that these are preserved by the representable functors $K(A,-)\colon K \to CAT$, and that K admits cotensor products $[X,B]$ for each finite category X and each B in K. It turns out to be sufficient to demand the existence of the cotensor product $[\underline{2},B]$ where $\underline{2}$ is the arrow category $0 \to 1$; the existence of the other $[X,B]$ then follows. If we replace "all finite limits" above by "all pullbacks" we get the <u>representable</u>

2-categories of Gray [11] or of Street's paper [25] in this volume; so a finitely complete 2-category is a representable one with a terminal object preserved by the $K(A,-)$.

3.4 We henceforth suppose that K admits the construction of algebras. We can express (3.13) as follows: there is a t-algebra $u^t\colon B^t \to B$; and the functor sending an arrow $p\colon A \to B^t$ to the t-algebra $u^t p$, and sending the 2-cell $\pi\colon p \Rightarrow p'$ to the algebra morphism $u^t\pi$, is an isomorphism $K(A,B^t) \to \mathcal{A}\ell g(A,t)$. It is easy to check that the t-algebra $t\colon B \to B$ arises thus from the arrow $f^t\colon B \to B^t$, and that the monad $u^t f^t$ generated by $f^t \dashv u^t$ is t itself.

Proposition 3.3. Let $f' \dashv u'\colon A \to B$. *Then the arrows* $p\colon A \to B^t$ *rendering commutative*

(3.14)

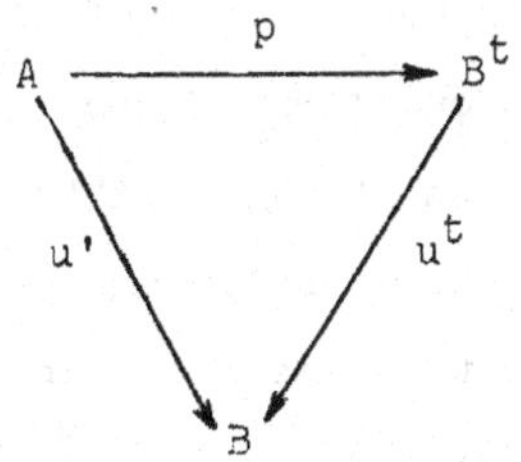

are in bijection with t-*actions on* u', *and hence by Proposition* 3.2 *with monad-maps* $\tau\colon t \Rightarrow u'f'$. □

If the monad $u'f'$ is t itself and $\tau = 1$, the p in (3.14) is called the *canonical comparison arrow*; the adjunction $f' \dashv u'$ is said to be *monadic* (*weakly monadic*) if p is an isomorphism (an equivalence). In fact it is easily seen that monadicity is a property of u' itself, and does not depend on the choice of the left adjoint f' to u'.

If in Proposition 3.3 we let t' be a second monad on B and take $f' \dashv u'$ to be the adjunction $f^{t'} \dashv u^{t'}\colon B^{t'} \to B$, we get:

Proposition 3.4. There is a bijection between arrows $p\colon B^{t'} \to B^t$ *rendering commutative*

(3.15)

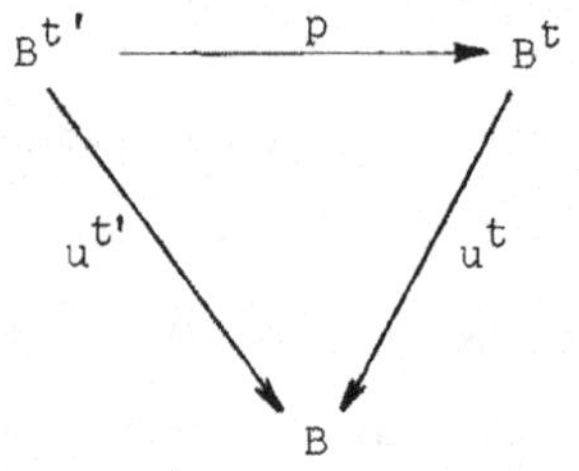

and monad maps τ: $t \Rightarrow t'$. □

We write B^τ for the p in (3.15) corresponding to τ, and call it an *algebraic map* between the algebra objects $B^{t'}$ and B^t. It is clear that (3.15) internalizes (3.7).

We can regard Proposition 3.4 as asserting that $t \mapsto u^t$, $\tau \mapsto B^\tau$ embeds the dual of the category of monads on B fully in the category K_0/B of objects over B; the canonical comparison arrow provides a reflexion into this subcategory of those u in K_0/B which have left adjoints. This is one form of the "semantics-structure" duality. For further details on the general theory of monads, and in particular on *distributive laws*, we refer again to Street [23].

3.5 Lawvere in [18] used the name "equational doctrine" for a monad in 2-*CAT* on *CAT*. Here we use the name *doctrine* (or 2-*monad*) for any monad in 2-*CAT*.

For us, then, a doctrine on a 2-category K is a 2-functor D: $K \to K$ with, for its unit and its multiplication, 2-natural transformations j: $1 \to D$ and m: $D^2 \to D$, satisfying on the nose the equations (3.1) (with D,j,m for t,η,μ).

Various relaxations are possible. Some things we should like to be doctrines (cf. [15] below) have the equalities in (3.1) replaced by isomorphisms. What Zöberlein [27] calls "doctrines" not only have isomorphisms for equalities in (3.1), but have j and m only *lax* natural transformations. The *lax monads* of Bunge [3] are weaker still: D only a lax 2-functor and mere morphisms in (3.1). To avoid complications we stick to the strict doctrines, and hope that there is a nice coherence theorem that will allow the results in our papers below to be applied at least to the "pseudo" case.

We also take D-<u>algebra</u> here in the strict sense: an object A of K (or more generally a 2-functor A with codomain K) with an action $n\colon DA \to A$ satisfying (3.2) (with A,n for s,ν). However Street considers lax algebras in [25] below, and defines them there; and Kelly considers them in relation to strict algebras in [16] and [17] below. When K = CAT, we also use "D-category" for "D-algebra".

For morphisms of D-algebras, on the other hand, the lax ones are the usual ones in nature, as we said in §1.5. We therefore depart from the nomenclature of §3.1, and define, for D-algebras A,A', a D-<u>morphism</u> $F\colon A \to A'$ to be a pair $(f,\bar{f})$ where $f\colon A \to A'$ is an arrow in K and $\bar{f}$ is a 2-cell

(3.16)
$$\begin{array}{ccc} DA & \xrightarrow{\;n\;} & A \\ {\scriptstyle Df}\big\downarrow & \Rightarrow\ \bar{f} & \big\downarrow{\scriptstyle f} \\ DA' & \xrightarrow[\;n'\;]{} & A' \end{array}$$

satisfying the axioms

(3.17)
$$\begin{array}{ccccc} D^2A & \xrightarrow{\;mA\;} & DA & \xrightarrow{\;n\;} & A \\ {\scriptstyle D^2f}\big\downarrow & & {\scriptstyle Df}\big\downarrow & \overset{\Rightarrow}{\bar{f}} & \big\downarrow{\scriptstyle f} \\ D^2A' & \xrightarrow[\;mA'\;]{} & DA' & \xrightarrow[\;n'\;]{} & A' \end{array} \quad = \quad \begin{array}{ccccc} D^2A & \xrightarrow{\;Dn\;} & DA & \xrightarrow{\;n\;} & A \\ {\scriptstyle D^2f}\big\downarrow & \overset{\Rightarrow}{D\bar{f}} & {\scriptstyle Df}\big\downarrow & \overset{\Rightarrow}{\bar{f}} & \big\downarrow{\scriptstyle f} \\ D^2A' & \xrightarrow[\;Dn'\;]{} & DA' & \xrightarrow[\;n'\;]{} & A', \end{array}$$

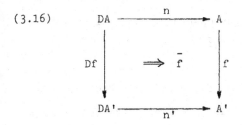

(3.18) 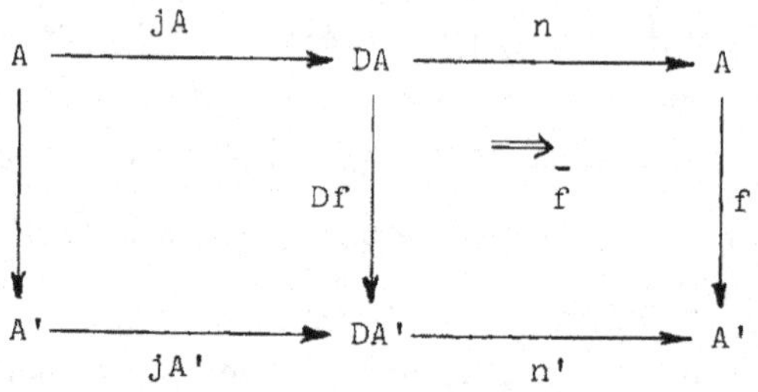

= identity.

In the case K = CAT we also say "D-functor" for "D-morphism". We call the D-morphism F <u>strong</u> if $\bar{f}$ is an isomorphism; we call it <u>strict</u> when $\bar{f} = 1$, so that $fn = n'.Df$. The strict D-morphisms, then, are the "morphisms of D-algebras" of §3.1; we also write f for the strict D-morphism $F = (f,1)$.

If we reverse the sense of $\bar{f}$ in (3.16), and also in the axioms (3.17) and (3.18), we get what we call an <u>op-D-morphism</u>. Clearly if $F = (f,\bar{f})$ is a strong D-morphism, then $(f,\bar{f}^{-1})$ is a strong op-D-morphism.

D-algebras and D-morphisms form a category under the operation of vertical pasting of diagrams like (3.16). We get subcategories with the same objects by restricting to strong or to strict D-morphisms. We now make these into 2-categories.

For D-morphisms $F,G: A \to A'$ we define a D-2-cell $\alpha: F \Rightarrow G$ to be a 2-cell $\alpha: f \Rightarrow g$ satisfying

(3.19) 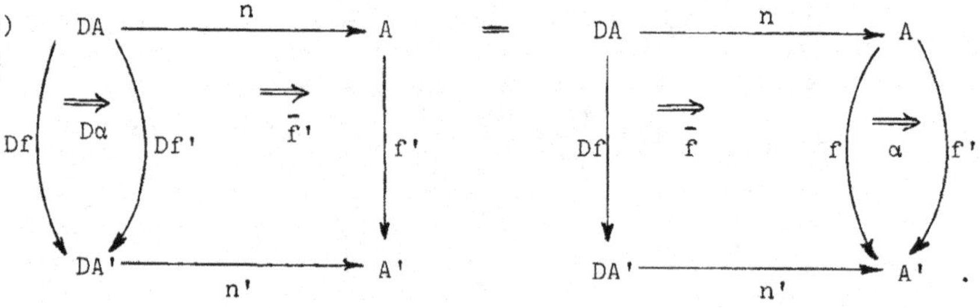

In the case $K = CAT$ we also say "D-natural transformation" for "D-2-cell". With the obvious laws of composition, D-algebras, D-morphisms, and D-2-cells form a 2-category D-$\mathcal{A}\ell g$, which we also write as D-CAT in the case $K = CAT$ or as D-$\mathcal{C}at$ when $K = \mathcal{C}at$.

We denote by D-$\mathcal{A}\ell g_*$ the sub-2-category in which only the strict D-morphisms are considered; it is this that is the object of algebras (here 2-category of algebras) K^D in the sense of §3.3. We say of K^D that it is doctrinal, or 2-monadic, over K (to emphasise that we mean monadic in 2-CAT, and not just in CAT).

It is not our intention here to go beyond definitions: some more substantive relations between D-morphisms and strict D-morphisms are examined below by Kelly in [16].

Examples of D-CAT are: monoidal categories, monoidal functors, and monoidal natural transformations; strict monoidal categories with arbitrary monoidal functors and monoidal natural transformations; symmetric monoidal categories, symmetric monoidal functors, and monoidal natural transformations; categories bearing a monad, with the monad functors and monad functor transformations of Street [23]; categories with assigned finite coproducts, arbitrary functors, and arbitrary natural transformations; categories bearing two monoidal structures $\otimes$ and $\oplus$ and having a distributivity of $\otimes$ over $\oplus$, with appropriate morphisms and 2-cells. With $K = CAT^2$ the objects of D-$\mathcal{A}\ell g$ may be pairs of monoidal categories with a monoidal functor between them; with $K = CAT^{|A|}$ for a category A, the objects of D-$\mathcal{A}\ell g$ may be lax functors $A \to CAT$, and the morphisms lax natural transformations (cf. Street [24]).

On the other hand the category of symmetric monoidal closed categories, and of morphisms preserving all the structure on the nose - the internal-hom as well as $\otimes$ and I - is not doctrinal over CAT. Indeed it is only a category, not a 2-category: there seems to be no natural definition of 2-cell. It is monadic over CAT, but the monad on CAT is only a functor, not a 2-functor.

3.6 Because in the doctrine case we have D-$\mathcal{A}\ell g$ as well as D-$\mathcal{A}\ell g_*$ = K^D, some new questions arise about the matter in §3.2 and §3.4 above.

We are using D-$\mathcal{A}\ell g$ to denote the D-algebras in the primary sense: actual objects of K acted on by D, and not 2-functors $A \to K$ acted on by D. Write U: D-$\mathcal{A}\ell g \to K$ for the forgetful 2-functor. Then it is not true in the sense of §3.5 that D acts on U: there is not a 2-natural transformation DU → U, but rather an op-lax-natural-transformation DU $\rightsquigarrow$ U. So a 2-functor E: $A \to$ D-$\mathcal{A}\ell g$ corresponds to a functor G = UE: $A \to K$ together with an action of D on G only in the weak sense that the action DG $\rightsquigarrow$ G is only op-lax-natural. Those G: $A \to K$ with honest actions of D correspond of course to the 2-functors $A \to$ D-$\mathcal{A}\ell g_*$, or to those 2-functors $A \to$ D-$\mathcal{A}\ell g$ that factorize through D-$\mathcal{A}\ell g_*$. Perhaps, therefore, the right thing to do is to allow both the n and the f in (3.16), when A is not an object of K but a 2-functor $A \to K$, to be op-lax; certainly the definitions work equally well if we do so, and reduce to the given ones when $A = \underline{1}$; but we want to stay as far from laxity as we can, and we don't think we shall need this extra generality in our papers below. So we shall pursue it no further here.

The same kind of observation, of course, applies to Proposition 3.4. A map d: D → D' of doctrines, in the sense of § 3.2, not only gives a 2-functor d-$\mathcal{A}\ell g_*$ = K^d satisfying

(3.20)

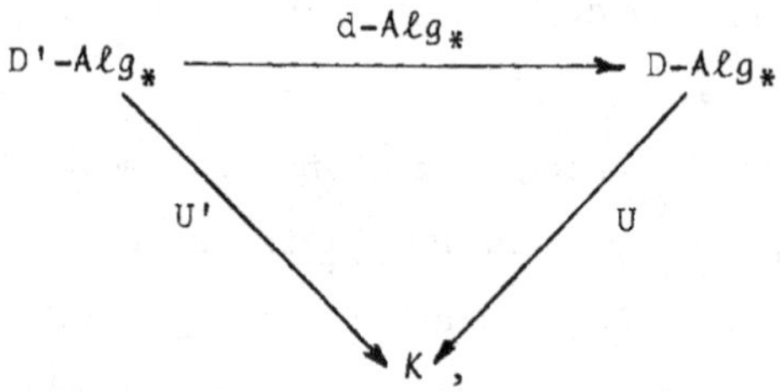

but also an evident 2-functor d-$\mathcal{Alg}$ satisfying

(3.21)
$$\begin{array}{ccc} D'\text{-}\mathcal{Alg} & \xrightarrow{\ d\text{-}\mathcal{Alg}\ } & D\text{-}\mathcal{Alg} \\ & {\scriptstyle U'}\searrow \quad \swarrow{\scriptstyle U} & \\ & K & \end{array},$$

where of course we are using U in an extended as well as a restricted sense. While, however, the only P: D'-$\mathcal{Alg}_*$ → D-$\mathcal{Alg}_*$ with UP = U' are the d-$\mathcal{Alg}_*$ for a 2-natural d: D → D' that is a doctrine map, there are many more P: D'-$\mathcal{Alg}$ → D-$\mathcal{Alg}$ with UP = U'; they correspond to op-lax-natural d: D $\rightsquigarrow$ D' satisfying the appropriate axioms. Again we hope to avoid these.

While we are at this level, we recall that the algebraic 2-functor d-$\mathcal{Alg}_*$ in (3.20) has a left adjoint in suitable cases. If K is 2-complete, so is D'-$\mathcal{Alg}_*$ = $K^{D'}$, and d-$\mathcal{Alg}_*$ preserves not only 2-limits but also cotensor products. It follows from Proposition 4.1 of Kelly [14] that the 2-functor d-$\mathcal{Alg}_*$ has a 2-left-adjoint if the underlying functor has a left adjoint. By Proposition 1.5.1 of Manes [20], this is so if the category D'-$\mathcal{Alg}_*$ has coequalizers of reflexive pairs. By Satz 10.3 of Gabriel-Ulmer [9], this is so when the underlying category K_0 of K is locally presentable (e.g. K = $\mathcal{Cat}$, K = $\mathcal{Cat}^2$, K = $\mathcal{Cat}/\Lambda$, etc.) and when D' has a <u>rank</u> in their sense - moreover they show that for a locally presentable K_0 the various reasonable definitions of having a rank coincide. We have just not thought out the question of whether d-$\mathcal{Alg}$ has a left adjoint in this situation; for example, j-$\mathcal{Alg}$: D-$\mathcal{Alg}$ → 1-$\mathcal{Alg}$ is just U: D-$\mathcal{Alg}$ → K; does it have a left adjoint?

We have in the present case the notion of a <u>modification of doctrine maps</u> δ: d ⇒ $\bar{d}$: D → D'; namely a modification δ: d ⇒ $\bar{d}$ satisfying

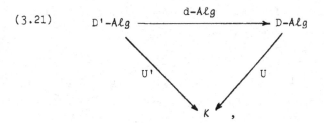

(3.22) $\quad j' = \delta j, \; m'.\delta\delta = \delta.m.$

Here $\delta\delta$ denotes neither the vertical composite $\delta\cdot\delta$ nor the horizontal composite $\delta*\delta$ of modifications, but the common value of

(3.23)
$$DD \;\overset{Dd}{\underset{D\bar{d}}{\Downarrow D\delta}}\; DD' \;\overset{dD'}{\underset{\bar{d}D'}{\Downarrow \delta D'}}\; D'D' \;=\; DD \;\overset{dD}{\underset{\bar{d}D}{\Downarrow \delta D}}\; D'D \;\overset{D'd}{\underset{D'\bar{d}}{\Downarrow D'\delta}}\; D'D,$$

which coincide by (1.1) and (1.2); notation in a 3-category presents frightful problems.

If δ is such a modification of doctrine maps, and if A is a D'-algebra with action $n'\colon D'A \to A$, then A has two D-algebra structures

$$n\colon DA \xrightarrow[dA]{} D'A \xrightarrow[n']{} A, \qquad \bar{n}\colon DA \xrightarrow[\bar{d}A]{} D'A \xrightarrow[n']{} A.$$

It is easily verified that $(1, n'.\delta A)$ is a D-morphism $(A,\bar{n}) \to (A,n)$:

(3.24)
$$\begin{array}{ccccc} DA & \xrightarrow{\bar{d}A} & D'A & \xrightarrow{n'} & A \\ {\scriptstyle 1}\downarrow & \overset{\Longrightarrow}{\scriptstyle \delta A} & \downarrow{\scriptstyle 1} & & \downarrow{\scriptstyle 1} \\ DA & \xrightarrow[dA]{} & D'A & \xrightarrow[n']{} & A\,. \end{array}$$

It follows that δ induces a 2-natural transformation $\delta\text{-}\mathcal{A\ell g}$ rendering commutative

(3.25)
$$\begin{array}{ccc} D'\text{-}\mathcal{A\ell g} & \overset{\bar{d}\text{-}\mathcal{A\ell g}}{\underset{d\text{-}\mathcal{A\ell g}}{\Downarrow \delta\text{-}\mathcal{A\ell g}}} & D\text{-}\mathcal{A\ell g} \\ & {\scriptstyle U'}\searrow \quad \swarrow{\scriptstyle U} & \\ & K\,, & \end{array}$$

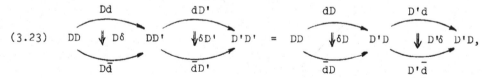

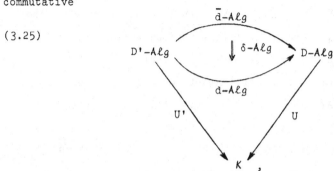

in the sense that $U.\delta\text{-}\mathcal{A}\ell g$ = identity. We leave the reader to prove:

Proposition 3.5. *Any 2-natural transformation* κ *rendering commutative*

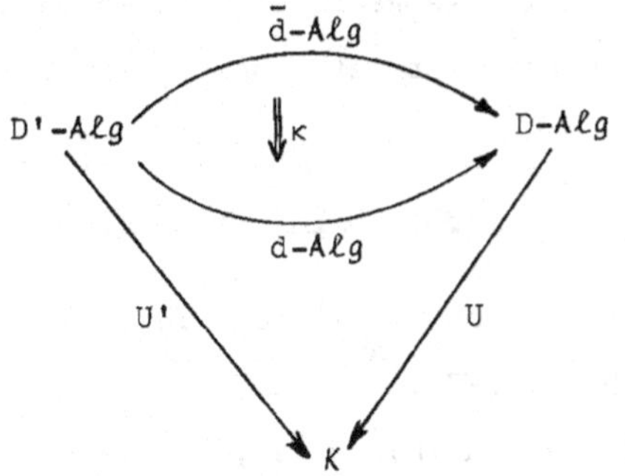

is $\delta\text{-}\mathcal{A}\ell g$ *for a unique modification* $\delta: d \Rightarrow \bar{d}$ *of doctrine maps.* □

It follows that equivalent doctrines D,D', in the sense that there are doctrine maps d: D → D', e: D' → D with $d.e \cong 1$ and $e.d \cong 1$ by modifications of doctrine maps, give equivalent 2-categories D-$\mathcal{A}\ell g$ and D'-$\mathcal{A}\ell g$; since the modifications here are isomorphisms, so that the D-morphism (3.24) is strong, even D-$\mathcal{A}\ell g_{**}$ and D'-$\mathcal{A}\ell g_{**}$ are equivalent, where D-$\mathcal{A}\ell g_{**}$ is the sub-2-category of D-$\mathcal{A}\ell g$ where only the *strong* D-morphisms are taken. However it is by no means the case that D-$\mathcal{A}\ell g_*$ and D'-$\mathcal{A}\ell g_*$ are equivalent.

As an example, let a D-algebra be a monoidal category A, with $\otimes: A\times A \to A$ and $I: \underline{\underline{1}} \to A$; let a D'-category be a category A with, for each $n \geq 0$, functors $\overset{n}{\otimes}: A^n \to A$, and with coherent isomorphisms $\overset{n}{\otimes}(\overset{m_1}{\otimes},\ldots,\overset{m_n}{\otimes}) \cong \overset{m}{\otimes}$, where $m = m_1 + \ldots + m_n$. A D'-algebra A gives a D-algebra A when we set $\otimes = \overset{2}{\otimes}$, $I = \overset{0}{\otimes}$; strict morphisms go into strict morphisms, so this comes from a doctrine map d: D → D'. A D-algebra A gives a D'-algebra A on setting $\overset{0}{\otimes} = I$, $\overset{1}{\otimes} = 1$, $\overset{n}{\otimes} = \otimes(1, \overset{n-1}{\otimes})$; again strict morphisms go to strict morphisms, so this comes from a doctrine map e: D' → D. The composite D-$\mathcal{A}\ell g$ → D'-$\mathcal{A}\ell g$ → D-$\mathcal{A}\ell g$ is the identity, whence e.d = 1; the composite D'-$\mathcal{A}\ell g$ → D-$\mathcal{A}\ell g$ → D'-$\mathcal{A}\ell g$ is clearly isomorphic to 1, whence d.e ≅ 1, and D and D' are equivalent. However the composite D'-$\mathcal{A}\ell g_*$ → D-$\mathcal{A}\ell g_*$ → D'-$\mathcal{A}\ell g_*$ is not isomorphic to 1; the original and final D'-algebra structures on A are such that 1: A → A is a strong, but not a strict, isomorphism between them.

BIBLIOGRAPHY

[1] J. Benabou, Introduction to bicategories, Lecture Notes in Math. 47 (1967), 1-77.

[2] M.C. Bunge, Bifibration induced adjoint pairs, Lecture Notes in Math. 195 (1971), 70-122.

[3] M.C. Bunge, Coherent extensions and relational algebras (preprint, March 1973).

[4] B.J. Day and G.M. Kelly, Enriched functor categories, Lecture Notes in Math. 106 (1969), 178-191.

[5] E.J. Dubuc, Kan extensions in enriched category theory, = Lecture Notes in Math. 145 (1970).

[6] C. Ehresmann, Catégories structurées, Ann. Sci. École Norm. Sup. 80 (1963), 349-425.

[7] C. Ehresmann, Catégories et structures (Dunod, Paris, 1965).

[8] S. Eilenberg and G.M. Kelly, Closed categories, Proc. Conf. on Categorical Algebra(La Jolla 1965.) (Springer, New York, 1966).

[9] P. Gabriel and F. Ulmer, Lokal präsentierbare Kategorien, = Lecture Notes in Math. 221 (1971).

[10] J.W. Gray, The categorical comprehension scheme, Lecture Notes in Math. 99 (1969), 242-312.

[11] J.W. Gray, Report "The meeting of the Midwest Category Seminar in Zürich, August 24-30, 1970", Lecture Notes in Math. 195 (1971), 248-255.

[12] J.W. Gray, Formal category theory, to appear in Lecture Notes in Math.

[13] A. Grothendieck, Catégories fibrées et descente, Séminaire de Géométrie Algébrique, Institut des Hautes Études Scientifiques, Paris (1961).

[14] G.M. Kelly, Adjunction for enriched categories, Lecture Notes in Math. 106 (1969), 166-177.

[15] G.M. Kelly, On clubs and doctrines (in this volume).

[16] G.M. Kelly, Coherence theorems for lax algebras and for distributive laws (in this volume).

[17] G.M. Kelly, Doctrinal adjunction (in this volume).

[18] F.W. Lawvere, Ordinal sums and equational doctrines, Lecture Notes in Math. 80 (1969), 141-155.

[19] F.E.J. Linton, Autonomous categories and duality of functors, Journal of Algebra 2 (1965), 315-349.

[20] E. Manes, A triple miscellany: some aspects of the theory of algebras over a triple (Dissertation, Wesleyan University, 1967).

[21] P.H. Palmquist, The double category of adjoint squares, Lecture Notes in Math. 195 (1971), 123-153.

[22] J.E. Roberts, A characterization of initial functors, Journal of Algebra 8 (1968), 181-193.

[23] R. Street, The formal theory of monads, J. Pure and Applied Algebra, 2 (1972), 149-168.

[24] R. Street, Two constructions on lax functors, Cahiers de Topologie et Géometrie Différentielle XIII, 3 (1972), 217-264.

[25] R. Street, Fibrations and Yoneda's lemma in a 2-category, (in this volume).

[26] R. Street, Elementary cosmoi, (in this volume).

[27] V. Zöberlein, Doktrinen auf 2-Kategorien (Manuscript, Math. Inst. der Univ. Zürich, 1973).

FIBRATIONS AND YONEDA'S LEMMA IN A 2-CATEGORY

by

Ross Street

Our purpose is to provide within a 2-category a conceptual proof of a set-free version of the Yoneda lemma using the theory of fibrations. In doing so we carry many definitions of category theory into a 2-category and prove in this more general setting results already familiar for *CAT*.

The La Jolla articles of Lawvere [5] and Gray [2] have strongly influenced this work. Both articles are written in styles which allow easy transfer into a 2-category. However, they also freely use the fact that *CAT* is cartesian closed, a luxury we do not allow ourselves.

The 2-category is required to satisfy an elementary completeness condition amounting to the existence of 2-pullbacks and comma objects. This relates the 2-category closely to a 2-category of category objects in a category. Such considerations appear in §1 and were considered by Gray [3].

Fibrations over B appear in §2 as pseudo algebras for a 2-monad on the 2-category of objects over B. This 2-monad is of a special kind distinguished by Kock [4]. We define lax algebras and lax homomorphisms for general 2-monads and provide alternative descriptions of pseudo algebras and lax homomorphisms for the special 2-monads. We are able then to give an equivalent definition of fibration generalizing the setting for the Chevalley criterion of Gray [2] p 56.

In order to eliminate the need for our 2-category to be cartesian closed in the remainder of our work we are led to introduce an extra variable; we must consider bifibrations from A to B rather than fibrations over B. A particular class of spans from A to B, called covering spans, is introduced in §3. As with their analogue in topology, covering spans are bifibrations. Furthermore, any arrow of spans between covering spans is a homomorphism. In the case of *CAT*, bifibrations correspond to category-valued functors and the last sentence reflects the fact that covering spans correspond to those functors which are discrete-category-valued; that is, set-valued. With this interpretation of covering spans as set-valued functors, we see that Corollary 16 is a generalization of the Yoneda lemma of category theory.

The concept of Kan extension of functors is one of the most fruitful concepts of category theory, and the definition just begs translation into a 2-category. This has already been used to some extent (see [6] and [7]). But the Kan extensions of functors which occur in practice are all pointwise (using the terminology of Dubuc [1]). Using comma objects we define pointwise extensions in a 2-category in §4. Note that, in general, for the 2-category *V-Cat*, this definition does not agree with Dubuc's; ours is too strong (we hope to remedy this by passing to some related 2-category). For *V=Set* and *V*=**2**, the definitions do agree; for *V=AbGp* and *V=Cat*, they do not. The closing section gives some applications of the Yoneda lemma and fibration theory to pointwise extensions illustrating their many pleasing properties.

§1. Representable 2-categories.

Let A denote a category. A *span* from A to B in A is a diagram (u_0, S, u_1):

$$\begin{array}{ccc} & S & \\ {\scriptstyle u_0}\swarrow & & \searrow{\scriptstyle u_1} \\ A & & B \end{array} .$$

When no confusion is likely, we abbreviate (u_0, S, u_1) to S; then the *reverse span* (u_1, S, u_0) is abbreviated to S^*. Also we identify an arrow $u: A \longrightarrow B$ with the span $(1, A, u)$ from A to B. An arrow of spans $f: (u_0, S, u_1) \longrightarrow (u_0', S', u_1')$ is a commutative diagram

$$\begin{array}{ccccc} & & S & & \\ & {\scriptstyle u_0}\swarrow & \downarrow{\scriptstyle f} & \searrow{\scriptstyle u_1} & \\ A & & & & B \\ & {\scriptstyle u_0'}\nwarrow & \downarrow & \nearrow{\scriptstyle u_1'} & \\ & & S' & & \end{array}$$

Let $SPN(A,B)$ denote the category of spans from A to B and their arrows.

When A has pullbacks, a span (u_0, S, u_1) from A to B and a span (v_0, T, v_1) from B to C have a *composite span* $(u_0\hat{v}_0, T\circ S, v_1\hat{u}_1)$ from A to C where the following square is a pullback.

$$\begin{array}{ccc} T\circ S & \xrightarrow{\hat{u}_1} & T \\ {\scriptstyle \hat{v}_0}\downarrow & & \downarrow{\scriptstyle v_0} \\ S & \xrightarrow[u_1]{} & B \end{array}$$

If $f: S \longrightarrow S'$ is an arrow of spans from A to B and $g: T \longrightarrow T'$ is an arrow of spans from B to C then the arrow $g\circ f: T\circ S \longrightarrow T'\circ S'$ induced on pullbacks is an arrow of spans.

An *opspan* from A to B in A is a span from A to B in A^{op}; however, arrows of opspans are arrows of diagrams in A.

Suppose A has pullbacks. A *category object* $\underset{\sim}{A}$ in A consists of the following data from A:

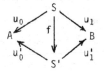

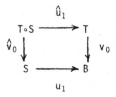

- an object A_0;
- a span (d_0,A_1,d_1) from A_0 to A_0;
- arrows of spans $i:(1,A_0,1) \longrightarrow (d_0,A_1,d_1)$,

  $c:(d_0\hat{d}_0,A_1 \circ A_1,d_1\hat{d}_1) \longrightarrow (d_0,A_1,d_1)$;

such that the following diagrams commute

$$\begin{array}{ccccc} A_1 & \xrightarrow{1\circ i} & A_1\circ A_1 & \xleftarrow{i\circ 1} & A_1 \\ & {\scriptstyle 1}\searrow & \downarrow{\scriptstyle c} & \swarrow{\scriptstyle 1} & \\ & & A_1 & & \end{array} \qquad \begin{array}{ccc} A_1\circ A_1\circ A_1 & \xrightarrow{1\circ c} & A_1\circ A_1 \\ {\scriptstyle c\circ 1}\downarrow & & \downarrow{\scriptstyle c} \\ A_1\circ A_1 & \xrightarrow[c]{} & A_1 \end{array} .$$

A *functorial arrow* $\underset{\sim}{f}:\underset{\sim}{A} \longrightarrow \underset{\sim}{B}$ consists of an arrow $f_0:A_0 \longrightarrow B_0$ and an arrow of spans $f_1:(f_0d_0,A_1,f_0d_1) \longrightarrow (d_0,B_1,d_1)$ such that the following commutes

$$\begin{array}{ccccc} A_0 & \xrightarrow{i} & A_1 & \xleftarrow{c} & A_1\circ A_1 \\ {\scriptstyle f_0}\downarrow & & \downarrow{\scriptstyle f_1} & & \downarrow{\scriptstyle f_1\circ f_1} \\ B_0 & \xrightarrow[i]{} & B_1 & \xleftarrow[c]{} & B_1\circ B_1 \end{array} .$$

If $\underset{\sim}{f},\underset{\sim}{f}':\underset{\sim}{A} \longrightarrow \underset{\sim}{B}$ are functorial arrows, a *transformation* from $\underset{\sim}{f}$ to $\underset{\sim}{f}'$ is an arrow of spans $n:(f_0,A_0,f_0') \longrightarrow (d_0,B_1,d_1)$ such that the following diagram commutes

$$\begin{array}{ccc} A_1 & \xrightarrow{(nd_1)\circ f_1} & B_1\circ B_1 \\ {\scriptstyle f_1'\circ(nd_0)}\downarrow & & \downarrow{\scriptstyle c} \\ B_1\circ B_1 & \xrightarrow[c]{} & B_1 \end{array} .$$

With the natural compositions we obtain a 2-category $CAT(A)$ of category objects in A.

A category object $\underset{\sim}{A}$ in A is determined up to isomorphism by the contravariant category-valued functor on A which assigns to each object X of A the category whose source and target functions are $A(X,d_0)$, $A(X,d_1)$: $A(A,A_1) \longrightarrow A(X,A_0)$ and whose identities and composition are determined by the functions $A(X,i)$, $A(X,c)$. Indeed, we have described the object function of a 2-fully-faithful 2-functor

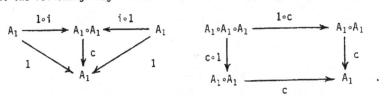

$$CAT(A) \longrightarrow [A^{op}, CAT] \ .$$

Henceforth we work in a 2-category K. By "span" we shall mean "span in the category K_0".

A *comma object* for the opspan (r,D,s) from A to B is a span $(d_0, r/s, d_1)$ from A to B together with a 2-cell

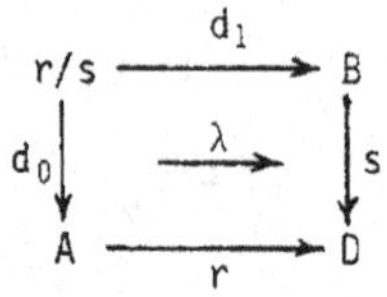

satisfying the following two conditions

- for any span (u_0,S,u_1) from A to B, composition with λ yields a bijection

between arrows of spans f and 2-cells σ;

- given 2-cells ξ,η such that the two composites

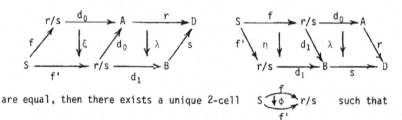

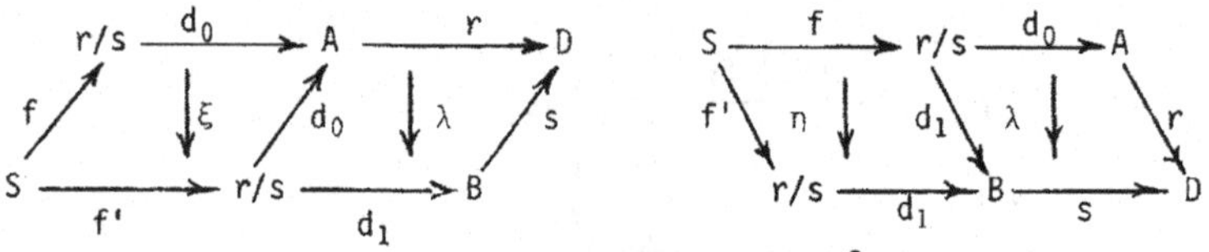

are equal, then there exists a unique 2-cell $S \underset{f'}{\overset{f}{\Downarrow \phi}} r/s$ such that

$\xi = d_0\phi$, $\eta = d_1\phi$.

In non-elementary terms, r/s is defined by a 2-natural isomorphism

$$K(S,r/s) \cong K(S,r)/K(S,s),$$

where the expression on the right hand side is the usual comma category of the functors $K(S,r)$, $K(S,s)$.

The comma object of the identity opspan $(1,A,1)$ from A to A is denoted by ΦA. It is defined by a 2-natural isomorphism

$$K(S,\Phi A) \cong K(S,A)^{\mathbf{2}} \quad ,$$

and so is the cotensor in K of the category $\mathbf{2}$ with the object A. When ΦA exists for each object A and when K has 2-pullbacks we say that K is a *representable 2-category* (Gray [3] uses "strongly representable").

<u>*Example.*</u> If A has pullbacks then $K = CAT(A)$ is a representable 2-category. //

Opcomma objects in K are comma objects in K^{op}. In a 2-category which is both representable and oprepresentable, Φ has a left 2-adjoint Ψ and any limit which exists in K_0 is automatically a 2-limit in K.

<u>*Proposition 1.*</u> *In a representable 2-category each opspan has a comma object.*

<u>*Proof.*</u> The formula is $r/s = s^* \circ \Phi D \circ r$. //

In a representable 2-category, an identity 2-cell $A \underset{1}{\overset{1}{\Downarrow 1}} A$ corresponds to an arrow $i : A \longrightarrow \Phi A$, and the composite 2-cell

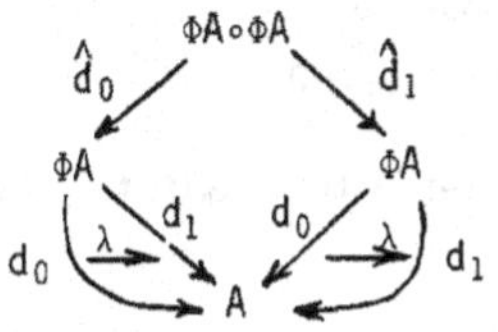

corresponds to an arrow $\Phi A \circ \Phi A \xrightarrow{c} \Phi A$. For each arrow $f : A \longrightarrow B$, the 2-cell $\Phi A \underset{d_1}{\overset{d_0}{\Downarrow \lambda}} A \xrightarrow{f} B$ corresponds to an arrow $\Phi f : \Phi A \longrightarrow \Phi B$.

<u>*Proposition 2.*</u> *In a representable 2-category the following results hold.*

(a) For each object A, *the arrows* i, c *enrich* $d_0, d_1 : \Phi A \longrightarrow A$ *to a category object* $\underset{\sim}{A}$ *in* K_0.

(b) For each arrow $f : A \longrightarrow B$, *the pair of arrows* $f, \Phi f$ *constitute a functorial arrow* $\underset{\sim}{f} : \underset{\sim}{A} \longrightarrow \underset{\sim}{B}$.

(c) For each 2-cell $A \underset{f'}{\overset{f}{\Downarrow \sigma}} B$, *the corresponding arrow* $\underset{\sim}{\sigma} : A \longrightarrow \Phi B$ *is a transformation from* $\underset{\sim}{f}$ *to* $\underset{\sim}{f}'$.

(d) The assignment

defines a 2-functor from K *to* $CAT(K_0)$.

<u>*Proof*</u>. (a) For each object X, $|K(X,A)^2| \rightrightarrows |K(X,A)|$ are the source and target functions for the category $K(X,A)$; so $K_0(X,\Phi A) \rightrightarrows K_0(X,A)$ are the source and target functions for a category, functorially in X. So $\Phi A \rightrightarrows A$ carries the structure of a category object in K_0. It is readily checked that this structure agrees with that of the proposition.

(b) For each X, $(K_0(X,f), K_0(X,\Phi f))$ corresponds to the functor $K(X,f)\colon K(X,A) \longrightarrow K(X,B)$.

(c) Similarly, $K_0(X,\underset{\sim}{\sigma})$ corresponds to the natural transformation $K(X,\sigma)\colon K(X,f) \longrightarrow K(X,f')$.

(d) What we have shown is that the composite

$$K \longrightarrow CAT(K_0) \longrightarrow [K_0{}^{op}, CAT]$$

is the Yoneda embedding, a well-known 2-functor. It follows that the first arrow is a 2-functor.//

<u>§2. Lax algebras and fibrations</u>

Suppose D is a 2-monad on a 2-category C and let $i\colon 1 \longrightarrow D$, $c\colon DD \longrightarrow D$ denote the unit and multiplication. A *lax* D-*algebra* consists of an object E, an arrow $c\colon DE \longrightarrow E$ and 2-cells

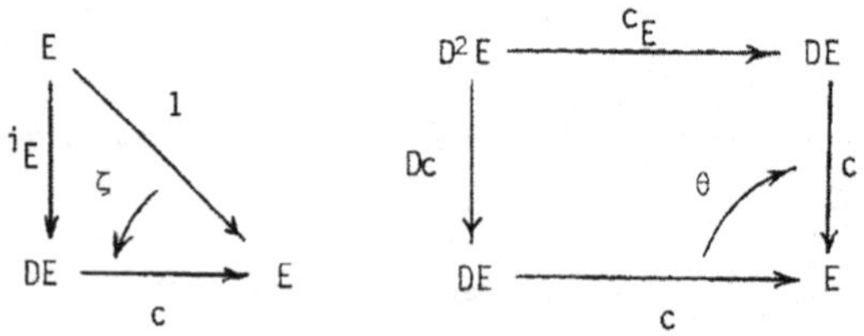

in the 2-category C such that the composites

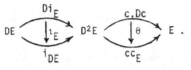

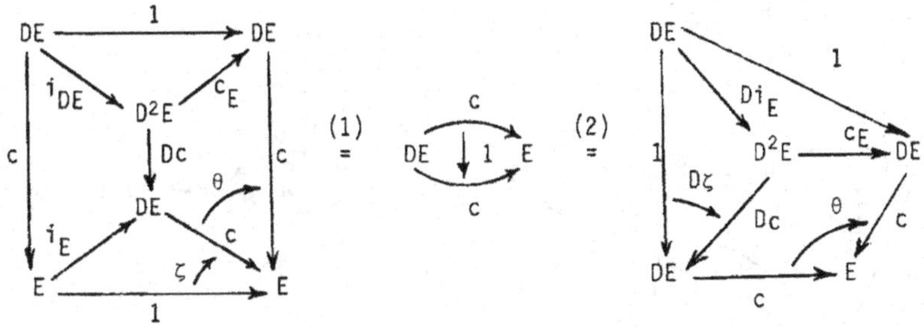

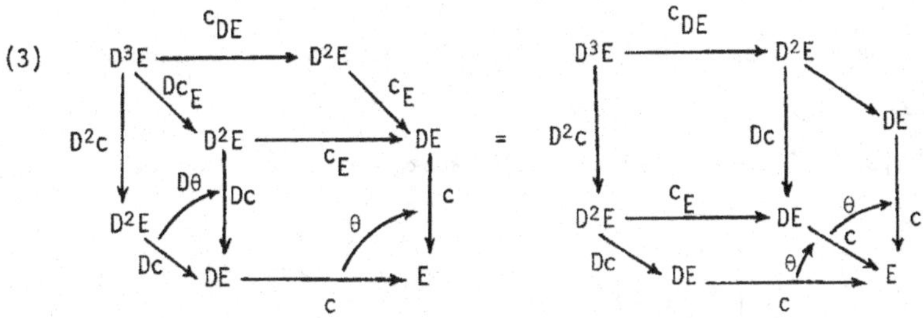

are equal as indicated. A *pseudo* D-*algebra* is a lax D-algebra in which ζ,θ are isomorphisms. A *normalized* lax D-algebra has ζ an identity 2-cell. A D-*algebra* is a lax D-algebra with both ζ,θ identities. Of course, for any E in C, DE with $c_E : D^2E \longrightarrow DE$ is the free D-algebra on E.

Kock [4] has distinguished those 2-monads D with the property that $c \dashv iD$ in the 2-functor 2-category $[C,C]$ with identity counit. Then the identity modification $D \underset{1}{\overset{c.Di}{\Downarrow 1}} D$ corresponds under the adjunction to a modification $D \underset{iD}{\overset{Di}{\Downarrow \iota}} D^2$. Suppose E is a lax D-algebra such that ζ is an isomorphism with inverse $\bar{\zeta}$, and consider the composite

$$DE \underset{i_{DE}}{\overset{Di_E}{\Downarrow \iota_E}} D^2E \underset{cc_E}{\overset{c.Dc}{\Downarrow \theta}} E .$$

On the one hand, $\theta\iota_E = (cc_E\iota_E)(\theta.Di_E) = \theta.Di_E = c.D\bar{\zeta}$.

On the other hand, $\theta\iota_E = (\theta i_{DE})(c.Dc.\iota_E) = (\bar{\zeta}c)(c.Dc.\iota_E)$.

So we have the equality

(4) 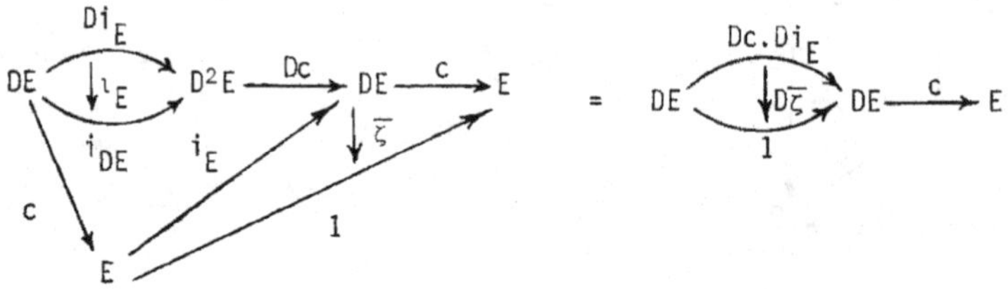

The next proposition generalizes slightly some of Kock's results; he considers the normalized case.

Proposition 3. *Suppose* D *is a 2-monad with the Kock property and suppose the 2-cell*

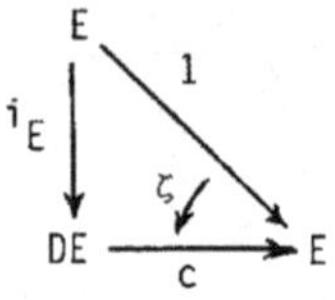

is an isomorphism with inverse $\overline{\zeta}$ *satisfying equality (4). Then:*

(a) $\overline{\zeta}$ *is the counit for an adjunction* $c \dashv i_E$ *with unit given by the composite*

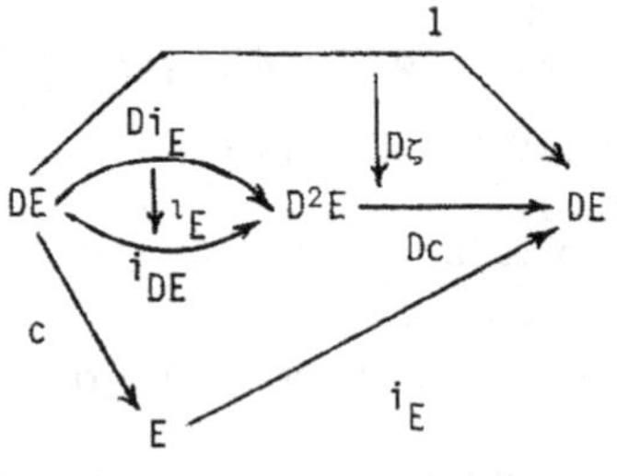

(b) the 2-cell $D^2E \overset{c.Dc}{\underset{c.c_E}{\underset{\Downarrow\theta}{\longrightarrow}}} E$ *corresponding under adjunction to the identity 2-cell* $E \overset{i_{DE}.i_E}{\underset{Di_E.i_E}{\underset{1\Downarrow}{\longrightarrow}}} D^2E$ *is unique with the property that the equality* (1) *holds;*

(c) this 2-cell θ *enriches* E, c, ζ *with the structure of pseudo* D*-algebra.*

Proof. (a) Let π denote the composite 2-cell displayed in *(a)*. Equality (4) gives $\bar{\zeta}c.c\pi = 1$. Since the composite $\iota_E i_E$ is the identity, we also have $i_E\bar{\zeta}.\pi i_E = 1$.

(b) Let τ denote the composite

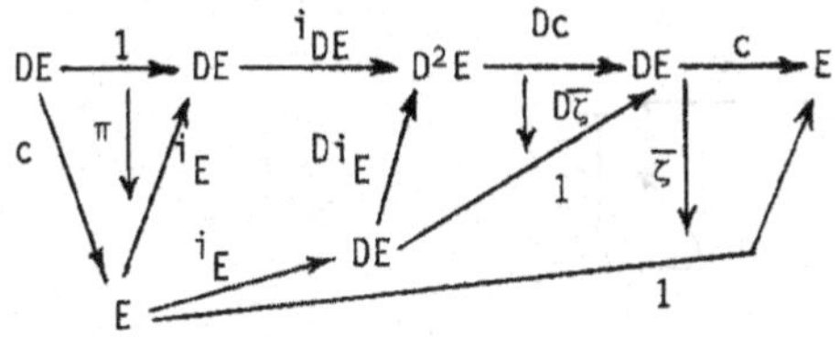

Then the 2-cell θ described in *(b)* is the composite

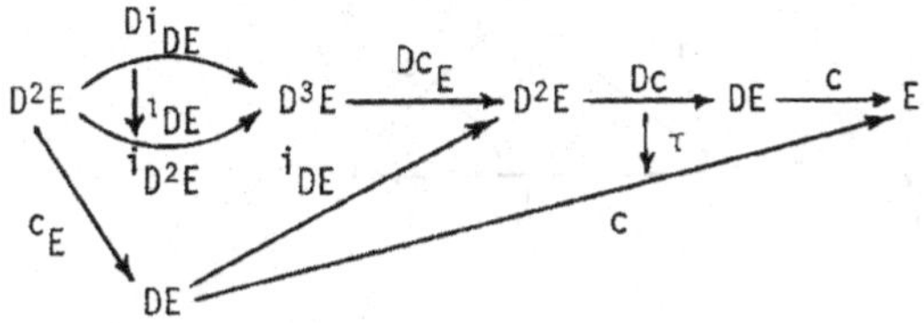

The 2-naturality of $i : 1 \longrightarrow D$ implies the equality $i_E\bar{\zeta}c = D\bar{\zeta}.Dc.i_{DE}$, from which it easily follows that $\tau = \bar{\zeta}c$. Using this and the equations $c_E i_{DE} = 1$, $\iota_{DE} i_{DE} = 1$, we deduce the equality (1).

To prove uniqueness, suppose θ satisfies (1). The 2-cell corresponding to θ under adjunction is the composite

$$i_{DE}i_E \xrightarrow{(\text{unit})i_{DE}i_E} Di_E.i_Ec.Dc.i_{DE}i_E \xrightarrow{Di_E.i_E\theta i_{DE}i_E} Di_E.i_Ecc_Ei_{DE}i_E$$

$$\xrightarrow{Di_E.i_E(\text{counit})} Di_E.i_E.$$

So (1) implies that this composite is independent of θ. For one such θ the composite is the identity, so the composite is the identity for all such θ. So θ is unique.

(c) Clearly θ is an isomorphism, so it remains to show that θ satisfies (2) and (3). Equality (2) follows from the equations $c_E.Di_E = 1$, $c.Dc.Dc_E.\iota_{DE}.Di_E = c.Dc.Dc_E.Di_{DE}.\iota_E = c.Dc.\iota_E$, $\tau = \overline{\zeta}c$ and (4). By the naturality of "replacing arrows by their right adjoints", equality (3) holds since identity 2-cells appear in the squares of the transformed equality.//

A *lax homomorphism* of lax D-algebras from E to E' consists of an arrow $f:E \longrightarrow E'$ and a 2-cell

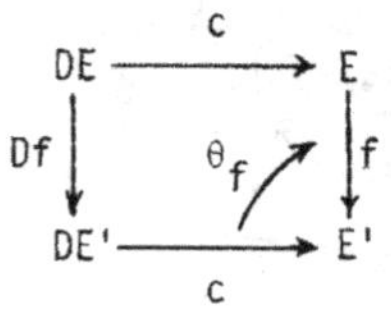

in C such that the composites

(5) 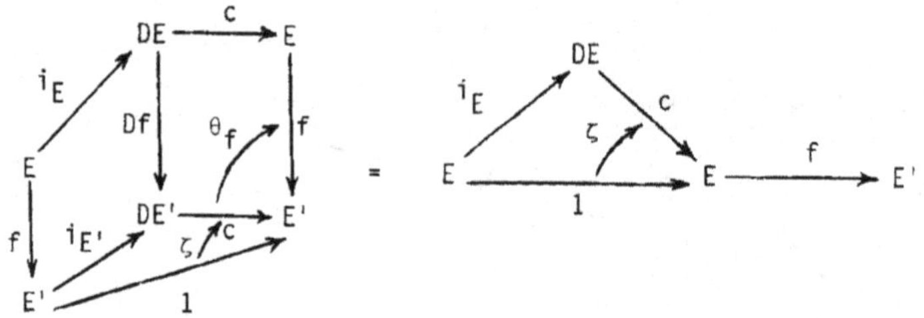

(6) 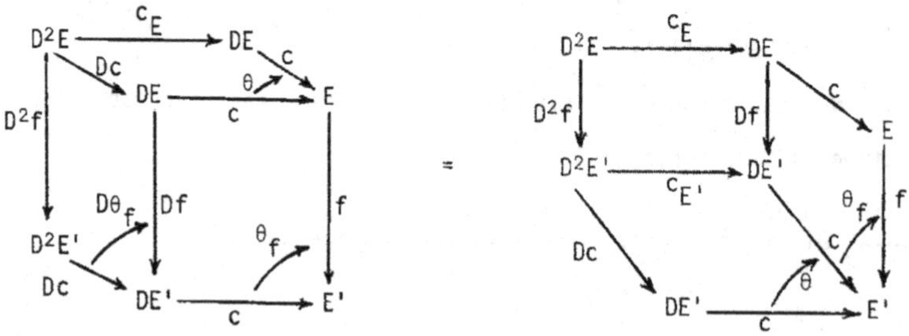

are equal as indicated. A lax homomorphism f is called a *pseudo homomorphism* when θ_f is an isomorphism, and is called a *homomorphism* when θ_f is an identity.

Proposition 4. *Suppose* D *is a 2-monad with the Kock property and suppose* $f:E \longrightarrow E'$ *is an arrow between pseudo* D*-algebras. Then the 2-cell* $\theta_f : c.Df \longrightarrow fc$ *which corresponds under adjunction to the identity 2-cell* $1:Df.i_E \longrightarrow i_{E'}.f$ *is unique with the property that equality* (5) *holds. Furthermore, this* θ_f *enriches* f *with the structure of lax homomorphism.*

Proof. Suppose θ_f is as explained in the proposition. Equality (5) holds since both the 2-cells $ci_{E'}f \longrightarrow f$ correspond to the identity 2-cell $i_{E'}f \xrightarrow{1} i_{E'}f$ under adjunction (recall that $\bar{\zeta}$ is the counit for $c \dashv i_{E'}$). On the other hand, suppose θ_f satisfies (5). Then θ_f corresponds under adjunction to the composite

$$Df.i_E \xrightarrow{\pi.Df.i_E} i_{E'}c.Df.i_E \xrightarrow{i_{E'}\theta_f i_E} i_{E'}fci_E \xrightarrow{i_{E'}f\zeta} i_{E'}f \quad ,$$

which is independent of θ_f by (5); so θ_f is unique. Finally, θ_f satisfies (6) since both the 2-cells $c.Dc.D^2f \longrightarrow fcc_E$ correspond under adjunction to the identity 2-cell $D^2f.i_{DE'}i_E = i_{DE'}.i_{E'}f \xrightarrow{1} Di_{E'}.i_{E'}f$ (recall that $\theta : c.Dc \longrightarrow cc_E$ corresponds to $1:i_{DE}i_E \longrightarrow Di_E.i_E$). //

For convenience we henceforth work in a representable 2-category *K*.

Proposition 5. *Suppose* $f:A \longrightarrow B$ *is an arrow with a right adjoint* u, *counit* ε *and unit* η. *For any arrow* $g:C \longrightarrow B$, *the arrow* $v:C \longrightarrow f/g$ *corresponding to the 2-cell* εg *is a right adjoint for* $d_1 : f/g \longrightarrow C$ *with counit the identity and unit* $\beta : 1 \longrightarrow vd_1$ *defined by the equations*

$$d_0\beta = u\lambda.\eta d_0 \quad , \quad d_1\beta = 1.$$

Proof. Using $\varepsilon f.f\eta = 1$, we see that the two composite 2-cells

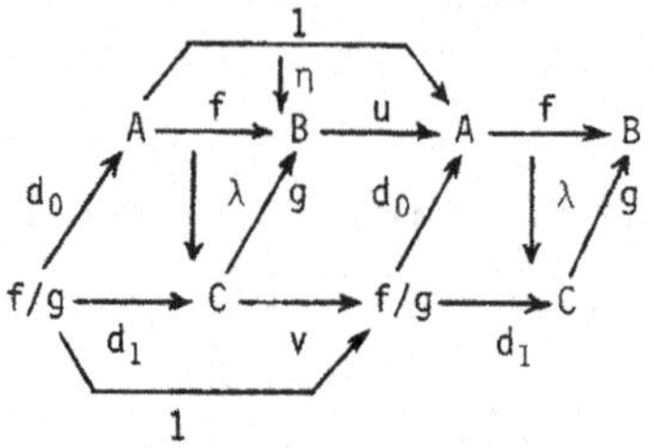

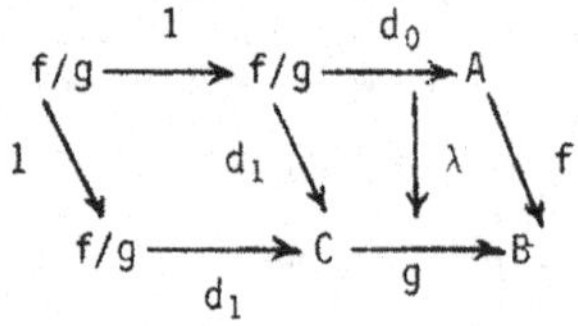

are equal; so there exists a unique 2-cell β as asserted. Using $u\varepsilon.\eta u = 1$, we also see that $\beta v = 1$. So β is a unit for $d_1 \dashv v$ with identity counit.//

<u>*Corollary 6.*</u> *For any arrow* $p:E \longrightarrow B$, *the arrow* $d_0:p/B \longrightarrow E$ *has a left adjoint* i_p *with unit the identity. Explicitly,* i_p *is the unique arrow whose composite with* λ *is the identity 2-cell* $E \overset{p}{\underset{p}{\Downarrow 1}} B$.

<u>*Proof.*</u> Since $1:B \longrightarrow B$ has a left adjoint, a dual of the proposition yields the result.//

<u>*Corollary 7.*</u> *An arrow* $f:A \longrightarrow B$ *has a right adjoint if and only if the arrow* $d_1:f/B \longrightarrow B$ *has a right adjoint. In this case there is a right adjoint for* d_1 *with counit the identity.*

<u>*Proof.*</u> If $d_1 \dashv v$ we can compose with $i_f \dashv d_0$ of Corollary 6 to obtain $f = d_1 i_f \dashv d_0 v$. The converse and the last sentence follow directly from Proposition 5.//

Corollary 4 applied to $p = 1_B$ gives i as left adjoint for $d_0:\Phi B \longrightarrow B$. The unit of this adjunction is the identity and the counit $\Phi B \overset{id_0}{\underset{1}{\Downarrow \iota_0}} \Phi B$ is the 2-cell defined by the equations $d_0\iota_0 = 1$, $d_1\iota_0 = \lambda$. Dually, $d_1:\Phi B \longrightarrow B$ has i as right adjoint with counit the identity and unit $\Phi B \overset{1}{\underset{id_1}{\Downarrow \iota_1}} \Phi B$ defined by $d_0\iota_1 = \lambda$, $d_1\iota_1 = 1$. Using the 2-pullback property of the square

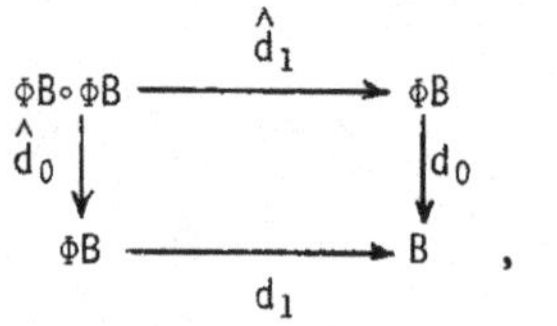

we see that $d_1\iota_0 = 1 = d_0\iota_1$ imply the existence of a unique 2-cell $\Phi B \overset{1\circ i}{\underset{i\circ 1}{\Downarrow \iota}} \Phi B\circ\Phi B$

such that $\hat{d}_0\iota = \iota_0$, $\hat{d}_1\iota = \iota_1$.

Proposition 8. *(a) The composite 2-cell*

$$\Phi B \overset{1\circ i}{\underset{i\circ 1}{\Downarrow \iota}} \Phi B\circ\Phi B \xrightarrow{\;c\;} \Phi B$$

is the identity 2-cell $\Phi B \overset{1}{\underset{1}{\Downarrow 1}} \Phi B$.

(b) The composite 2-cell

$$B \xrightarrow{\;i\;} \Phi B \overset{1\circ i}{\underset{i\circ 1}{\Downarrow \iota}} \Phi B\circ\Phi B$$

is the identity 2-cell $B \overset{i\circ i}{\underset{i\circ i}{\Downarrow 1}} \Phi B\circ\Phi B$.

(c) The arrow $c:\Phi B\circ\Phi B \longrightarrow \Phi B$ *is left adjoint to* $i\circ 1$ *with counit the identity and with unit given by the composite*

$$\begin{array}{c} \Phi B\circ\Phi B \overset{1\circ i\circ 1}{\underset{i\circ 1\circ 1}{\Downarrow \iota\circ 1}} \Phi B\circ\Phi B\circ\Phi B \xrightarrow{\;1\circ c\;} \Phi B\circ\Phi B \\ \text{(top: } \Phi B\circ\Phi B \xrightarrow{1} \Phi B\circ\Phi B\text{; bottom: } \Phi B\circ\Phi B \xrightarrow{c} \Phi B \xrightarrow{i\circ 1} \Phi B\circ\Phi B) \end{array}$$

Proof. (a) This follows from the calculations

$$d_0c\iota = d_0\hat{d}_0\iota = d_0\iota_0 = 1, \quad d_1c\iota = d_1\hat{d}_1\iota = d_1\iota_1 = 1.$$

(b) This follows from the calculations

$$\hat{d}_0\iota i = \iota_0 i = 1_i \ , \quad \hat{d}_1\iota i = \iota_1 i = 1_i .$$

(c) Using (a), we have

$$c(1\circ c)(\iota\circ 1) = c(c\circ 1)(\iota\circ 1) = c(c\iota\circ 1) = 1_c ;$$

and using (b), we have

$$(1\circ c)(\iota\circ 1)(i\circ 1) = (1\circ c)(\iota i\circ 1) = (1\circ c)1_{i\circ i\circ 1} = 1_{i\circ c(i\circ 1)} = 1_i .$$

So $c \dashv i\circ 1$ with counit and unit as stated.//

Let K_B denote the comma object of the opspan

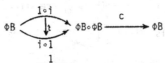

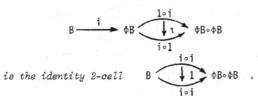

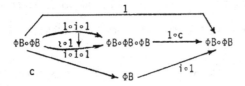

$$\begin{array}{ccc} K & & 1 \\ & {\scriptstyle 1}\searrow \quad \swarrow{\scriptstyle B} & \\ & K & \end{array}$$

in 2-*CAT*. So K_B is the 2-category whose objects are pairs (E,p) where $p:E \longrightarrow B$ is an arrow in K, whose arrows $f:(E,p) \longrightarrow (E',p')$ are arrows $f:E \longrightarrow E'$ in K such that $p'f = p$, and whose 2-cells $(E,p) \overset{f}{\underset{g}{\Downarrow\sigma}} (E',p')$ are 2-cells $E \overset{f}{\underset{g}{\Downarrow\sigma}} E'$ in K such that $p'\sigma = 1_p$. We often write E for (E,p).

Let $L:K_B \longrightarrow K_B$ denote the 2-functor given by:

$$(E,p) \overset{f}{\underset{g}{\Downarrow\sigma}} (E',p') \longmapsto (p/B,d_1) \overset{f/B}{\underset{g/B}{\Downarrow\sigma/B}} (p'/B,d_1);$$

or, in other words, $L(E,p) = (\Phi B \circ p, d_1 \hat{p})$, $Lf = 1 \circ f$, $L\sigma = 1 \circ \sigma$. Let $i:1 \longrightarrow L$, $c:L^2 \longrightarrow L$ denote the 2-natural transformations with (E,p)-components

$$E \xrightarrow{i \circ 1} \Phi B \circ p \ , \quad \Phi B \circ \Phi B \circ E \xrightarrow{c \circ 1} \Phi B \circ E \ .$$

The diagrams which say that $\underset{\sim}{B}$ (see Proposition 2(a)) is a category object precisely say that L is a 2-monad on K_B with unit i and multiplication c. Moreover, Proposition 8 shows that L has the Kock property so that Propositions 3 and 4 apply.

An arrow $p:E \longrightarrow B$ is called a *0-fibration over* B when (E,p) supports the structure of pseudo L-algebra. The 0-fibration is called *split* when (E,p) supports the structure of an L-algebra.

<u>*Proposition 9.*</u> *(Chevalley criterion). The arrow* $p:E \longrightarrow B$ *is a 0-fibration over* B *if and only if the arrow* $\bar{p}:\Phi E \longrightarrow p/B$ *corresponding to the 2-cell*

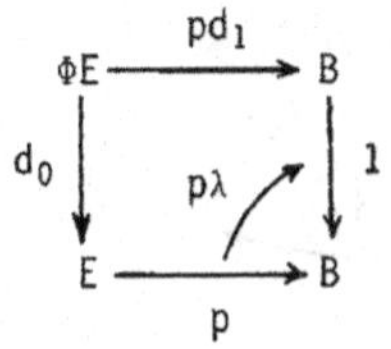

has a left adjoint with unit an isomorphism.

Proof. Suppose (E,p) is a pseudo L-algebra. The counit of Corollary 6 is readily seen to be

$$p/B = \Phi B\circ E \;\underset{1}{\overset{(\mathrm{id}_0)\circ 1}{\Downarrow \iota_0\circ 1}}\; \Phi B\circ E = p/B \;;$$

this 2-cell corresponds to an arrow $k: p/B \longrightarrow \Phi(p/B)$. Let ℓ be the composite $p/B \xrightarrow{k} \Phi(p/B) \xrightarrow{\Phi c} \Phi E$. One readily verifies that $\bar{p}\ell = L(c\iota_E)$. Let $p/B \;\underset{\bar{p}\ell}{\overset{1}{\Downarrow \eta}}\; p/B$ denote the 2-cell $L\zeta$; it is an isomorphism. Let $\Phi E \;\underset{1}{\overset{\ell\bar{p}}{\Downarrow \varepsilon}}\; \Phi E$ denote the unique 2-cell satisfying $d_0\varepsilon = \bar{\zeta}d_0$, $d_1\varepsilon = (\bar{\zeta}d_1)(c\bar{p}\iota_1)$. (Note that $\bar{p}i = i_E$). By applying d_0, d_1 to $\bar{p}\varepsilon.\eta\bar{p}$ it is readily seen that $\bar{p}\varepsilon.\eta\bar{p} = 1$. Also $d_0(\varepsilon\ell.\ell\eta) = 1$ is immediate. To complete the proof that $\ell \dashv \bar{p}$ with counit ε and unit η, we must show that $d_1(\varepsilon\ell.\ell\eta) = 1$. But $d_1(\varepsilon\ell.\ell\eta) = (\bar{\zeta}d_1\ell)(c\bar{p}\iota_1\ell)(cL\zeta)$.

From the calculations

$$d_0\bar{p}\iota_1\ell = d_0\iota_1\ell = \lambda\ell = \lambda.\Phi c.k = c\lambda k = c(\iota_0\circ 1) = c(\hat{d}_0\circ 1)(\iota\circ 1) = d_0(1\circ c)(\iota\circ 1),$$

$$d_1\bar{p}\iota_1\ell = \Phi p.d_1\iota_1\ell = 1_{\Phi(pc)k} = 1_{d_1k} = 1_{d_1} = d_1(\iota_1\circ 1) = d_1(\hat{d}_1\circ 1)(\iota\circ 1) = d_1(1\circ c)(\iota\circ 1),$$

we deduce that $\bar{p}\iota_1\ell = (1\circ c)(\iota\circ 1) = Lc.\iota_E$. So, by condition (4), we have $d_1(\varepsilon\ell.\ell\eta) = (\bar{\zeta}c)(c.Lc.\iota_E)(cL\zeta) = 1$.

Conversely, suppose $\ell \dashv \bar{p}$ with counit ε and isomorphism unit η. Since $\ell \dashv \bar{p}$ with counit ε and $d_1 \dashv i$ with counit 1, we have $d_1\ell \dashv \bar{p}i = i_E$ with counit $d_1\varepsilon i: d_1\ell pi \longrightarrow d_1 i = 1$. So put c equal to the composite $p/B \xrightarrow{\ell} \Phi E \xrightarrow{d_1} E$ and $\bar{\zeta} = d_1\varepsilon i$. It is readily checked that the composite

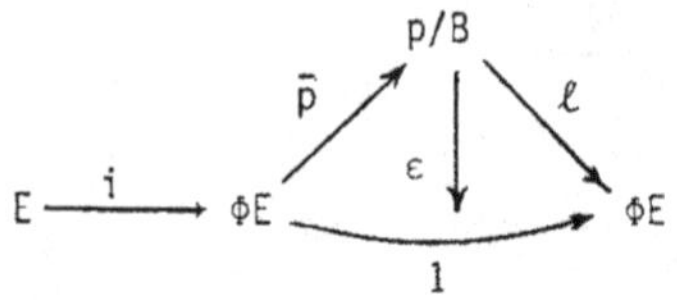

is an isomorphism with inverse the composite

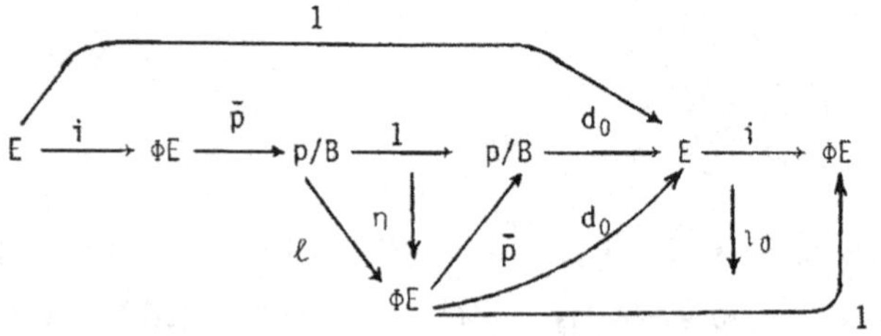

So $\bar{\zeta}$ is an isomorphism. The equality (4) for Proposition 3 follows easily. So E is a pseudo L-algebra.//

Compare the above proposition with Gray [2] p.56; so we have related the definition of 0-fibration here with the definition of opfibration in [2] when K = *Cat*. Notice that the unit of the adjunction $\ell \dashv \bar{p}$ for Gray is not just an isomorphism but an identity. It is worth pointing out the reason for this since we will need the observation in the next paper. A 0-fibration will be called *normal* when there is a normalized pseudo L-algebra structure on it. In *Cat* every 0-fibration is normal, but in other 2-categories this need not be the case. In the proof of the Chevalley criterion, if ζ is an identity then so is η. So, *for a normal* 0-*fibration,* $\bar{p}:\Phi E \longrightarrow p/B$ *has a left adjoint with unit an identity.*

For any arrow $g:B' \longrightarrow B$, "pulling back along g" is a 2-functor $g^*:K_B \longrightarrow K_{B'}$;

for each E in K_B, the diagram

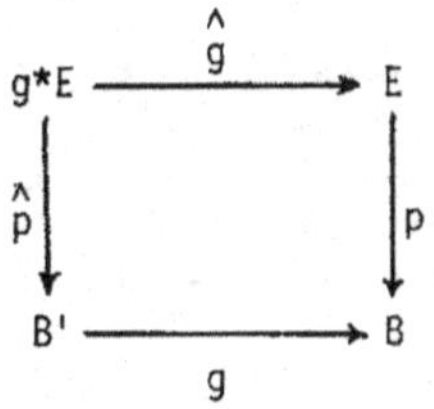

is a pullback. The composite 2-cell

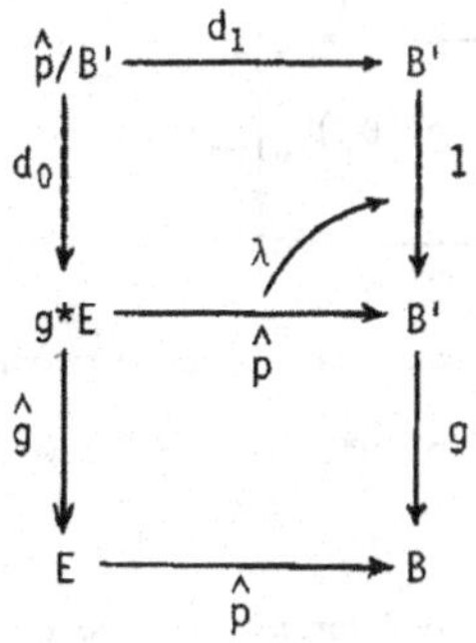

induces an arrow of spans $\check{g}_E$:

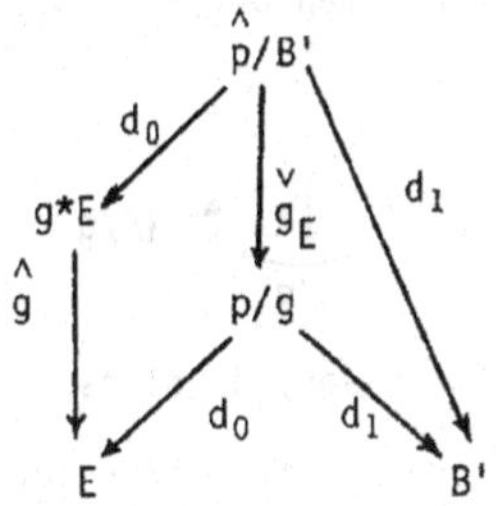

for each E in K_B. One readily checks that $\check{g}_E$, $E \in K_B$ are the components of a 2-natural transformation $\check{g}$:

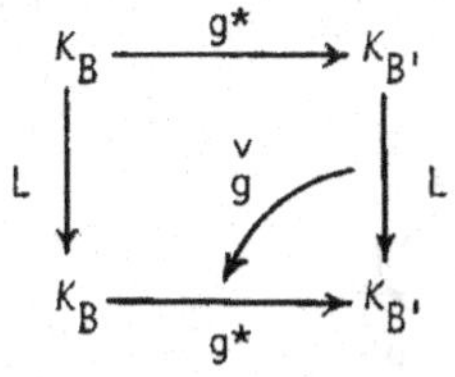

Indeed, in the language of Street [6], the pair $(g^*,\check{g})$ is a monad functor from

(K_B, L) to $(K_{B'}, L)$ in the 2-category 2-CAT.

Proposition 10. *Suppose* $g: B' \longrightarrow B$ *is an arrow in* K. *For each lax* L*-algebra* E, *the arrow*

$$Lg^*E \xrightarrow{\check{g}_E} g^*LE \xrightarrow{g^*(c)} g^*E$$

enriches g^*E *with the structure of lax* L*-algebra. For each lax homomorphism* $f: E \longrightarrow E'$ *of lax* L*-algebras, the 2-cell*

$$\begin{array}{ccccc} Lg^*E & \xrightarrow{\check{g}_E} & g^*LE & \xrightarrow{g^*(c)} & g^*E \\ {\scriptstyle Lg^*E}\downarrow & & \downarrow{\scriptstyle g^*(Lf)} & \nearrow{\scriptstyle g^*(\theta_f)} & \downarrow{\scriptstyle g^*(f)} \\ Lg^*E & \xrightarrow[\check{g}_{E'}]{} & g^*LE' & \xrightarrow[g^*(c)]{} & g^*E' \end{array}$$

enriches $g^*(f): g^*E \longrightarrow g^*E'$ *with the structure of lax homomorphism. If* E *is a pseudo* L*-algebra or an* L*-algebra then so is* g^*E. *If* f *is a pseudo homomorphism or a homomorphism then so is* $g^*(f)$.//

Corollary 11. *The pullback of a (split) 0-fibration along any arrow is a (split) 0-fibration.*//

Let $R: K_A \longrightarrow K_A$ denote the 2-functor given by:

$$(E,q) \underset{g}{\overset{f}{\Downarrow\sigma}} (E',q') \longmapsto (A/q, d_0) \underset{A/g}{\overset{A/f}{\Downarrow A/\sigma}} (A/q', d_0).$$

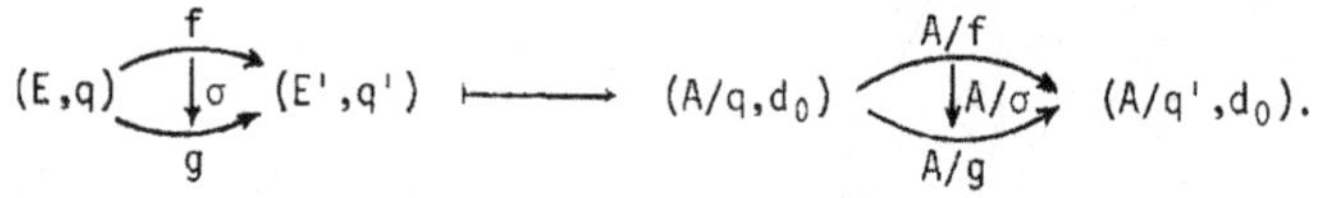

There is a 2-monad structure on R and the theory develops as for L; just replace K by K^{co}. An arrow $q: E \longrightarrow A$ is called a *1-fibration over* A when (E,q) supports the structure of pseudo R-algebra.

Note that the category $SPN(A,B)$ of spans from A to B becomes a 2-category by taking as 2-cells the 2-cells σ of K as in the diagram

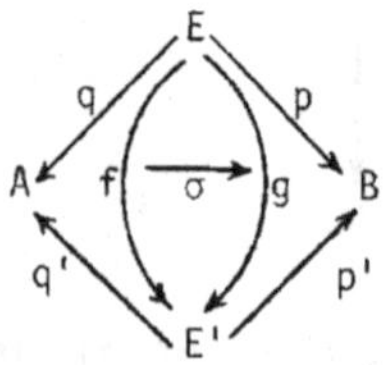

where $q'\sigma = 1_q$, $p'\sigma = 1_p$. Let $M:SPN(A,B) \longrightarrow SPN(A,B)$ denote the 2-functor given by:

$$\left(E \underset{g}{\overset{f}{\rightrightarrows}} E' \;\; \sigma\!\downarrow \right) \longmapsto \left(\Phi B \circ E \circ \Phi A \underset{1\circ g\circ 1}{\overset{1\circ f\circ 1}{\rightrightarrows}} \Phi B \circ E' \circ \Phi A \;\; \downarrow 1\circ\sigma\circ 1 \right).$$

This 2-functor supports the structure of 2-monad too; the unit $i:1 \longrightarrow M$ and multiplication $c:MM \longrightarrow M$ have as components

$$E \xrightarrow{i\circ 1\circ i} \Phi B\circ E\circ \Phi A \quad \text{and} \quad \Phi B\circ \Phi B\circ E\circ \Phi A\circ \Phi A \xrightarrow{c\circ 1\circ c} \Phi B\circ E\circ \Phi A\ .$$

A span (q,E,p) for A to B is called a *bifibration from* A *to* B when it supports the structure of pseudo M-algebra. A *split bifibration* is an M-algebra.

Results on L-algebras and R-algebras can be transferred to M-algebras via the following result. The corresponding statement for lax algebras is left to the reader.

Proposition 12. *Suppose* E *is a span from* A *to* B. *The* M-*algebra structures* $c:\Phi B\circ E\circ \Phi A \longrightarrow E$ *are in bijective correspondence with pairs of arrows of spans* $c_L:\Phi B\circ E \longrightarrow E$, $c_R:E\circ \Phi A \longrightarrow E$ *such that* c_L, c_R *are* L-*algebra*, R-*algebra structures on* E *related by the condition that*

$$\begin{array}{ccc} ME & \xrightarrow{1\circ c_R} & LE \\ {\scriptstyle c_L\circ 1}\big\downarrow & & \big\downarrow{\scriptstyle c_L} \\ RE & \xrightarrow[c_R]{} & E \end{array}$$

commutes; the bijection is determined by

$$c_L = (\Phi B\circ E \xrightarrow{1\circ 1\circ i} \Phi B\circ E\circ \Phi A \xrightarrow{c} E)$$

$$c_R = (E\circ \Phi A \xrightarrow{i\circ 1\circ 1} \Phi B\circ E\circ \Phi A \xrightarrow{c} E)$$

$$c = c_L(1\circ c_R) = c_R(c_L\circ 1).$$

Furthermore, an arrow of spans is a homomorphism of M-*algebras if and only if it is a homomorphism of both the corresponding* L-*algebras and the corresponding* R-*algebras.* //

Combining this with Corollary 11 and the dual for 1-fibrations we have:

Corollary 13. *For any arrows* $f: A' \longrightarrow A$, $g: B' \longrightarrow B$, *each (split) bifibration* E *from* A *to* B *induces a (split) bifibration* $g^* \circ E \circ f$ *from* A' *to* B'.//

There is a more general composition of bifibrations which we will not need. If E is a bifibration from A to B and F a bifibration from B to C then the bifibration $F \otimes E$ from A to C can be defined by the usual "tensor product of bimodules" coequalizer, provided this coequalizer exists and is preserved by certain pullbacks.

§3. Yoneda's Lemma within a 2-category.

Again we work in a representable 2-category K.

A *covering span* is defined to be a span which is the comma object of some opspan.

Theorem 14. *Any covering span is a split bifibration. Any arrow of spans between covering spans is a homomorphism.*

Proof. Any comma object r/s is a composite $s^* \circ \Phi D \circ r$. But ΦD is the value of M at the identity span of D; so ΦD is a free split bifibration. So r/s is a split bifibration by Corollary 13.

Suppose $f: r/s \longrightarrow u/v$ is an arrow of spans from A to B. We must prove that f commutes with the M-algebra structures on r/s and u/v. By Proposition 12, it suffices to show that f commutes with the L-algebra and R-algebra structures separately. By duality, it suffices to show that f commutes with just the L-algebra structures.

The L-algebra structure $c: \Phi B \circ (r/s) \longrightarrow r/s$ comes from that of ΦD via the commutative square

$$\begin{array}{ccc} \Phi B \circ (r/s) & \xrightarrow{\Phi s \circ \underset{\sim}{\lambda}} & \Phi D \circ \Phi D \\ {\scriptstyle c}\downarrow & & \downarrow{\scriptstyle c} \\ r/s & \xrightarrow[\underset{\sim}{\lambda}]{} & \Phi D \end{array} .$$

Equivalently, note that $\Phi B \circ (r/s)$ is the comma object of the opspan (r, D, sd_0) from A to ΦB (composed with $d_1: \Phi B \longrightarrow B$) since we have the pullback

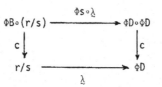

$$\begin{array}{ccc} r/sd_0 & \xrightarrow{\hat{d}_1} & \Phi B \\ {\scriptstyle \hat{d}_0}\downarrow & & \downarrow{\scriptstyle d_0} \\ r/s & \xrightarrow[d_1]{} & B \end{array}$$

and $c:r/sd_0 \longrightarrow r/s$ corresponds to the composite 2-cell

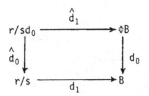

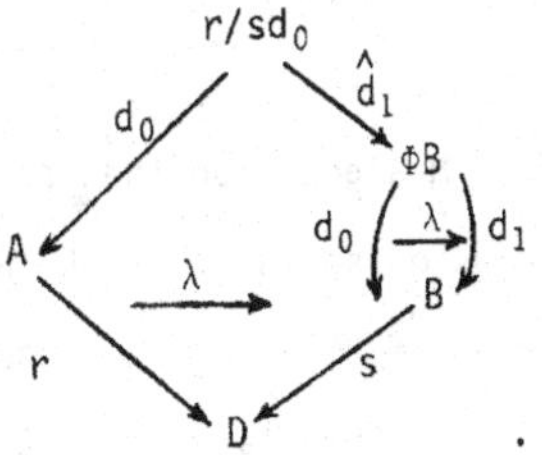

.

The main trick of the proof is to introduce the 2-cell $\alpha : \hat{d}_0 \Rightarrow c : r/sd_0 \to r/s$ defined by $d_0\alpha = 1_{d_0}$, $d_1\alpha = \lambda\hat{d}_1$; of course, we also have such an α for u/v. The arrow $L(f)$ is defined by the commutative diagram

$$\begin{array}{ccccc} r/s & \xleftarrow{\hat{d}_0} & r/sd_0 & \xrightarrow{\hat{d}_1} & \Phi B \\ {\scriptstyle f}\downarrow & & \downarrow{\scriptstyle L(f)} & & \downarrow{\scriptstyle 1} \\ u/v & \xleftarrow[\hat{d}_0]{} & u/vd_0 & \xrightarrow[\hat{d}_1]{} & \Phi B \end{array} .$$

The calculations

$$d_0\alpha L(f) = 1_{d_0}L(f) = 1_{d_0} = d_0\alpha = d_0 f\alpha$$
$$d_1\alpha L(f) = \lambda\hat{d}_1 L(f) = \lambda\hat{d}_1 = d_1\alpha = d_1 f\alpha$$

show that the following composites are equal

$$\left(r/sd_0 \underset{c}{\overset{\hat{d}_0}{\Downarrow\alpha}} r/s \xrightarrow{f} u/v\right) = \left(r/sd_0 \xrightarrow{L(f)} u/vd_0 \underset{c}{\overset{\hat{d}_0}{\Downarrow\alpha}} u/v\right).$$

So $c.L(f) = fc$, which proves that f is a homomorphism.//

Let *COV*(A,B) denote the full subcategory *SPN*(A,B) whose objects are the covering spans. Let *SPL*(A,B) denote the category of algebras for the monad M on *SPN*(A,B); it is the category of split bifibrations from A to B and their homomorphisms (up to equivalence).

Corollary 15. *The inclusion functor* COV(A,B)⟶ SPN(A,B) *factors through the underlying functor* SPL(A,B)⟶ SPN(A,B).//

Corollary 16 (Yoneda lemma). *Suppose* f:A⟶ B *is an arrow and* E *is a covering span from* A *to* B. *Composition with the arrow of spans* i_f:f⟶ f/B *yields a bijection between arrows of spans from* f/B *to* E *and arrows of spans from* f *to* E.

Proof. Note that f/B is the free M-algebra on the span f from A to B. This gives a bijection between arrows of spans f⟶ E and homomorphisms f/B⟶ E. But by Theorem 14, any arrow of spans f/B⟶ E is a homomorphism.//

Remark. Take K = CAT and A = $\mathbb{1}$ in Corollary 16. Covering spans E from $\mathbb{1}$ to B functorially correspond to functors e from B into some category of sets. An arrow f: $\mathbb{1}$⟶ B is just an object b of B. The bijection of the corollary becomes the usual bijection between natural transformations B(b,-)⟶ e and elements of eb obtained by evaluating at 1_b.

The following special case of Corollary 16 appears in Gray [3].

Corollary 17. *The functor* $K(A,B)^{op}$⟶ SPN(A,B) *given by*

is fully faithful.

Proof. The definition of comma objects gives the bijection

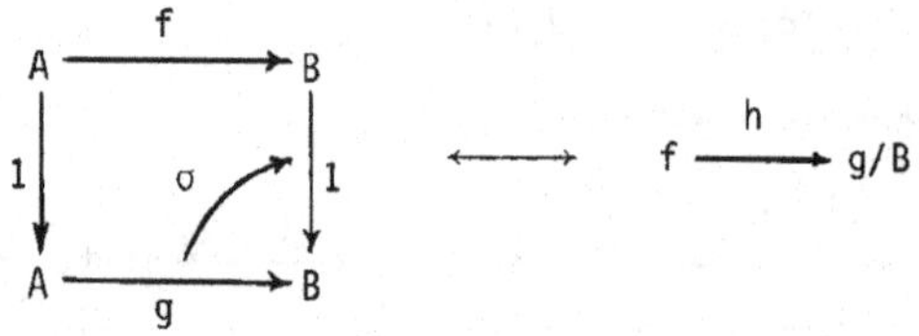

between 2-cells σ and arrows of spans h. The Yoneda lemma provides the bijection between such h and arrows of spans f/A⟶ g/B.//

§4. Pointwise extensions.

Recall the definition of left extension in a 2-category (see [6]).

Proposition 18. *There is a bijection between 2-cells*

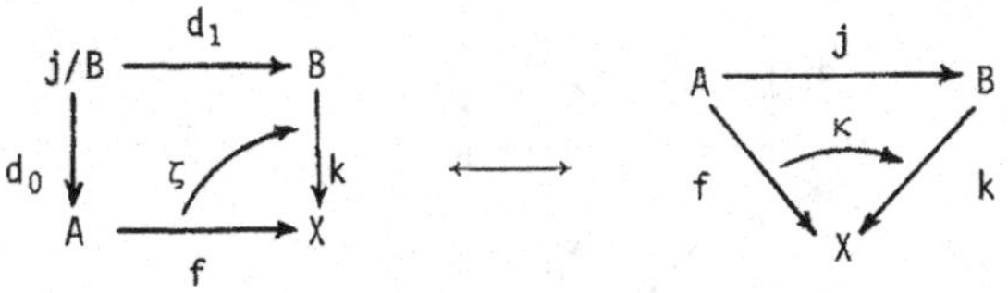

obtained by composition with i_j*. The 2-cell* κ *exhibits* k *as a left extension of* f *along* j *if and only if the corresponding* ζ *exhibits* k *as a left extension of* fd_0 *along* d_1*.*

Proof. By definition of comma objects there is a bijection between 2-cells ζ and arrows of spans $j/B \longrightarrow f/k$, and a bijection between 2-cells κ and arrows of spans $j \longrightarrow f/k$. The first sentence of the proposition now follows by the Yoneda lemma. For any arrow $\ell : B \longrightarrow X$, there are bijections

$$f \longrightarrow \ell j \quad \longleftrightarrow \quad j \longrightarrow f/\ell \quad \longleftrightarrow \quad j/B \longrightarrow f/\ell$$
$$\longleftrightarrow \quad j/B \longrightarrow \Phi X \quad \longleftrightarrow \quad fd_0 \longrightarrow \ell d_1$$

where the first, third and fourth are from the definition of comma object and the second is by Yoneda.//

The 2-cell

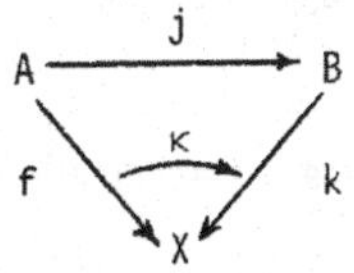

is said to exhibit k as a *pointwise left extension* of f along j when, for each arrow $g : C \longrightarrow B$, the composite 2-cell

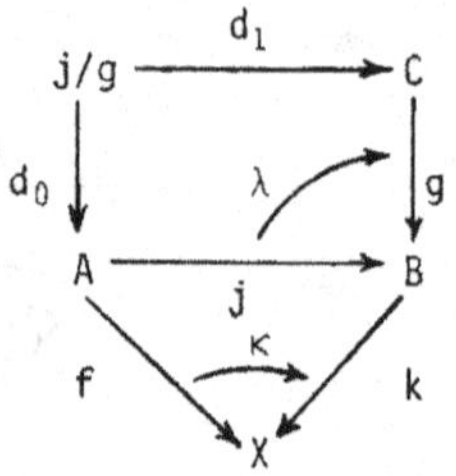

exhibits kg as a left extension of fd_0 along d_1.

Taking $g = 1_B$ in this definition we obtain the following corollary to the last proposition.

Corollary 19. *A pointwise left extension is a left extension.*//

Remark. When $K = CAT$, the pointwise left extensions are precisely those given by the formula

$$kb = \varinjlim (j/b \xrightarrow{d_0} A \xrightarrow{f} X).$$

To see this, take $C = \mathbb{1}$, $g = b$ and note that left extension along $j/b \longrightarrow \mathbb{1}$ is direct limit.

Recall that left extensions along an arrow with a right adjoint always exist and are obtained by composing with the right adjoint. The following result is a direct corollary of Proposition 5.

Proposition 20. *Any left extension along an arrow which has a right adjoint is pointwise.*//

Proposition 21. *An arrow* $f:B \longrightarrow A$ *is a left adjoint to* $u:A \longrightarrow B$ *if and only if there is an isomorphism* $f/A \cong B/u$ *of spans from* A *to* B.

Proof. Using the Yoneda lemma, we have bijections

$$1 \xrightarrow{\eta} uf \longleftrightarrow f \longrightarrow B/u \longleftrightarrow f/A \xrightarrow{m} B/u$$

$$fu \xrightarrow{\varepsilon} 1 \longleftrightarrow u^* \longrightarrow f/A \longleftrightarrow B/u \xrightarrow{n} f/A \quad .$$

It is readily checked that η, ε are a unit and counit for an adjunction $f \dashv u$ if and only if the corresponding m, n are mutually inverse isomorphisms.//

An arrow $j:A \longrightarrow B$ is said to be *fully faithful* when, given any 2-cell $C \underset{jv}{\overset{ju}{\Downarrow \tau}} B$, there exists a unique 2-cell $C \underset{v}{\overset{u}{\Downarrow \sigma}} A$ such that τ is the composite $C \underset{v}{\overset{u}{\rightrightarrows}} A \xrightarrow{j} B$ (with $\sigma: u \Rightarrow v$). It is readily seen that j is fully faithful if and only if the arrow of spans $\Phi A \longrightarrow j/j$ corresponding to $j\lambda$ is an isomorphism.

Proposition 22. *If* $j:A \longrightarrow B$ *is fully faithful and if the 2-cell*

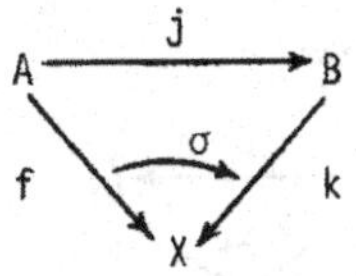

exhibits k *as a pointwise left extension of* f *along* j*, then* σ *is an isomorphism.*

Proof. Since k is a pointwise left extension and $\Phi A \longrightarrow j/j$ is an isomorphism, the composite 2-cell

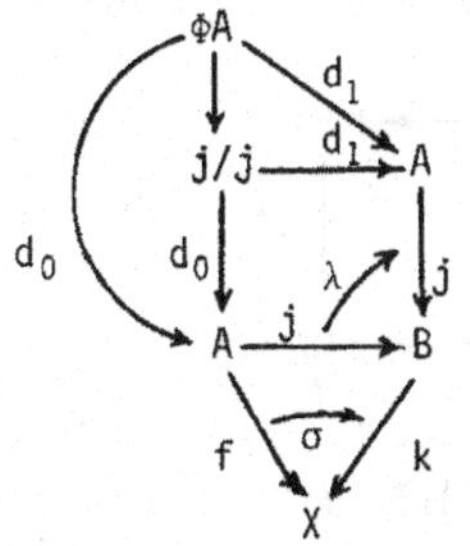

exhibits kj as a left extension of fd_0 along d_1. By Proposition 18, the corresponding 2-cell

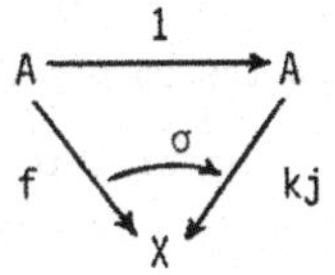

exhibits kj as a left extension of f along 1_A. But also the identity 2-cell exhibits f as a left extension of f along 1_A. So σ is an isomorphism.//

For a 0-fibration $p:E \longrightarrow B$ and arrow $b:G \longrightarrow B$, we denote by E_b the pullback of b along p.

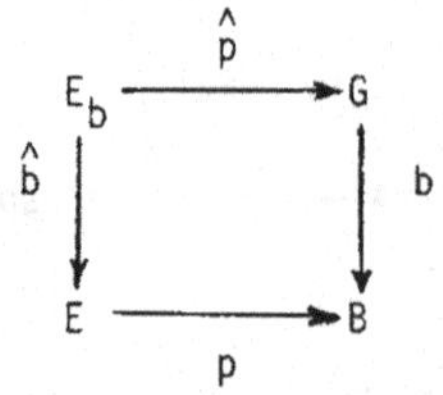

Proposition 23. *Suppose in the diagram*

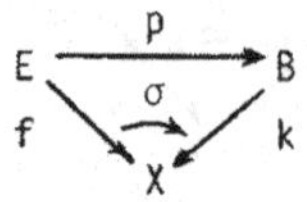

that p *is a normal* 0-*fibration. The* 2-*cell* σ *exhibits* k *as a pointwise left extension of* f *along* p *if and only if, for each arrow* $b:G \longrightarrow B$, *the* 2-*cell* $\sigma\hat{b}$ *exhibits* kb *as a left extension of* $f\hat{b}$ *along* $\hat{p}$.

Proof. The following square is readily seen to be a pullback.

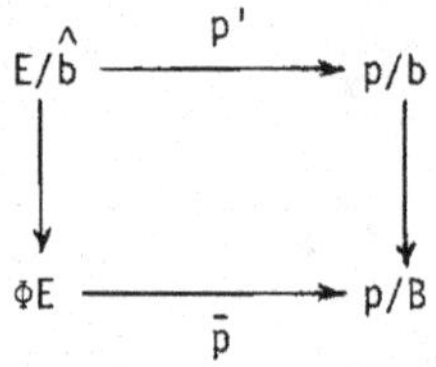

Since p is a normal 0-fibration, $\bar{p}$ has a left adjoint with unit an identity (Chevalley criterion). This property is preserved by pullback: so p' has a left adjoint ℓ' with unit an identity. The arrow $d_1:E/\hat{b} \longrightarrow E_b$ has a right adjoint $i_{\hat{b}}:E_b \longrightarrow E/\hat{b}$ (dual of Corollary 6). So the composite $d_1\ell':p/b \longrightarrow E_b$ has a right adjoint $p'i_{\hat{b}}$. Let η denote the unit of this adjunction. One readily checks the equations

$$\hat{b} = d_0 p' i_b\ , \quad \hat{p} d_1 \ell' = d_1 \quad \text{and} \quad p d_0 \eta = \lambda\ .$$

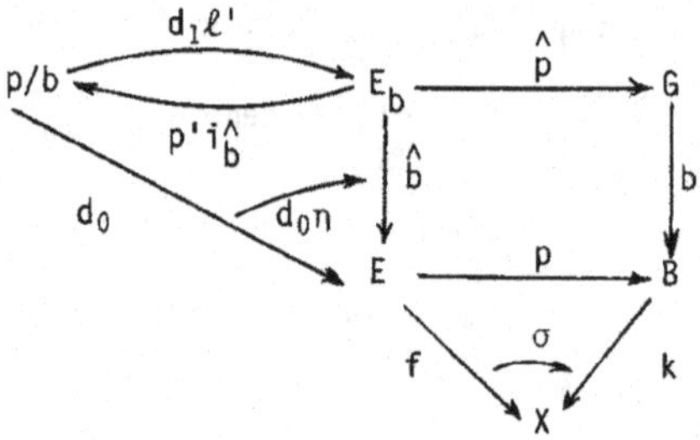

So $fd_0\eta$ exhibits $\hat{fb}$ as a left extension of fd_0 along $d_1\ell'$. It follows that the composite 2-cell

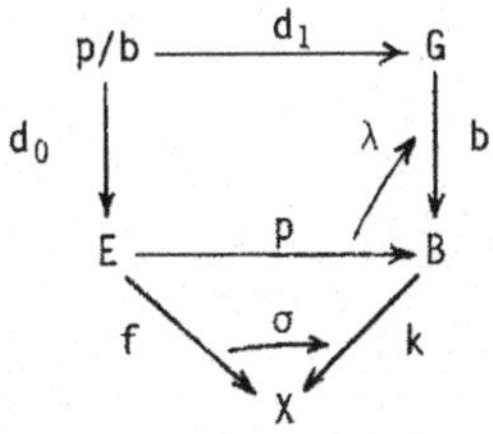

exhibits kb as a left extension of fd_0 along d_1 if and only if $\sigma\hat{b}$ exhibits kb as a left extension of $\hat{fb}$ along $\hat{p}$.//

Proposition 24. *Suppose in the diagram*

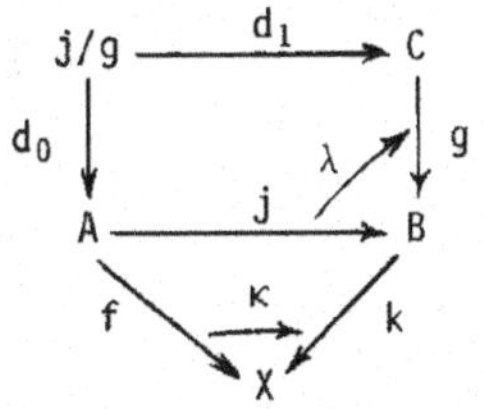

that κ *exhibits* k *as a pointwise left extension of* f *along* j. *Then the composite 2-cell exhibits* kg *as a pointwise left extension of* fd_0 *along* d_1.

Proof. Take $b:G \longrightarrow C$. The following square is a pullback.

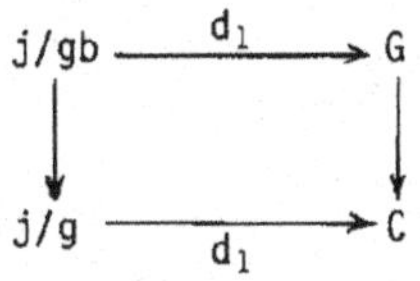

.

If this is mounted on the top of the diagram of the proposition we obtain the diagram

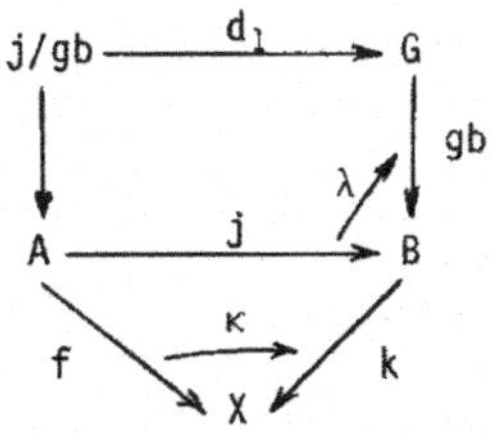

;

and this composite 2-cell does exhibit kgb as a left extension of fd_0 along d_1 (from the pointwise property of κ). By Proposition 12 and Theorem 14 we have that $d_1:j/g \longrightarrow C$ is a normal 0-fibrations (indeed, split). So Proposition 23 applies with $p = d_1:j/g \longrightarrow C$ to yield the result.//

Bibliography

[1] E.J. Dubuc, *Kan extensions in enriched category theory.* Lecture Notes in Math. 145 (1970) 1-173.

[2] J.W. Gray, *Fibred and cofibred categories.* Proc. Conference on Categorical Alg. at La Jolla (Springer, 1966) 21-83.

[3] J.W. Gray, *Report on the meeting of the Midwest Category Seminar in Zurich.* Lecture Notes in Math. 195 (1971) 248-255.

[4] A. Kock, *Monads for which structures are adjoint to units.* Aarhus University Preprint Series 35 (1972-73) 1-15.

[5] F.W. Lawvere, *The category of categories as a foundation for mathematics.* Proc. Conference on Categorical Algebra at La Jolla (Springer, 1966) 1-20.

[6] R.H. Street, *The formal theory of monads.* Journal of Pure and Applied Algebra 2 (1972) 149-168.

[7] R.H. Street, *Two constructions on lax functors.* Cahiers de topologie et géométrie différentielle XIII (1972) 217-264.

ELEMENTARY COSMOI I

by

Ross Street

The theory of categories enriched in some base closed category V, is couched in set-theory; some of the interesting results even require a hierarchy of set-theories. Yet there is a sense in which the results themselves are of an elementary nature. It seems reasonable then to ask which are the essential elementary results on which the rest of the theory depends. In unpublished joint work with R. Walters, an axiom system was developed which amounts to Theorems 6 and 7 of the present paper restated in terms of the representation arrow. We were able to deduce a great deal of the desired theory. One model for this system is provided by the 2-category V-Cat of small V-enriched categories together with the 2-functor $P:(V\text{-}Cat)^{coop} \to V\text{-}Cat$ given by $PA = [A^{op}, V]$ (= the V-enriched category of V-functors from A^{op} to V) where V is an appropriate small full subcategory of V.

In the case $V = SET$, $V = Set$, there is a universal property of the presheaf construction P which is more fundamental than the axioms mentioned above. With size considerations aside this universal property amounts to, for each category A, a pseudo-natural equivalence between the category of functors from B to PA and the category of covering spans from A to B. Generalizing to a representable 2-category K, we obtain the definition of an elementary precosmos as presented in this paper, the adjective "elementary" is dropped for brevity. (Strictly the universal property only determines $P:K^{coop} \to K$ as a pseudo functor, so we further ask that there should be a choice of P on arrows which makes it a 2-functor.) A cosmos is a precosmos for which P has a left 2-adjoint.

Our use of the word "cosmos" is presumptuous. To J. Benabou the word means "bicomplete symmetric monoidal closed category", such categories V being rich enough so that the theory of categories enriched in V develops to a large extent just as the theory of ordinary categories. It is not modifying this meaning much to apply the term to the constuction P of V-valued V-enriched presheaves for such a V, together with whatever structure is needed to make P well defined. However, we frankly do not know how a cosmos in this sense in general gives an example of an elementary cosmos in the sense of the present paper. The problem amounts to a well-known one in the theory of V-enriched categories concerning the relationship between comma objects and pointwise kan extensions. If we naively take K to be the 2-category V-*Cat* then the pointwise left kan extensions given by the coend formula (see Dubuc [6]) are not always pointwise left extensions in K in the sense of the previous paper [22]; the comma objects in K are just not right for extension purposes with a general V. We conjecture that there is some variant of V-*Cat* which is an elementary cosmos and provides fuller information on V-enriched categories (see Linton [19] p 228).

Despite this degree of ignorance, we believe there is good reason for presenting our work in its present form. Although we do have proofs for many of our results (we mention in particular Theorem 35) in the joint work with Walters, the proofs of the present paper are shorter and simpler. Further, our work can be regarded as a different approach to the elementary theory of the (2-) category of categories emphasising the role of the set-valued presheaf construction (compare Lawvere [14]). Also the (pre-) ordered objects in any elementary topos provide an example of an elementary cosmos; in particular, the 2-category of ordered-set-valued sheaves on a site is a cosmos. This observation contributes to topos theory in that our theory puts the techniques of adjoint arrows, kan extensions, comma objects, completeness, etc, at our disposal to examine the ever-present ordered objects in a topos. The 2-category of category-valued sheaves on a site is most probably a cosmos. Finally, we repeat the hope that enriched categories can be shown to fit into our present framework and mention that in a forthcoming paper we will show that *Cat*-enriched categories (that is, 2-categories) do fit in by expanding to double categories.

The notations and results of [12] and [22] are freely used throughout this paper.

Table of Contents

1. Internal attributes.

Let K be a representable 2-category. For objects A, B, recall that the category $SPN(A,B)$ of spans from A to B supports a monad M which assigns to each span S the span MS obtained as the inverse limit of the diagram

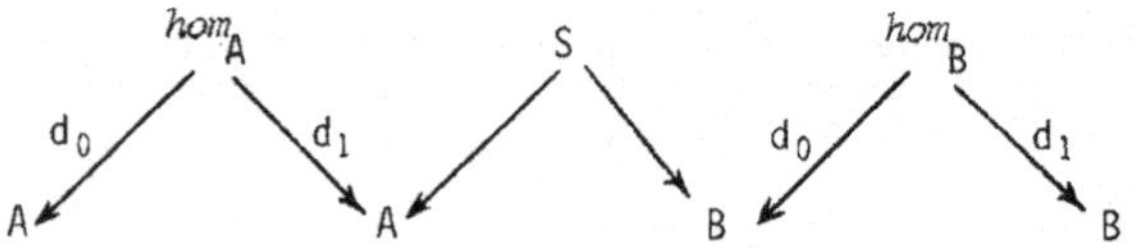

where, in this paper, we denote the comma object $A/A = \Phi A$ by hom_A. We freely call the M-algebras split bifibrations from A to B, although a split bifibration is strictly the underlying span of an M-algebra.

Before giving the next definition we must describe a pseudo functor

$$SPL\colon K^{coop} \times K^{op} \longrightarrow CAT.$$

For each pair of objects A, B, $SPL(A,B)$ is the category of split bifibrations from A to B. For each pair of arrows $f\colon A' \to A$, $g\colon B' \to B$, the functor

$$g^{*}\circ - \circ f\colon SPN(A,B) \longrightarrow SPN(A',B')$$

lifts in a canonical way (see Corollary 13 of the last paper) to a functor

$$SPL(f,g) = g^{*}\circ - \circ f\colon SPL(A,B) \longrightarrow SPL(A',B').$$

For 2-cells $A' \underset{f}{\overset{h}{\Downarrow\sigma}} A$, $B' \underset{k}{\overset{g}{\Downarrow\tau}} B$, the natural transformation $SPL(\sigma,\tau)\colon SPL(f,g) \longrightarrow SPL(h,k)$ is defined as follows. For each E in $SPL(A,B)$ with action $c\colon ME \longrightarrow E$, the component $SPL(\sigma,\tau)_E = \tau^{*}\circ E\circ\sigma$ is the composite

$$g^{*}\circ E\circ f \xrightarrow{1\circ i_E\circ 1} g^{*}\circ ME\circ f = (B/g)\circ E\circ (f/A) \xrightarrow{(B/\tau)\circ 1\circ (\sigma/A)} (B/k)\circ E\circ (h/A) = k^{*}\circ ME\circ h \xrightarrow{1\circ c\circ 1} k^{*}\circ E\circ h.$$

The routine verifications required to prove that SPL is a pseudo functor are left to the reader.

Suppose we have a 2-functor $P\colon K^{coop} \longrightarrow K$ and, for each object A in K, a split bifibration $\in_A$ from A to PA. These data are said to *endow* K *with attributes* when the functors

$$K(B,PA) \xrightarrow{\{A,B|-\}} SPL(A,B)$$

$$B \underset{k}{\overset{h}{\underset{\downarrow\sigma}{\rightrightarrows}}} PA \longmapsto (h^{*}\circ\in_A \xrightarrow{\sigma^{*}\circ\in_A} k^{*}\circ\in_A)$$

are fully faithful and form the components of a pseudo natural transformation. Pseudo naturality of $\{A,B|-\}$ in B is automatic, but in A it means that $(Pu)^{*}\circ\in_A \cong \in_{A'}\circ u$ naturally in $u\colon A \to A'$. Split bifibrations isomorphic to ones of the form $\{A,B|h\}$ for some $h\colon B \longrightarrow PA$ are called *attributes from* A *to* B. We call PA the *object of attributes of type* A.

A pair of arrows $a\colon X \longrightarrow A$, $b\colon Y \longrightarrow B$ is said to be *admissible with respect to a split bifibration* E from A to B when $b^{*}\circ E\circ a$ is an attribute from X to Y. Then there is an arrow $E(a,b)\colon Y \longrightarrow PX$ defined uniquely up to isomorphism by the condition

$$\{X,Y|E(a,b)\} = b^{*}\circ E\circ a \ .$$

For 2-cells $\alpha\colon a' \longrightarrow a$, $\beta\colon b \longrightarrow b'$, where both the pairs a, b and a', b' are admissible with respect to E, we can define a 2-cell $E(\alpha,\beta)\colon E(a,b) \longrightarrow E(a',b')$ so that the condition of the last sentence becomes natural. Then "E" becomes a pseudo natural transformation in the following sense.

Proposition 1. *If the pair of arrows* $a\colon X \longrightarrow A$, $b\colon Y \longrightarrow B$ *is admissible with respect to a split bifibration* E *from* A *to* B *then, for all arrows* $u\colon H \longrightarrow X$, $v\colon K \longrightarrow Y$, *the pair* au, bv *is admissible with respect to* E *and there is a natural isomorphism*

$$E(au,bv) \cong Pu.E(a,b).v \ .$$

Proof. The following isomorphisms are all natural:

$$(Pu.E(a,b).v)^{*}\circ\in_H \cong v^{*}\circ E(a,b)^{*}\circ(Pu)^{*}\circ\in_H \cong v^{*}\circ(E(a,b)^{*}\circ\in_X)\circ u \cong$$
$$v^{*}\circ(b^{*}\circ E\circ a)\circ u \cong (bv)^{*}\circ E\circ(au) \cong E(au,bv)^{*}\circ\in_H \ . \ /\!/$$

We say that an opspan $X \xrightarrow{f} A \xleftarrow{g} Y$ is *admissible* when the pair f,g is admissible with respect to hom_A; or in other words, when f/g is an attribute from X to Y. As a special case of our above notation, we then have an arrow $hom_A(f,g)\colon Y \longrightarrow PX$ satisfying the condition

$$\{X,Y|hom_A(f,g)\} \cong f/g \ .$$

Call an arrow $f\colon X \longrightarrow A$ *admissible* when the opspan $f,1_A$ is admissible; call f *coadmissible* when the opspan $1_A,f$ is admissible. Call an object A *legitimate* when hom_A is an attribute. The following is an immediate consequence of Proposition 21 of the last paper.

Theorem 2. An arrow $u\colon A \longrightarrow B$ *is a right adjoint for an admissible arrow* $f\colon B \longrightarrow A$ *if and only if there is an isomorphism*

$$hom_A(f,1) \cong hom_B(1,u). \ /\!/$$

Recall that an object G is said to be *orthogonal* to an arrow $f\colon A \longrightarrow B$ when the functor

$$K(G,f)\colon K(G,A) \longrightarrow K(G,B)$$

is an isomorphism. This is an elementary condition since a functor is an isomorphism when it is bijective on objects and fully faithful. A class G of objects of K is said to be *strongly generating* when, given $f\colon A \longrightarrow B$, if each G in G is orthogonal to f then f is an isomorphism.

When the arrows $j\colon A \longrightarrow B$, $f\colon A \longrightarrow X$ are each admissible, we have the following string of bijections.

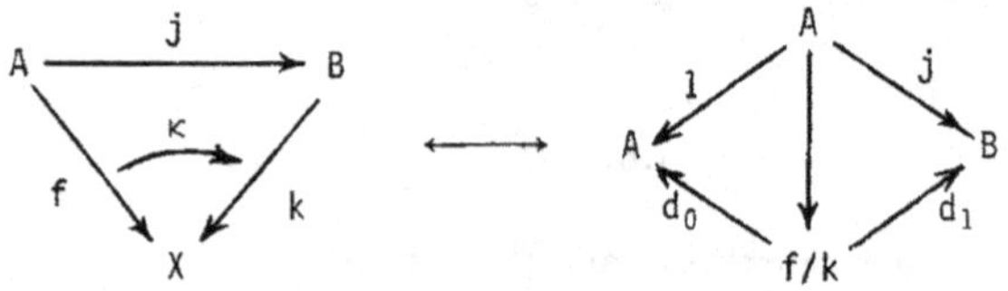

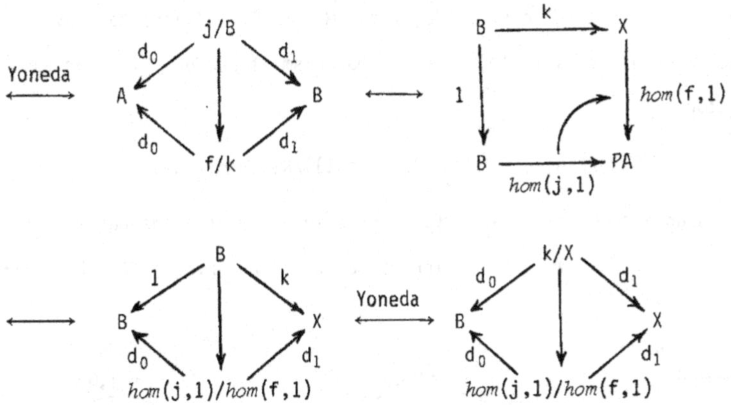

So that 2-cells κ correspond to arrows of spans $k/X \longrightarrow \mathit{hom}(j,1)/\mathit{hom}(f,1)$.

Theorem 3. *Suppose* $\mathcal{G}$ *is a strongly generating class of objects and suppose that* f, j *are each admissible in the diagram*

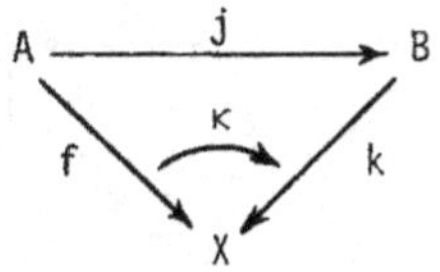

Then the following three conditions are equivalent:

(a) the 2-cell κ *exhibits* k *as a pointwise left extension of* f *along* j;

(b) the arrow of spans $k/X \longrightarrow \mathit{hom}_B(j,1)/\mathit{hom}_X(f,1)$ *corresponding to* κ *is an isomorphism;*

(c) for each object G *in* $\mathcal{G}$ *and each arrow* $b\colon G \longrightarrow B$, *the composite 2-cell*

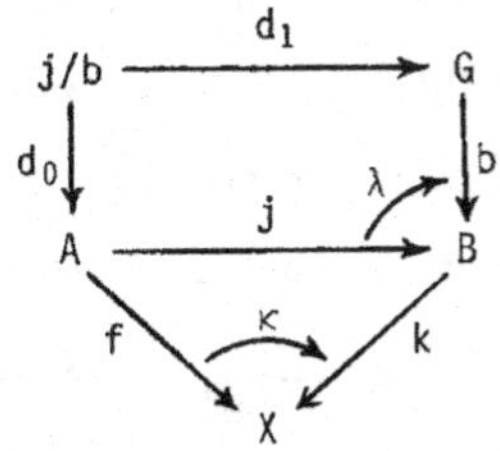

exhibits kb *as a left extension of* fd_0 *along* d_1.

Proof. Since (a) is a special case of (c) with $G = K$, it suffices to prove (b) ⇔ (c). The arrow of spans in (b) is an isomorphism if and only if, for each G in G, the functor

$$(*) \qquad K(G,k)/K(G,X) \longrightarrow K(G,hom(j,1))/K(G,hom(f,1))$$

induced by κ is an isomorphism. This functor is an arrow of spans between the categories $K(G,B)$ and $K(G,X)$. Given arrows $b: G \longrightarrow B$, $x: G \longrightarrow X$, there are bijections

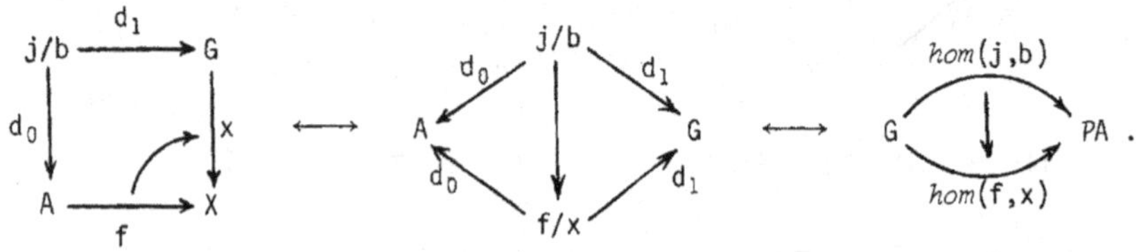

So we obtain a function

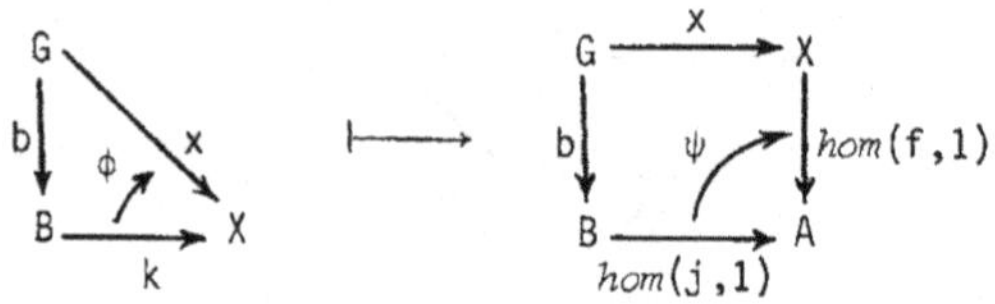

by first pasting ϕ on the right hand side of the 2-cell in (c) and then tracing through the bijections of the last sentence. This function is readily seen to be the object function of the functor (*). So the object function of (*) is a bijection if and only if "pasting on the right hand side of the 2-cell in (c)" is a bijection for all b,x; that is, if and only if property (c) holds. In this case it is readily seen that $(\beta,\xi): (b,\phi,x) \longrightarrow (b',\phi',x')$ is an arrow of $K(G,k)/K(G,X)$ if and only if $(\beta,\xi): (b,\psi,x) \longrightarrow (b',\psi',x')$ is an arrow of $K(G,hom(j,1))/K(G,hom(f,1))$ where ϕ,ϕ' go to ψ,ψ' respectively under the bijection. So (*) is an isomorphism of categories if and only if (c) holds.//

Corollary 4. Suppose $j: A \longrightarrow B$, $f: A \longrightarrow X$, $k: B \longrightarrow X$ *are all admissible. Then* k *is a left extension of* f *along* j *if and only if there is an isomorphism*

$$hom_X(k,1) \cong hom_{PA}(hom_B(j,1),hom_X(f,1)). \;//$$

Remark. For $K = CAT$ and $PA = [A^{op}, S]$ where S is a category of sets, the condition (b) of Theorem 3 amounts to the formula

$$X(kb,x) \cong PA(B(j-,b),X(f-,x))$$

from which all the properties of left extensions can be proved. The coend formula

$$kb = \int^{a} B(ja,b) \otimes fa$$

is an easy consequence ([5] p 187 and [20] p236). Also $G = \{1\}$ is strongly generating, so from (c) we have that the formula

$$kb = \underrightarrow{lim}(j/b \xrightarrow{d_0} A \xrightarrow{f} X)$$

actually determines the pointwise left extension.//

§2. Precosmoi.

A *precosmos* is a representable 2-category endowed with attributes satisfying the property that any arrow of spans between attributes is a homomorphism of split bifibrations. Theorem 14 of the last paper implies that those attributes which are covering spans automatically have this property; we shall see that in some cases the property implies that attributes are covering spans. Of course, as in the last paper, we have the consequence that, for an admissible arrow $f: A \longrightarrow B$ and any attribute E from A to B, composition with $i_f: A \longrightarrow f/B$ yields a bijection between arrows of spans

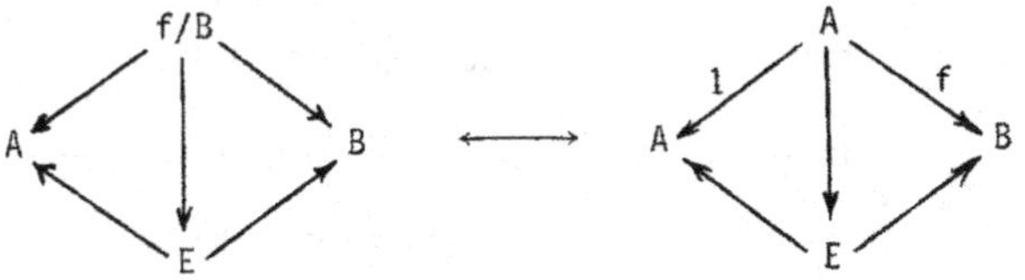

There is a dual consequence for coadmissible arrows $u: B \longrightarrow A$.

Theorem 5. *If* $a: K \longrightarrow A$ *is coadmissible then* $hom_A(1,a)$ *is admissible and there is an isomorphism*

$$hom_{PA}(hom_A(1,a),1) \cong Pa.$$

Proof. We shall show that the attribute $\in_A \circ a \cong (Pa)^* \circ \in_K$ has the universal property of the comma object $hom_A(1,a)/PA$ from which the result follows. Note that, for any span (u,S,v) from K to PA, we have $(hom(1,a)u)^* \circ \in_A \cong A/au$. There are bijections

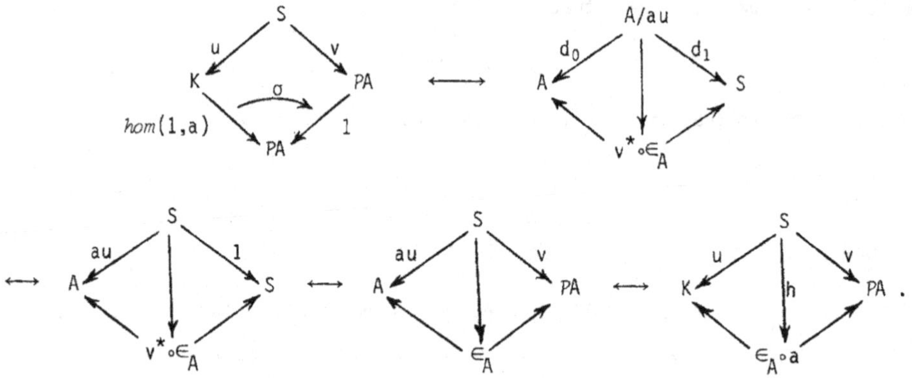

The composite bijection $\sigma \longleftrightarrow h$ is natural in spans S so that $\in_A \circ a$ has the desired comma property.//

If the pair $1_A, b$ is admissible with respect to the split bifibration E from A to B then we have an isomorphism

$$hom_{PA}(hom_A(1,a),E(1,b)) \cong E(a,b)$$

for coadmissible a (combine Proposition 1 and Theorem 5). This is very reminiscent of the Yoneda lemma of category theory (see Mac Lane [20] p.61)

Suppose $f: C \longrightarrow X$ is any arrow and $A \xrightarrow{u} C \xleftarrow{v} B$ is an admissible opspan such that the opspan $A \xrightarrow{fu} X \xleftarrow{fv} B$ is also admissible. The canonical arrow of spans $u/v \longrightarrow fu/fv$ (induced by the 2-cell $f\lambda$) corresponds to a 2-cell

$$\chi^f = \chi^f_{u,v}: hom_C(u,v) \longrightarrow hom_X(fu,fv),$$

called *the effect of* f *on homs.*

<u>*Theorem 6.*</u> *Suppose* f: A ⟶ B, g: C ⟶ A *are arrows such that both* g *and* fg *are admissible. Then the 2-cell*

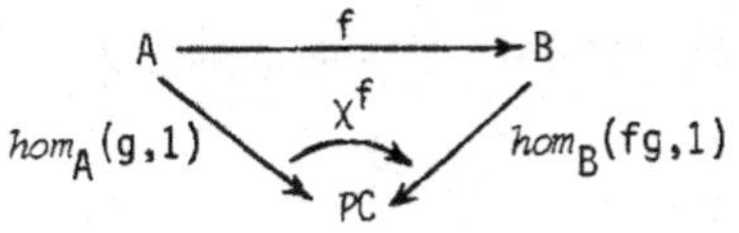

exhibits $hom_B(fg,1)$ *as a left extension of* $hom_A(g,1)$ *along* f.

<u>*Proof.*</u> For any arrow h: B ⟶ PC we have the following bijections between arrows of spans

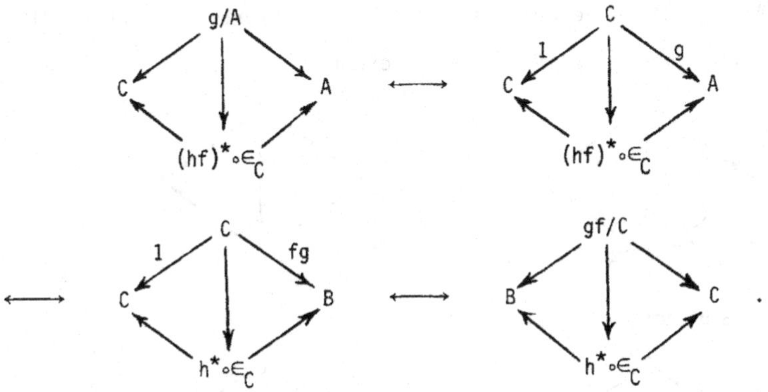

The composite bijection corresponds to a bijection between 2-cells

as required.//

<u>*Theorem 7.*</u> *Suppose* f: A ⟶ B *is an admissible arrow. For all coadmissible arrows* g: C ⟶ A, *the 2-cell*

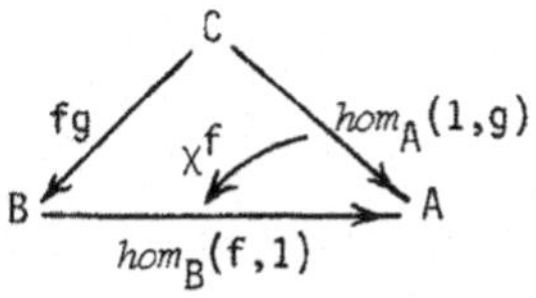

exhibits fg *as a left lifting of* $hom_A(1,g)$ *through* $hom_B(f,1)$. *When* A *is legitimate,* $hom_B(f,1)$ *is characterized up to isomorphism by this property.*

Proof. Suppose an arrow $h: B \longrightarrow PA$ is given. In order that a 2-cell $hom_A(1,g) \longrightarrow hfg$ should have the left lifting property, it should establish a bijection

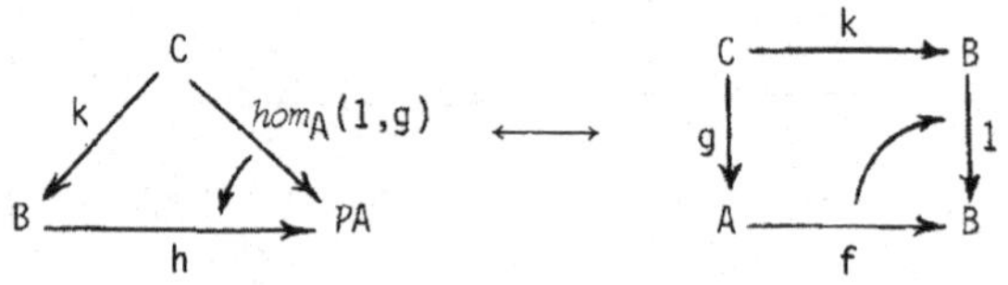

By applying $\{A,C|-\}$ to the left hand side and by the property of comma objects on the right hand side, this corresponds to a bijection

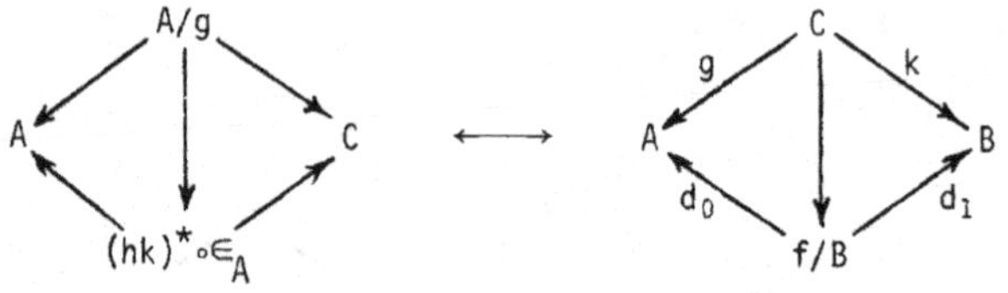

which amounts to a bijection

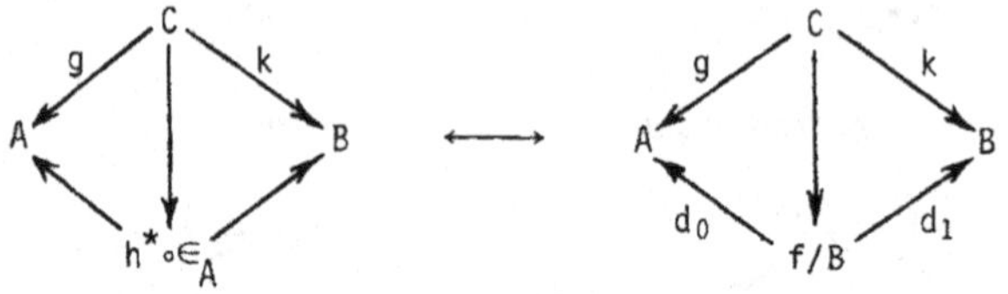

If $h = hom_B(f,1)$ then $h^* \circ \in_A \cong f/B$ so we certainly have such a bijection. On the other hand, if A is legitimate then the latter bijection for all spans (g,C,k) implies $h^* \circ \in_A \cong f/B$; so $h \cong hom_B(f,1)$.//

The uniqueness clause in the last theorem is an observation of R.F.C. Walters.

§3. The representation arrow.

For each legitimate object A, the arrow $y_A = hom_A(1,1): A \longrightarrow PA$ (defined by $\{A,A|y_A\} \cong hom_A$) is called the *representation arrow* of A. It was the search for the proof of the next result which led us to the Yoneda lemma of the last paper. We now obtain it from Theorem 5 by taking $a = 1_A$.

Theorem 8. *If* A *is legitimate then* y_A *is admissible and there is an isomorphism*

$$hom_{PA}(y_A,1) \cong 1.$$

Indeed, there is an isomorphism

$$\{A,B|h\} \cong y_A/h$$

which is natural in $h: B \longrightarrow PA$. //

Corollary 9. *If* A *is legitimate then each attribute out of* A *is a covering span.* //

A precosmos will be called *uniform* when the attributes are precisely the covering spans. So a precosmos is uniform if and only if each object is legitimate.

We now proceed to prove properties of the representation arrow which are familiar in category theory. Note that, if A is legitimate, any pair of arrows $f: X \longrightarrow A$, $g: Y \longrightarrow A$ is admissible and, by Proposition 1,

$$hom_A(f,g) \cong Pf.y_A.g .$$

So, with suitable legitimacy conditions, Theorems 6 and 7 can be stated as properties of the representation arrow. In joint work with Walters, these properties were taken as axioms and some results of the remainder of this section were proved (again with extra legitimacy assumptions). Many of the later results also follow from these axioms.

Recall that an arrow $f: A \longrightarrow C$ is *fully faithful* when the canonical arrow $hom_A \longrightarrow f/f$ is an isomorphism. We say that f is *dense* when the identity 2-cell exhibits 1_C as a pointwise left extension of f along f.

Theorem 10. *For legitimate* A, *the representation arrow* $y_A\colon A \longrightarrow PA$ *is fully faithful and dense.*

Proof. Combine the definition of y_A with Theorem 8 to yield isomorphisms

$$hom_A \ = \ \{A,A|y_A\} \ \cong \ y_A/y_A \ ;$$

so y_A is fully faithful. For denseness we must show that, for any arrow $k\colon B \longrightarrow PA$, the 2-cell

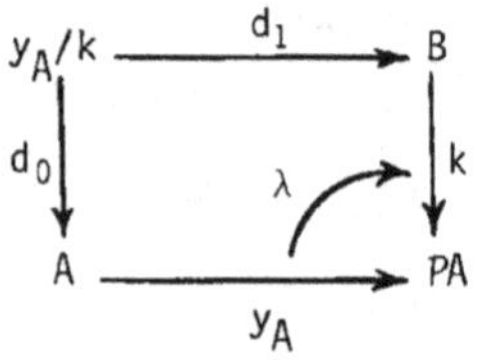

exhibits k as a left extension of $y_A d_0$ along d_1. So take $h\colon B \longrightarrow PA$. Then we have bijections:

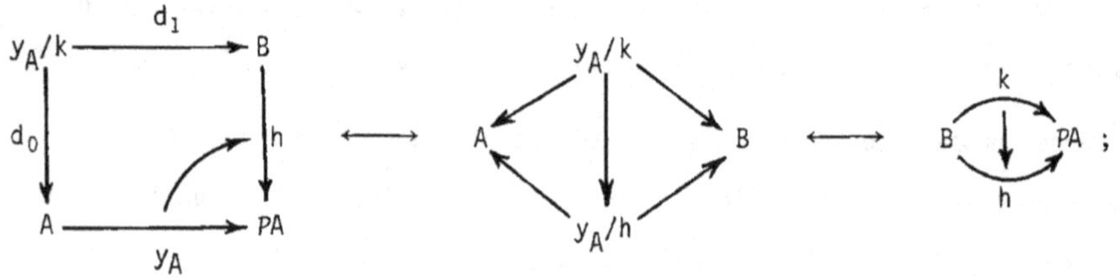

the first by the property of comma objects, the second by Theorem 8 and the fully faithfulness of $\{A,B|-\}$.//

Theorem 11. *Suppose* $f\colon A \longrightarrow B$ *is an arrow with* B *legitimate. The arrow* $Pf\colon PA \longrightarrow PB$ *has a right adjoint* $\forall f\colon PA \longrightarrow PB$ *if and only if* $hom_B(f,1)$ *is admissible. In this case,*

$$\forall f \ = \ hom_{PA}(hom_B(f,1),1).$$

Proof. Suppose $Pf \dashv \forall f$. By Theorem 8, $y_B/\forall f$ is an attribute from B to PA. But $y_B/\forall f \cong Pf.y_B/PA \cong hom(f,1)/PA$; so $hom(f,1)$ is admissible.

Suppose $hom(f,1)$ is admissible and put $\forall f = hom_{PA}(hom_B(f,1),1)$. By Theorem 6, the following 2-cells exhibit left extensions preserved by any arrow of the form Pg:

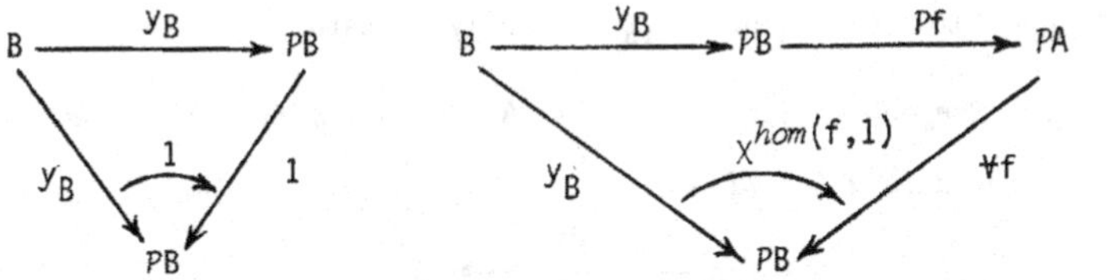

;

for the first of these note that $hom_{PB}(y_B,1) = 1$ by Theorem 8, and for the second note that $hom_B(f,1) = Pf.y_B$. It follows that there exists a unique 2-cell

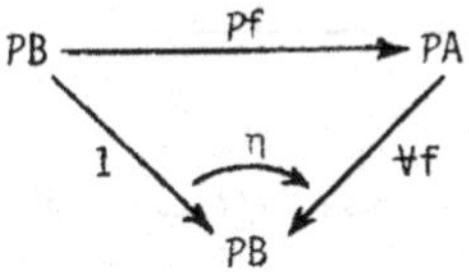

such that $\eta y_B = \chi^{hom(f,1)}$, and that this 2-cell is a left extension preserved by any Pg (and so by Pf). So $\forall f$ is a right adjoint for Pf with unit η.//

The arrow $\forall f$ is an internal expression of right extensions, and yet, as we saw, left extensions were used in the above proof. There is another relationship between $\forall$ and left extensions. Referring back to Corollary 4, suppose B is legitimate and $hom_B(j,1)$ is admissible, then the isomorphism of the corollary becomes

$$hom_X(k,1) \cong \forall j.hom_X(f,1).$$

If X is legitimate, this becomes

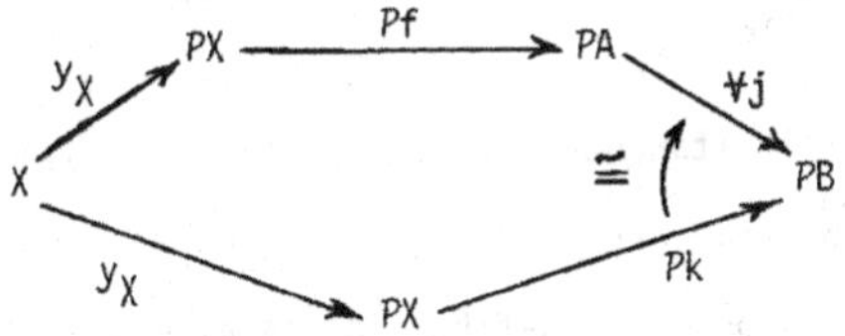

Theorem 12. If both A *and* PA *are legitimate then* $Py_A: P^2A \longrightarrow PA$ *is a left adjoint for* $y_{PA}: PA \longrightarrow P^2A$ *with counit an isomorphism.*

Proof. Take $f = y_A$ in Theorem 11. By Theorem 8 we have

$$\forall y_A = hom_{PA}(hom_{PA}(y_A,1),1) \cong hom_{PA}(1,1) = y_{PA}.$$

The counit is an isomorphism since y_{PA} is fully faithful.//

Proposition 13. *For any legitimate object* A *and any admissible opspan* $A \xrightarrow{f} X \xleftarrow{g} B$, *there is a 2-cell*

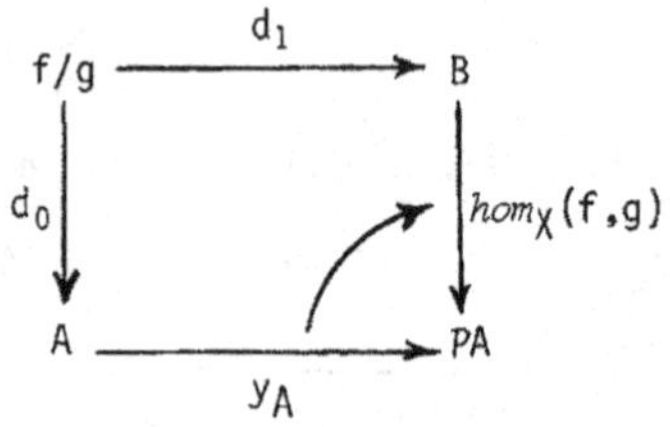

which exhibits $hom_X(f,g)$ *as a pointwise left extension of* $y_A d_0$ *along* d_1.

Proof. Consider the composite 2-cell

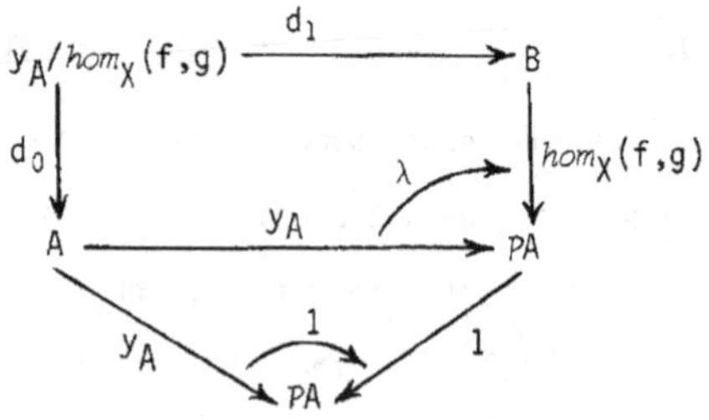

By Theorem 10, the lower triangle is a pointwise left extension; so, by Proposition 24 of the last paper, the composite exhibits $hom_X(f,g)$ as a *pointwise* left extension of $y_A d_0$ along d_1. Using Theorem 8 we have $y_A/hom_X(f,g) \cong hom_X(f,g)^* \circ \epsilon_A$ $\cong f/g$; so the result follows.//

Functors which are simultaneously 0- and 1-fibrations have been considered by Chevalley [4]. A 0-fibration over a category $\mathcal{B}$ corresponds to a pseudo functor $R: \mathcal{B}^{op} \longrightarrow CAT$; when the original 0-fibration is also a 1-fibration, the functor Rf has a right adjoint $\check{R}f$ for each arrow f in $\mathcal{B}$. Chevalley (and later Beck and Benabou-Roubaud [3] in their study of descent data) considered a compatibility condition on the fibration which, in terms of R, amounts to the following:

for each pullback
$$\begin{array}{ccc} P & \xrightarrow{k} & B \\ {\scriptstyle h}\downarrow & & \downarrow{\scriptstyle g} \\ A & \xrightarrow[f]{} & C \end{array}$$
in $\mathcal{B}$, the induced 2-cell
$$\begin{array}{ccc} RA & \xrightarrow{Rh} & RP \\ {\scriptstyle \check{R}h}\downarrow & \nearrow & \downarrow{\scriptstyle \check{R}g} \\ RC & \xrightarrow[Rf]{} & RB \end{array}$$

is an isomorphism.

This statement still makes sense verbatim in the case where $\mathcal{B}$ is a 2-category, but the result is *too strong*. There are two generalizations which present themselves in this case. The first is to require that f should be a 0-fibration in $\mathcal{B}$ (if $\mathcal{B}$ is a category this is no condition on f); an interesting combination of fibrations at two different levels! The second is to replace the pullback by a comma object

$$\begin{array}{ccc} f/g & \xrightarrow{d_1} & B \\ {\scriptstyle d_0}\downarrow & \overset{\lambda}{\nearrow} & \downarrow{\scriptstyle g} \\ A & \xrightarrow[f]{} & C \end{array}$$

in $\mathcal{B}$ (if $\mathcal{B}$ is a category this reduces to the pullback again). The two generalizations are closely related and we have both in a precosmos.

<u>*Theorem 14*</u>. *Suppose* $f\colon A \longrightarrow C$, $g\colon B \longrightarrow C$ *are arrows between legitimate objects. Then the 2-cell*

$$\begin{array}{ccc} PA & \xrightarrow{Pd_0} & P(f/g) \\ {\scriptstyle \forall f}\downarrow & \nearrow & \downarrow{\scriptstyle \forall d_1} \\ PC & \xrightarrow[Pg]{} & PB \end{array}$$

corresponding under adjunction to $P\lambda$, *is an isomosphism whenever* $\forall f$, $\forall d_1$ *exist.*

<u>*Proof*</u>. Proposition 13 yields that $hom_C(f,g)$ is a pointwise left extension of $y_A d_0$ along d_1. By Theorem 3, we have an isomorphism

$$hom_C(f,g)/PA \;\cong\; hom_B(d_1,1)/hom_{PA}(y_A d_0,1).$$

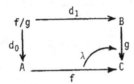

Now $hom_{PA}(y_A d_0,1) \;\cong\; Pd_0 . hom_{PA}(y_A,1) \;\cong\; Pd_0$; and also $hom_B(d_1,1)$ is admissible when $\forall d_1$ exists (Theorem 11). So the span on the right hand side of the displayed isomorphism above is an attribute and we have an isomorphism

$$hom_{PA}(hom_C(f,g),1) \cong hom_{P(f/g)}(hom_B(d_1,1),Pd_0).$$

Using Proposition 1 and Theorem 11, we obtain the result.//

Theorem 15. *Suppose* $p\colon E \longrightarrow B$ *is a normal 0-fibration,* $b\colon G \longrightarrow B$ *is an arrow, and* E, B, G *are legitimate. Then the 2-cell*

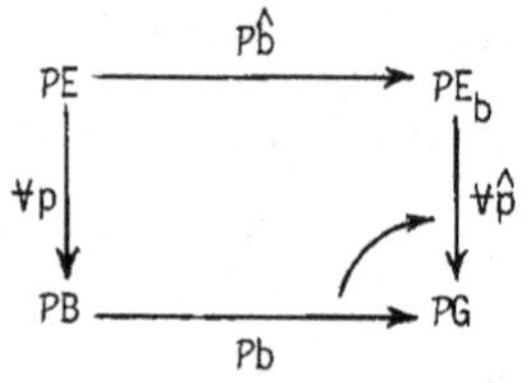

corresponding under adjunction to the identity 2-cell $P\hat{p}.Pb \longrightarrow P\hat{b}.Pp$*, is an isomorphism whenever* $\forall p$*,* $\forall\hat{p}$ *exist.*

Proof. From Proposition 13, the 2-cell χ exhibits $hom_B(p,1)$ as a pointwise left extension of y_E along p. By Proposition 23 of the last paper this implies that the composite

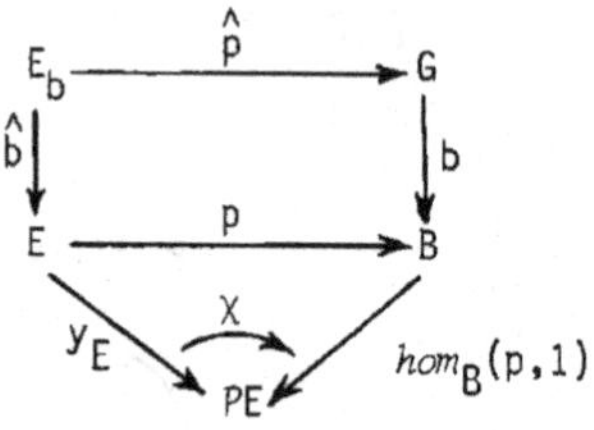

exhibits $hom_B(p,1)b$ as a pointwise left extension of $y_E\hat{b}$ along $\hat{p}$. By Theorem 3, we have an isomorphism

$$hom_B(p,b)/PE \cong hom_G(\hat{p},1)/hom_{PE}(y_E\hat{b},1).$$

But $hom_{PE}(y_E\hat{b},1) \cong P\hat{b}$, and $hom_G(\hat{p},1)$ is admissible when $\forall\hat{p}$ exists. So the span on the right hand side of the above displayed isomorphism is an attribute. This gives:

$$hom_{PE}(hom_B(p,b),1) \cong hom_{PE_b}(hom_G(\hat{p},1),P\hat{b}).$$

The result follows using Proposition 1 and Theorem 11.//

Remark. Consider the case where $K = CAT$. The isomorphism of Theorem 14 expresses internally the fact that right extensions of set-valued presheaves are pointwise. Of course, the arrows can be replaced by their left adjoints to obtain an isomorphism

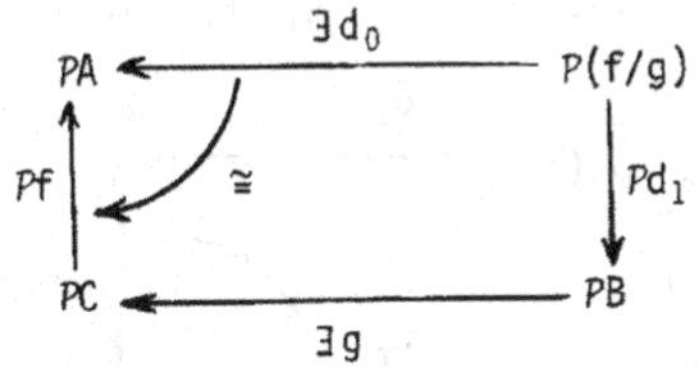

whenever $\exists g$, $\exists d_0$ exist. This is an internal expression of the pointwiseness of left extensions of set-valued presheaves. The particular case of Theorem 15 with $G = \mathbb{1}$ yields the well-known formula

$$(\forall p)F = \underleftarrow{lim}(E_b^{op} \longrightarrow E^{op} \longrightarrow Set)$$

for the right extension of a presheaf F along a 1-fibration $p^{op}: E^{op} \longrightarrow B^{op}$. The isomorphism obtained by replacing the arrows of the 2-cell in Theorem 15 by their left adjoints has a rather curious interpretation.//

Suppose A is a legitimate object. We now give the generalization of *colimit* needed for our work. When it exists, the pointwise left extension of $f: A \longrightarrow X$ along $y_A: A \longrightarrow PA$ is denoted by $lexf: PA \longrightarrow X$. The 2-cell exhibiting this extension is an isomorphism since y_A is fully faithful.

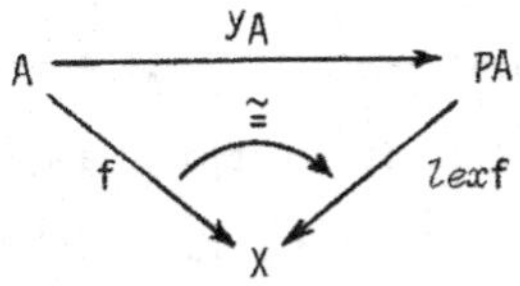

<u>*Theorem 16.*</u> *Suppose* A *is legitimate. The pointwise left extension* $lex\,f$ *of* $f\colon A \longrightarrow X$ *along* y_A *exists if and only if, for all admissible arrows* $j\colon A \longrightarrow B$, *the pointwise left extension* k *of* f *along* j *exists. In this case,* k *is the composite*

$$B \xrightarrow{\;hom_B(j,1)\;} PA \xrightarrow{\;lex\,f\;} X.$$

<u>*Proof.*</u> Since y_A is admissible (Theorem 8), "if" is clear. Suppose $lex\,f$ exists. By the pointwise property, for any $g\colon C \longrightarrow B$, the composite 2-cell

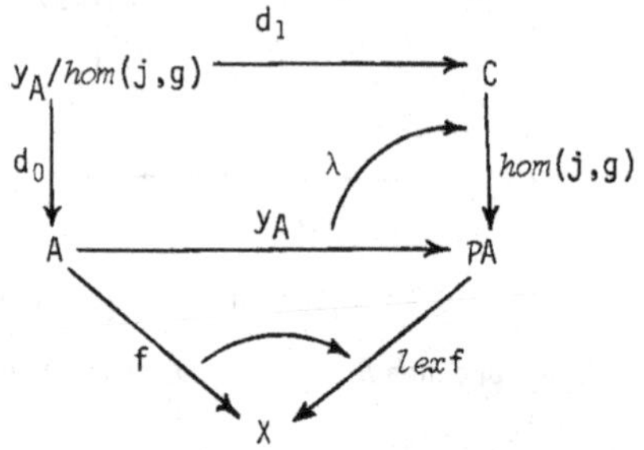

exhibits $kg = lex\,f\,.hom(j,g)$ as the left extension of fd_0 along d_1. But the span $y_A/hom(j,g)$ can be replaced by j/g (Theorem 8).//

<u>*Proposition 17.*</u> *Suppose* A *is legitimate and* $f\colon A \longrightarrow X$ *is admissible. If* $lex\,f$ *exists then it is a left adjoint for* $hom_X(f,1)$. *If* $hom_X(f,1)$ *has a left adjoint then* $lex\,f$ *exists.*

<u>*Proof.*</u> In order that f' should be a pointwise left extension of f along y_A, we should have $f'/X \cong hom(y_A,1)/hom(f,1) \cong PA/hom(f,1)$ by Theorem 3; and this is precisely the condition that $f' \dashv hom(f,1)$.//

The particular case when f is the identity arrow of the legitimate object A deserves special attention; then $hom(f,1) = y_A\colon A \longrightarrow PA$. An object A is said to be *absolutely cocomplete* when it is legitimate and $y_A\colon A \longrightarrow PA$ has a left adjoint $lex_A\colon PA \longrightarrow A$. If $A = PK$ for some legitimate K then Theorem 12 gives $lex_A = Py_K$; so A is absolutely cocomplete. For absolutely cocomplete objects we have the most favourable form of adjoint functor theorem.

Theorem 18. *Suppose* A *is absolutely cocomplete. An admissible arrow* $f: A \longrightarrow X$ *has a right adjoint if and only if* $\mathit{lex}f$ *exists and the canonical 2-cell* $\mathit{lex}f \longrightarrow f\,\mathit{lex}_A$ *is an isomorphism.*

Proof. If $f \dashv u$ then, composing with $\mathit{lex}_A \dashv y_A$, we have $f\,\mathit{lex}_A \dashv y_A u = \mathit{hom}(1,u) \cong \mathit{hom}(f,1)$. So $\mathit{lex}f$ exists and is isomorphic to $f\,\mathit{lex}_A$.

Suppose $f\,\mathit{lex}_A \cong \mathit{lex}f$. Theorem 16 applies to $1: A \longrightarrow A$ to yield $u = \mathit{lex}_A.\mathit{hom}(f,1)$ as the left extension of 1_A along f. It remains to prove that f preserves this extension. But $fu = f\,\mathit{lex}_A.\mathit{hom}(f,1) \cong \mathit{lex}f\,.\mathit{hom}(f,1)$, and the latter is the left extension of f along f (Theorem 16).//

Remark. For categories, the colimit of a functor $f: A \longrightarrow X$ is obtained from $\mathit{lex}f : PA \longrightarrow X$ by evaluating at the terminal presheaf. Recall that the colimit of the identity functor of A is the terminal object in A. It follows that absolutely cocomplete categories have terminal objects. The formula for $\mathit{lex}f$ in terms of colimits in X is:

$$(\mathit{lex}f)G = \int^{a} Ga \otimes fa.//$$

Since we have discussed the case where the representation arrow has a left adjoint, we digress briefly to point out some trivialities regarding the case where it has a right adjoint. An object A is called *degenerate* when it is legitimate and $y_A: A \longrightarrow PA$ has a right adjoint.

Proposition 19. *A degenerate object* A *has the following properties:*

(a) $y_A: A \longrightarrow PA$ *is an equivalence;*

(b) *each admissible arrow with source* A *has a right adjoint;*

(c) *each arrow* $B \longrightarrow PA$ *is isomorphic to one of the form* $\mathit{hom}_A(1,s)$ *for some* $s: B \longrightarrow A$.

Proof. Let t be a right adjoint for y_A. Recall Theorem 10. The counit $y_A t \longrightarrow 1$ is an isomorphism since y_A is dense, and the unit $1 \longrightarrow t y_A$ is an isomorphism since y_A is fully faithful. So $y_A: A \approx PA$. Now we can apply Theorem 2 to obtain that $u = t\, hom(f,1)$ is a right adjoint for admissible $f: A \longrightarrow B$. This proves (a) and (b). Using (a) we have an equivalence of categories $K(B,A) \approx K(B,PA)$; (c) follows from this and Theorem 8.//

§4. Extension systems.

Monoidal (= multiplicative) categories [1], [7] have been generalized by Benabou [2] to bicategories. We now make the corresponding generalization for closed categories.

An *extension system* E consists of the following data:

(i) *objects* $A, B, C, \ldots$;

(ii) for each pair A,B of objects, a category $E(A,B)$ whose objects are called *arrows* and whose arrows are called *2-cells* ;

(iii) for each object A, an arrow $y_A \in E(A,A)$;

(iv) for objects A, B, C, a functor

$$[\ ,\ ]: E(X,A)^{op} \times E(X,B) \longrightarrow E(A,B)\ ;$$

(v) for arrows $f \in E(X,A)$, $g \in E(X,B)$, $h \in E(X,C)$, 2-cells

$$\nu^h_{f,g}: [f,g] \longrightarrow [[h,f],[h,g]],\quad \chi^f: y_A \longrightarrow [f,f],\quad \omega_g: g \longrightarrow [y_X,g]\ ,$$

the latter an isomorphism;

such that the following axioms are satisfied:

ES1. $\nu^h_{f,g}$, χ^f, ω_g are natural in their subscripts and extraordinary natural in their superscripts;

ES2. the following diagrams commute

$$\begin{array}{ccc} [f,f] & \xrightarrow{\nu} & [[g,f],[g,f]] \\ & {}_{\chi}\nwarrow \quad \nearrow_{\chi} & \\ & y_A & \end{array} \qquad \begin{array}{ccc} [f,g] & \xrightarrow{\nu} & [[f,f],[f,g]] \\ & {}_{\omega}\searrow \quad \swarrow_{[\chi,1]} & \\ & [y_A,[f,h]] & \end{array} \qquad \begin{array}{ccc} [f,g] & \xrightarrow{\nu^{y_X}} & [[y_X,f],[y_X,g]] \\ & {}_{[1,\omega]}\searrow \quad \swarrow_{[\omega,1]} & \\ & [f,[y_X,g]] & \end{array}$$

$$\begin{array}{ccc} [f,g] & \xrightarrow{\qquad\nu^k\qquad} & [[k,f],[k,g]] \\ \downarrow{\scriptstyle \nu^h} & & \\ [[h,f],[h,g]] & & \downarrow{\scriptstyle [1,\nu^g]} \\ \downarrow{\scriptstyle \nu^{[h,k]}} & & \\ [[[h,k],[h,f]],[[h,k],[h,g]]] & \xrightarrow[{[\nu^h,1]}]{} & [[k,f],[[h,k],[h,g]]] \end{array} \quad ;$$

ES3. the composite function

$$E(X,A)(f,g) \xrightarrow{[-,g]} E(A,A)([g,g],[f,g]) \xrightarrow{E(A,A)(\chi^g,1)} E(A,A)(y_A,[f,g])$$

is a bijection.

Special cases. 1) An extension system with precisely one object is a *closed category*. This does not quite agree with the definition of closed category appearing in [7]. Reference to a category of sets has been eliminated as required for example by Lawvere [14] p12. Also, a monoidal category such that each of the functors $X \otimes -$ has a right adjoint is closed in our sense (compare this with [7] Theorem 5.8 p493). Note that, for any extension system E, $E(A,A)$ becomes a closed category.

2) A bicategory B in which all right extensions exist yields an extension system with $[f,h]$ taken as the right extension of h along f. Such a bicategory we call an *extensional bicategory* (also called "closed bicategory" by some authors).

3) Suppose E is an extension system such that, for all $f\colon D \longrightarrow A$, $k\colon A \longrightarrow C$, there exists an arrow $k \otimes f\colon D \longrightarrow C$ and a natural isomorphism

$$[k \otimes f,-] \cong [k,[f,-]].$$

Then E becomes an extensional bicategory with composition given by $\otimes$.

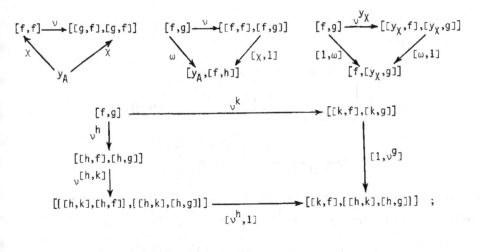

In a precosmos K, suppose the arrow $h: C \longrightarrow X$ is admissible. The composite 2-cell

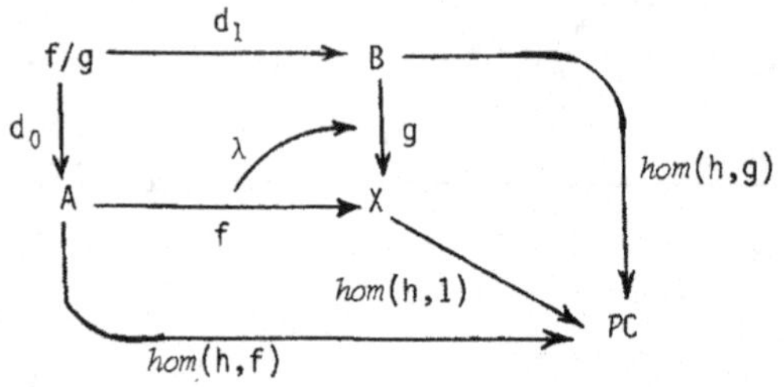

induces an arrow of spans $f/g \longrightarrow hom_X(h,f)/hom_X(h,g)$. Provided the source and target of the latter arrow are attributes, we obtain a 2-cell

$$\nu^g_{f,h}: hom_X(f,g) \longrightarrow hom_{PC}(hom_X(h,f), hom_X(h,g)).$$

The next two theorems can be proved using Theorems 6 and 7.

Theorem 20. *An extension system $\mathcal{P}rof$ is defined by the following data:*

(i) the objects are the legitimate objects A of K for which PA is legitimate;

(ii) $\mathcal{P}rof(A,B) = K(B,PA)$;

(iii) for $f \in \mathcal{P}rof(X,A)$, $g \in \mathcal{P}rof(X,B)$, *take* $[f,g] = hom_X(f,g)$;

(iv) y_A *is the representation arrow;*

(v) ν, χ *are as previously defined and* ω *is the isomorphism of Theorem 8.*//

Theorem 21. *For each object K of $\mathcal{P}rof$ and each legitimate object A of K, the data hom_A, ν, χ enrich the category $K(K,A)$ with the structure of a V_K-category, where V_K is $\mathcal{P}rof(K,K)$ with its closed category structure. Indeed, the 2-functor* $K(K,-): L \longrightarrow CAT$ *lifts to a 2-functor* $L \longrightarrow V_K$ - CAT.//

Freyd's tensor product of functors [8] p120 can be carried over to arrows in a precosmos. Given arrows $f: A \longrightarrow X$, $g: B \longrightarrow PA$, their *tensor product* $g \otimes f: B \longrightarrow X$, when it exists, is defined as the left extension as exhibited by a 2-cell:

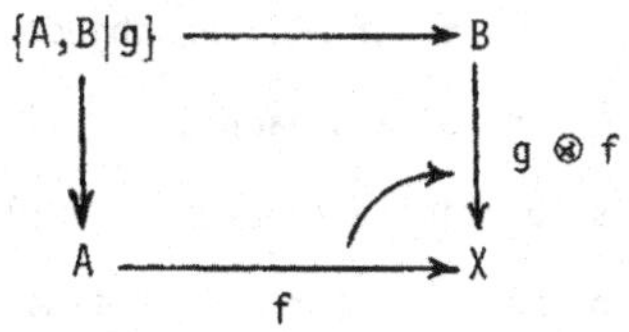

If $A \xrightarrow{f} X \xrightarrow{h} C$ is an admissible opspan then there is a bijection

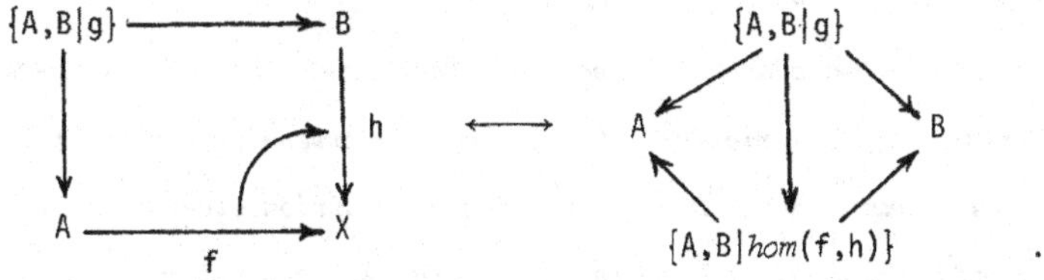

Using the left extension property of $g \otimes f$ on the left hand side and fully faithfulness of $\{A,B|-\}$ on the right, we obtain a natural bijection between 2-cells:

Proposition 22. *Suppose* A *is legitimate and* $f\colon A \longrightarrow X$ *is an arrow for which* lexf *exists. Then, for all* $g\colon B \longrightarrow PA$, $g \otimes f$ *exists and there is a natural isomorphism*

$$g \otimes f \;\cong\; (\mathit{lexf})g.$$

Furthermore, if X *is legitimate then, for all arrows* $h\colon C \longrightarrow X$, *there is a natural isomorphism*

$$\mathit{hom}_X(g \otimes f, h) \;\cong\; \mathit{hom}_{PA}(g, \mathit{hom}_X(f,h)).$$

Proof. By Theorem 8, $\{g|A,B\} = y_A/g$; so the first isomorphism follows from the pointwise property of lexf. When X is legitimate the left hand side of the second isomorphism exists, and

$$\{B,C|\mathit{hom}(g \otimes f,h)\} \;\cong\; g \otimes f/h \;\cong\; (\mathit{lexf})g/h \;\cong\; g/\mathit{hom}(f,1)h \;\cong\; g/\mathit{hom}(f,h)\ ;$$

so the arrow $\mathit{hom}(g \otimes f,h)$ has the defining property of $\mathit{hom}(g,\mathit{hom}(f,h))$.//

Let $\mathcal{G}$ denote a class of objects of $\mathcal{K}$. An object X of $\mathcal{K}$ is called *$\mathcal{G}$-cocomplete* when, for each G in $\mathcal{G}$ and each arrow $f\colon G \longrightarrow X$, $\mathit{lexf}\colon PG \longrightarrow X$ exists. If each G in $\mathcal{G}$ is legitimate Theorem 16 implies:

X is $\mathcal{G}$-cocomplete if and only if, for all admissible arrows $j\colon G \longrightarrow B$ with G in $\mathcal{G}$ and all arrows $f\colon G \longrightarrow X$, the pointwise left extension of f along j exists.

Theorem 23. *Let $\mathcal{G}$ be a class of objects of $\mathcal{P}rof$ such that PG is $\mathcal{G}$-cocomplete for each G in $\mathcal{G}$. The restriction $\mathcal{P}rof(\mathcal{G})$ of the extension system $\mathcal{P}rof$ to objects in $\mathcal{G}$ is an extensional bicategory with composition given by tensor product. In the notation of Theorem 21, if K is in $\mathcal{G}$ and A is legitimate and $\mathcal{G}$-cocomplete then $K(K,A)$ becomes a tensored V_K-category.*//

Suppose $\mathcal{G}$ is as in the last theorem only regard it as a full sub-2-category of $\mathcal{K}$, and let $I\colon \mathcal{G}^{co} \longrightarrow \mathcal{P}rof(\mathcal{G})$ denote the pseudo functor which is the identity on objects and which is given on hom-categories by the functors

$$K(A,B)^{op} \xrightarrow{\ hom(-,1)\ } K(B,PA) \;=\; \mathcal{P}rof(A,B)\ ;$$

the isomorphisms $I(gf) \cong Ig \otimes If$, $I1_A = y_A$ are canonical.

The following proposition extends a theorem of Benabou on profunctors to our setting.

Proposition 24. *Arrows in the image of I have right adjoints in $\mathcal{P}rof(\mathcal{G})$. Indeed, if $f\colon A \longrightarrow B$ is in $\mathcal{G}$ then $hom_B(f,1) \dashv hom_B(1,f)$ in $\mathcal{P}rof(\mathcal{G})$.*

Proof. The 2-cell $\chi'\colon hom(f,1) \longrightarrow hom(hom(1,f), hom(1,1))$ corresponds to a 2-cell $hom(f,1) \otimes hom(1,f) \longrightarrow hom(1,1) = y_B$. What we must show is that, for all arrows $g\colon B \longrightarrow C$ in $\mathcal{P}rof(\mathcal{G})$, the composite 2-cell

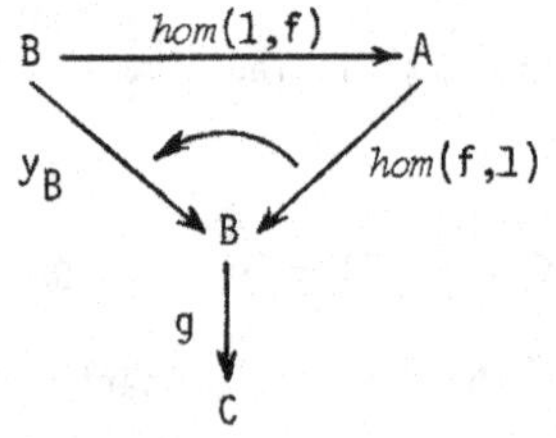

exhibits $g \otimes hom(f,1)$ as a right extension of $g \otimes y_B$ along $hom(1,f)$. But we are in an extensional bicategory so the right extension of $g \otimes y_B = g$ along $hom(1,f)$ is

$$hom(hom(1,f)g) \cong Pf.g \cong g \otimes hom(f,1),$$

as required.//

Lawvere [18] has viewed non-symmetric metric spaces as enriched categories and found a condition which can be stated in our context and reduces to Cauchy completeness for metric spaces. An object X of $\mathcal{P}rof(G)$ is called *Cauchy(-G)-complete* when each arrow $A \longrightarrow X$ in $\mathcal{P}rof(G)$ with a right adjoint is isomorphic to an arrow of the form $hom_X(f,1)$ (compare Proposition 19).

§5. Cosmoi.

Another way of expressing the precosmos condition is: for all objects A, B, the composite functor

$$K(B,PA) \xrightarrow{\{A,B|-\}} SPL(A,B) \xrightarrow{\text{und}} SPN(A,B)$$

is fully faithful. We now show that reflections with respect to this "inclusion" functor are just certain left extensions.

Theorem 25. *(Comprehension scheme). Suppose* (u,S,v) *is a span from* A *to* B *and that* $u: S \longrightarrow A$ *is coadmissible. The left extension* $k: B \longrightarrow PA$ *of* $hom_A(1,u)$ *along* v *exists precisely when the left adjoint of the functor*

$$\{A,B|-\}: K(B,PA) \longrightarrow SPN(A,B)$$

exists at the span (u,S,v).

Proof. For any arrow $h: B \longrightarrow PA$, there are bijections

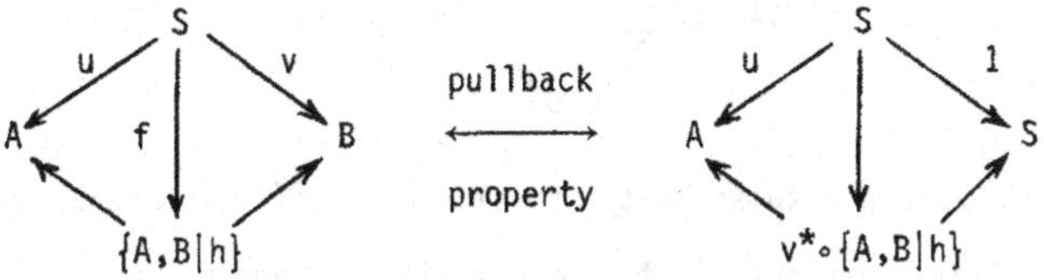

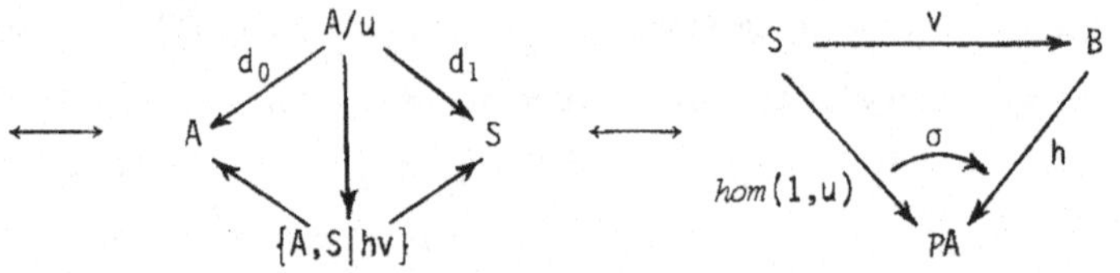

So we have the bijection between 2-cells σ and 2-cells $B \overset{k}{\underset{h}{\Downarrow \tau}} A$ if and only if we have the bijection between arrows of spans f and 2-cells τ.//

Proposition 26. Suppose the span (u,S,v) *from* A *to* B *has an opcomma object which is an admissible opspan. If* A *is legitimate then the left adjoint in the last theorem exists at* (u,S,v).

Proof. (Lax Doolittle). The reflection for the inclusion functor $COV(A,B) \longrightarrow SPN(A,B)$ is obtained by forming the comma object of the opcomma object. When A is legitimate, $Att(A,B) \subset COV(A,B)$. The condition of the theorem ensures that the reflection lands in $Att(A,B) \approx K(B,PA)$.//

Theorem 27. Suppose A *is legitimate and* $f\colon A \longrightarrow B$ *is coadmissible. The arrow* $Pf\colon PA \longrightarrow PB$ *has a left adjoint* $\exists f\colon PA \longrightarrow PB$ *if and only if the pointwise left extension of* $hom_B(1,f)$ *along* y_A *exists. In this case, there is an isomorphism*

$$\exists f \;\cong\; lexhom_B(1,f).$$

Proof. By Theorem 5, $hom(1,f)$ is admissible and $hom(hom(1,f),1) \cong Pf$. The result now follows from Proposition 17.//

Under the conditions of the last theorem, we have an isomorphism

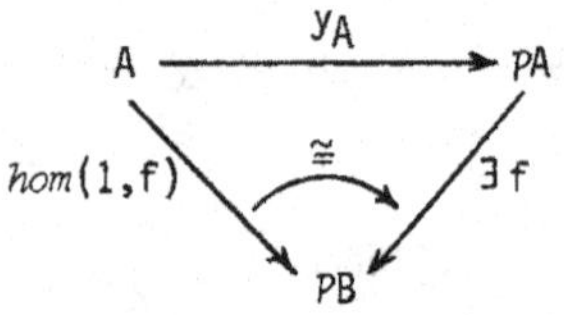

which exhibits $\exists f$ as a pointwise left extension of $hom(1,f)$ along y_A. (In the case where B is legitimate this is the more familiar diagram

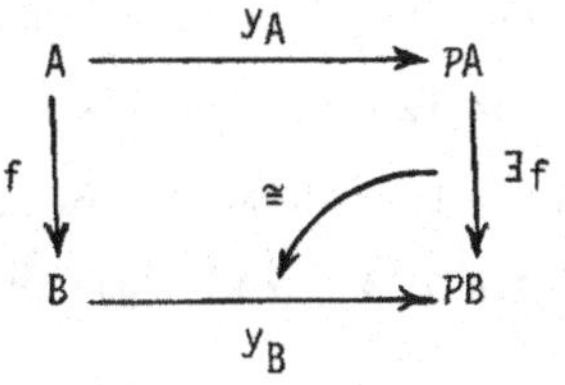

which expresses the pseudo naturality of $y: 1 \longrightarrow \exists$.)

Theorem 25 and Proposition 26 provide conditions under which certain left extensions exist. If these results are to be used to produce $\exists f$, we need to know that the extensions are pointwise (by Theorem 27). In a cosmos we shall see that, with size conditions, left extensions of arrows into objects of the form PK are necessarily pointwise. At best this approach requires opcomma objects in K (Proposition 26). In order to continue this approach and to present another approach which does not depend on opcomma objects (but does suffer from size problems in the non-uniform case) we require K to be a cosmos.

A *cosmos* is a precosmos for which the 2-functor $P: K^{coop} \longrightarrow K$ has a left 2-adjoint $P^*: K \longrightarrow K^{coop}$. This means of course that there is a 2-natural isomorphism

$$K(A, P^*B)^{op} \cong K(B, PA).$$

We denote the composite of this isomorphism with the functor $\{A,B|-\}$ by

$$\{A,B|-\}^*: K(A,P^*B)^{op} \longrightarrow SPL(A,B);$$

and the attribute $\{A,B|1_A\}^*$ from A to B when $A = P^*B$ is denoted by $\ni_B$. A split bifibration E from A to B in K becomes a split bifibration E^* from B to A in K^{co}, so that $P^*, \ni$ endow K^{co} with attributes. Indeed, K^{co} is also a cosmos. All previous precosmos theory dualizes.

It is consistent with our notation to write $E^*(a,b): X \longrightarrow P^*Y$ for the arrow corresponding to $E(a,b): Y \longrightarrow PX$ in the situation of Proposition 1. In particular, for an admissible opspan $X \xrightarrow{a} A \xrightarrow{b} Y$, we have $hom^*(a,b): X \longrightarrow P^*Y$ defined by $\{X,Y|hom^*(a,b)\}^* \cong a/b$.

An arrow $j: A \longrightarrow B$ will be called *lax-fibre small* when A and B are legitimate and there exists a strongly generating class G of legitimate objects such that, for all arrows $g: G \longrightarrow B$ with G in G, j/g is legitimate and both $hom_B^*(1,g): B \longrightarrow P^*G$ and $hom_A^*(1,d_0): A \longrightarrow P^*(j/g)$ are coadmissible.

Theorem 28. If $j: A \longrightarrow B$ *is lax-fibre small then any left extension along* j *of the form*

$$A \xrightarrow{\quad j \quad} B$$

with f, k, PK

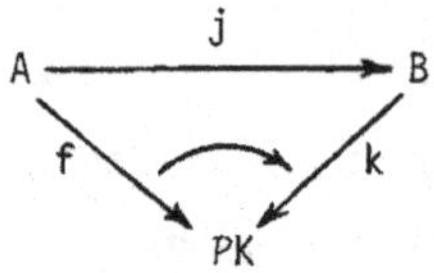

is pointwise.

Proof. Take G as above. We will prove that (c) of Theorem 3 is satisfied. By applying Theorems 11 and 14 in K^{co} we obtain an isomorphism

$$
\begin{array}{ccc}
P^*C & \xrightarrow{P^*d_1} & P^*(j/g) \\
{\scriptstyle \forall^*g}\downarrow & \cong & \downarrow{\scriptstyle \forall^*d_0} \\
P^*B & \xrightarrow[P^*j]{} & P^*A
\end{array}
$$

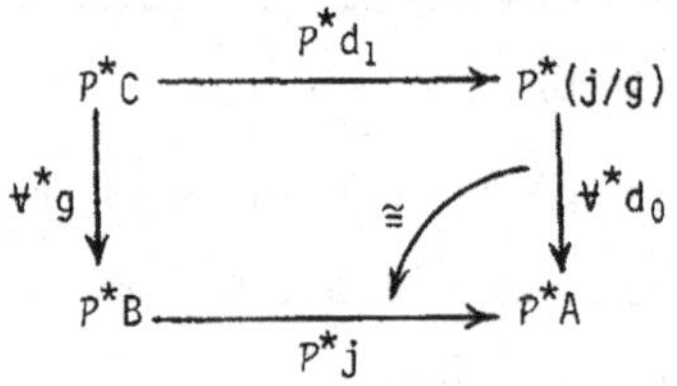

where $\forall^*g$, $\forall^*d_0$ are the left adjoints (in K; that is right adjoints in K^{co}) of P^*g, P^*d_0, respectively. Now apply $K(K,-)^{op}$ and use the adjunction $P^* \dashv P$ to obtain an isomorphism

$$
\begin{array}{ccc}
K(G,PK) & \xrightarrow{K(d_1,1)} & K(j/g,PK) \\
\downarrow & \cong & \downarrow \\
K(B,PK) & \xrightarrow[K(j,1)]{} & K(A,PK)
\end{array}
$$

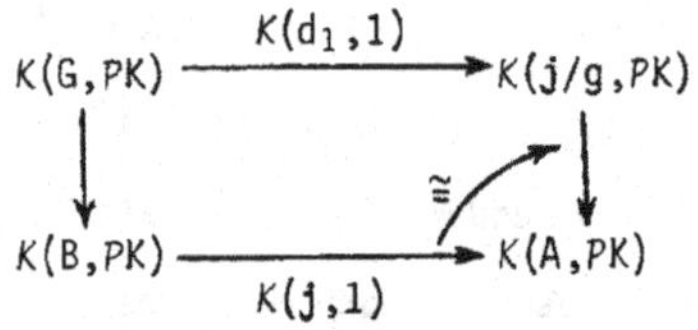

where the vertical arrows are the right adjoints of $K(g,1)$, $K(d_0,1)$. The diagram obtained by replacing the arrows by their left adjoints (when defined) also commutes up to isomorphism. But the value of the left adjoint to $K(j,1)$ at f is k. So the value of the left adjoint to $K(d_1,1)$ at fd_0 is isomorphic to kg. So kg is the left extension of fd_0 along d_1.//

An object A is called *small* when $y_A : A \longrightarrow PA$ is lax-fibre small. An arrow $f: A \longrightarrow B$ is called *kan* when A is small, B is legitimate, and, the opcomma object of the span (f,A,y_A) from B to PA exists and is an admissible opspan.

Corollary 29. *If* $f: A \longrightarrow B$ *is a kan arrow then* $Pf: PB \longrightarrow PA$ *has a left adjoint* f *and a right adjoint* $\forall f$.

Proof. Since PA is legitimate, $hom_B(f,1): B \longrightarrow PA$ is admissible; so $\forall f$ exists by Theorem 11. By Proposition 26, the left adjoint of the functor

$$\{B,PA|-\}: K(PA,PB) \longrightarrow SPN(B,PA)$$

exists at (f,A,y_A). By Theorem 25, the left extension $k: PA \longrightarrow PB$ of $hom_B(1,f)$ along y_A exists. By Theorem 28, this extension is pointwise. So, by Theorem 27, $k = \exists f$.//

Left extensions in a precosmos have been discussed. From duality present in a cosmos we obtain results about right extensions. The results about left extensions of arrows into objects of the form PK dualize to results about right extensions of arrows into objects of the form P^*K. More surprisingly, we can prove results about right extensions of arrows into PK.

Theorem 30. *Suppose* $j: A \longrightarrow B$ *is an arrow with* B *legitimate. If* $hom^*_B(1,j)$ *is coadmissible then each arrow* $f: A \longrightarrow PK$ *has a right extension along* j *preserved by any arrow of the form* $Pg: PK \longrightarrow PK'$. *(If* K^{op} *is representable then the converse of the last sentence holds.)*

Proof. By the dual of Theorem 11, $hom^*_B(1,j)$ coadmissible is equivalent to the existence of a left adjoint $\forall^* j$ to $P^*j: P^*B \longrightarrow P^*A$. From the 2-adjunction $P^* \dashv P$ we have a commutative square

$$\begin{array}{ccc} K(B,PK) & \xrightarrow{K(j,1)} & K(A,PK) \\ \cong\downarrow & & \downarrow\cong \\ K(K,P^*B)^{op} & \xrightarrow[K(1,P^*j)^{op}]{} & K(K,P^*A)^{op} \end{array} .$$

But $K(1,P^*j)^{op} \dashv K(1,\forall^*j)^{op}$, so $K(j,1)$ has a right adjoint. So each arrow $f\colon A \longrightarrow PK$ has a right extension along j. The preservation property follows from the naturality of the above square in K. We leave the converse to the reader.//

Corollary 31. Suppose $f\colon A \longrightarrow B$ *is an arrow such that* PA *is legitimate and* $hom^*_{PA}(1,Pf)$ *is coadmissible. Then* $Pf\colon PA \longrightarrow PB$ *has a left adjoint* $\exists f\colon PA \longrightarrow PB$.

Proof. Apply the theorem to the arrow Pf. The right extension of $1\colon PB \longrightarrow PB$ along Pf gives an arrow $\exists f\colon PA \longrightarrow PB$, and since this right extension is preserved by Pf we have $\exists f \dashv Pf$.//

Corollary 32. Suppose $f\colon A \longrightarrow B$ *is an arrow in a uniform cosmos. Then* $Pf\colon PB \longrightarrow PA$ *has a left adjoint* $\exists f$ *and a right adjoint* $\forall f$.//

§6. Universal constructions.

Whilst 2-CAT with $PA = [A^{op}, Cat]$ is not a cosmos in our sense, we do have pointwise left extensions in the sense of Dubuc ([6] with V = CAT) and representation arrows. So we can carry over the definition of "colimit" given in §3 to 2-categories themselves. Since $\mathbb{1}$ is a strong generator for 2-CAT, any pointwise left extension of an arrow into K along a representation arrow should be constructible from the tensor product of arrows $\ulcorner\Theta\urcorner\colon \mathbb{1} \longrightarrow [A^{op},Cat]$ with arrows $\Gamma\colon A \longrightarrow K$. If we identify 2-functors out of $\mathbb{1}$ with objects of the target 2-category, the definition of tensor product (see §4) amounts to the following.

Given 2-functors $\Theta\colon A^{op} \longrightarrow Cat$, $\Gamma\colon A \longrightarrow K$, the object $\Theta \otimes \Gamma$ of K is defined up to isomorphism by an isomorphism

$$K(\Theta \otimes \Gamma, K) \cong [A^{op},Cat](\Theta,[\Gamma-,K])$$

natural in K. Note that we could also ask for 2-naturality in K in which case we say $\Theta \circledast \Gamma$ is *2-enriched;* this is automatic when K is representable.

Examples. 1) If A is a category and Θ is the constant functor at $\mathbb{1}$ then $\Theta \otimes \Gamma = \underset{\longrightarrow}{\lim}\, \Gamma$; just the usual colimit of the functor $\Gamma: A \longrightarrow K$. So pushouts, coequalizers, coproducts all fall as special cases.

2) Take A to be the three object category with two non-identity arrows thus: $1 \longleftarrow 2 \longrightarrow 1'$. Let $\Theta: A^{op} \longrightarrow Cat$ denote the opspan $\mathbb{1} \xrightarrow{\delta_0} \mathbb{2} \xleftarrow{\delta_1} \mathbb{1}$ in Cat. A functor $\Theta: A \longrightarrow K$ is a span $A \xleftarrow{u} S \xrightarrow{v} B$ in K and $\Theta \circledast \Gamma$ is the opcomma object $u\backslash v$ of this span.

3) Let $\underline{\Delta}$ denote the simplicial category. Take A to be the 2-category with one object $*$ and $A(*,*) = \underline{\Delta}$. A 2-functor $\Gamma: A \longrightarrow K$ can be identified with a monad (A,s) in the 2-category K (using the language of [21]). So a 2-functor $\Theta: A^{op} \longrightarrow Cat$ is a monad in Cat. We take this monad Θ to be the monad called $A\underline{\Delta}$ by Lawvere in [15] pp150-1. Then

$$[A^{op},Cat](\Theta,[\Gamma-,K]) \cong [A^{op},Cat](\Theta,(K(A,K),K(s,1))) \cong K(A,K)^{K(s,1)} \cong K(A_s,K).$$

So $\Theta \otimes \Gamma$ is the construction A_s of kleisli algebras (= construction of algebras in K^{op}) for the monad (A,s) in K (again see [21]).

4) Take A to be the 2-category with two objects and non-identity arrows and 2-cells thus: $0 \underset{}{\overset{}{\Downarrow}} 1$. A 2-functor $\Gamma: A \longrightarrow K$ is just a 2-cell $A \underset{g}{\overset{f}{\Downarrow \sigma}} B$ in K. We leave it to the reader to find the 2-functor $\Theta: A^{op} \longrightarrow Cat$ (that is, a natural transformation) such that $\Theta \otimes \Gamma = B[\sigma^{-1}]$ with the following universal property:

- there is an arrow $B \longrightarrow B[\sigma^{-1}]$ such that the 2-cell $p\sigma$ is an isomorphism, and, if $B \longrightarrow K$ is an arrow such that $k\sigma$ is an isomorphism then k factors uniquely as $B \longrightarrow B[\sigma^{-1}] \longrightarrow K$.

We call $B[\sigma^{-1}]$ the *localization* of B at σ.

5) There is a dual notion of *cotensor* which generalizes "limit". As examples we obtain pullbacks, equalizers, products, comma objects, eilenberg-moore constructions and oplocalizations within a 2-category.//

Theorem 33. *A representable 2-category with a 2-terminal object admits the following constructions:*

(a) *finite limits;*

(b) *the construction of algebras;*

(c) *oplocalization.*

These constructions are all 2-enriched.

Proof. It is well known that pullbacks and a terminal object imply all finite limits. Part (b) is intimated by Gray [10] and a proof will appear in his forthcoming book [11]; we also discussed it in [12].

For part (c), take a 2-cell $A \underset{g}{\overset{f}{\rightrightarrows}} B$ with $\sigma : f \Rightarrow g$. Using limits and the category object structure on hom_B (= ΦB in [22] Proposition 2), one readily constructs an arrow $iso_B \longrightarrow hom_B$ such that, for all K, the full image of the composite

$$K(K, iso_B) \longrightarrow K(K, hom_B) \cong K(K,B)^{\mathbf{2}}$$

is precisely the full subcategory of $K(K,B)^{\mathbf{2}}$ consisting of the arrows $K \longrightarrow B$ which are isomorphisms. Then form the pullback

$$\begin{array}{ccc} P & \longrightarrow & iso_B \\ {\scriptstyle e}\downarrow & & \downarrow \\ A & \underset{\underset{\sim}{\sigma}}{\longrightarrow} & hom_B \end{array}$$

One readily checks that e is the universal arrow into A with the property that σe is an isomorphism.//

Familiar techniques prove the following.

Theorem 34. *Suppose* $\Theta: A^{op} \longrightarrow Cat$, $\Gamma: A \longrightarrow K$ *are 2-functors such that* $\Theta \otimes \Gamma$ *exists. If* $\Omega: K \longrightarrow H$ *is a 2-functor with a right 2-adjoint then*

$$\Theta \otimes (\Omega\Gamma) \cong \Omega(\Theta \otimes \Gamma).//$$

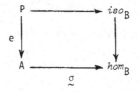

As promised in the introduction of [21] we shall prove the result which relates the eilenberg-moore construction to the internal sheaves of a certain type on the kleisli construction. Recall that ([12]§3.3), for a monad (A,s) in $\mathcal{K}$, the eilenberg-moore object E (= A^s) is defined by the condition that there is a diagram

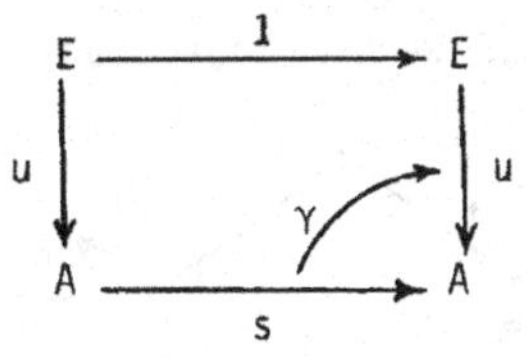

such that (u,γ) is the universal s-algebra. Dually, the kleisli object K (= A_s) is defined by the condition that there is a diagram

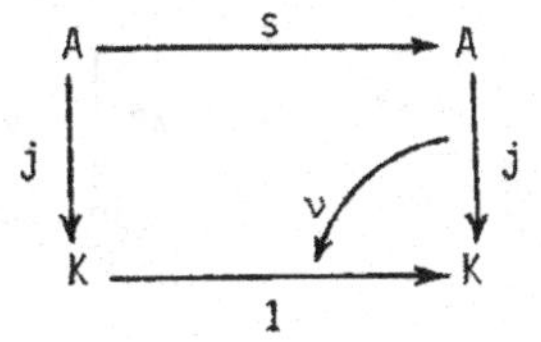

such that (j,ν) is the universal s-opalgebra. Note that (A,s) is a comonad and K is the co-eilenberg-moore object in $\mathcal{K}^{coop}$. So if P has a left 2-adjoint, Theorem 34 yields that PK is the co-eilenberg-moore object for the comonad (PA,Ps) in $\mathcal{K}$; that is,

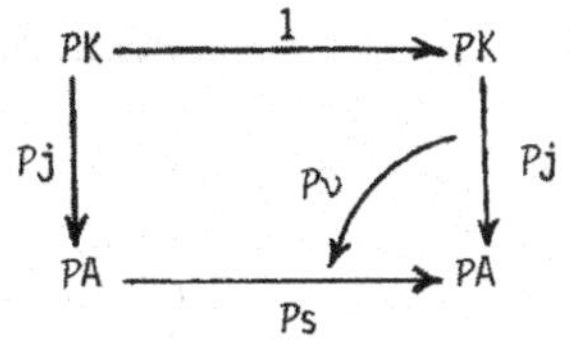

is the universal Ps-coalgebra.

<u>*Theorem 35.*</u> *Suppose* $\mathcal{K}$ *is a representable 2-category endowed with attributes such that* P: $\mathcal{K}^{coop} \longrightarrow \mathcal{K}$ *has a left 2-adjoint. Suppose* A *is legitimate and that* (A,s) *is a monad for which the kleisli object* K *exists in* $\mathcal{K}$. *An object* E *is the eilenberg-moore object for* (A,s) *if and only if there is a pullback*

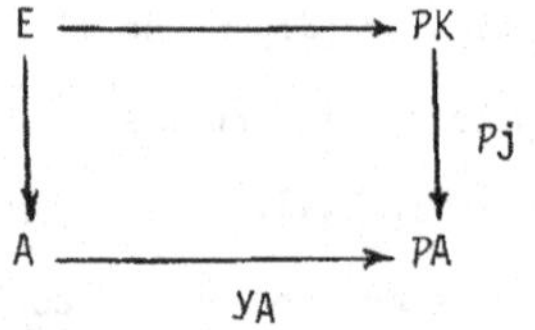

Proof. For any arrow $a: X \longrightarrow A$ we have bijections

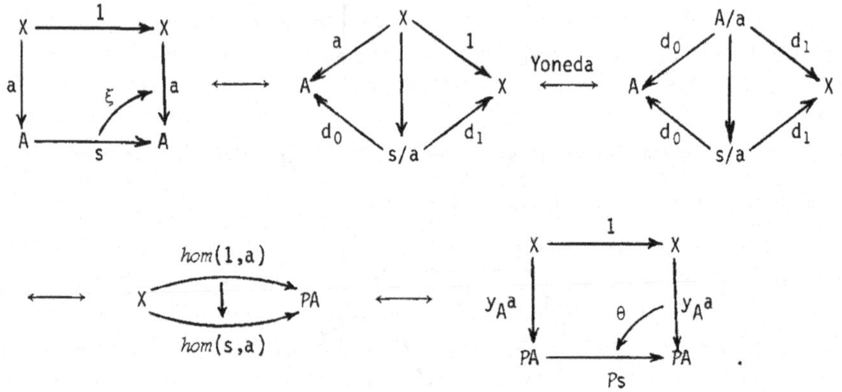

One readily sees that (a,ξ) is an s-algebra if and only if $(y_A a,\theta)$ is a Ps-coalgebra; the diagrams just translate naturally through the bijections. But $(Pj,P\nu)$ is the universal Ps-coalgebra, so in this case there exists a unique arrow $x: X \longrightarrow PK$ such that $Pj.x = y_A a$ and $P\nu.x = \theta$. In other words, we have a natural bijection

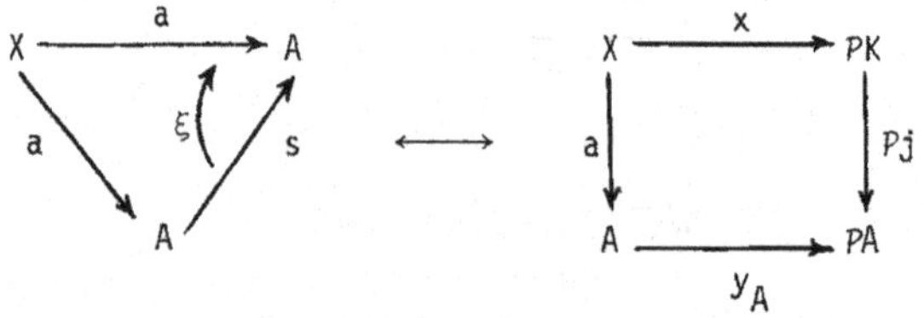

between such s-algebras and such commutative squares.//

The above theorem represents only one amongst many ways in which the various "limits" are related. One would like to obtain the kleisli construction from the eilenberg-moore since the latter does not require colimits. There are two approaches.

The first is to take the left adjoint $f: A \longrightarrow E$ to u and factor it $A \xrightarrow{j} K \xrightarrow{k} E$ where k is fully faithful and j is "bijective on objects"; but when do such factorizations exist? The second is a general one which applies to any "colimit". The monad (A,s) gives a comonad (PA,Ps) and we can form the eilenberg-moore object X for this comonad. Now we need a "recognition theorem" to tell us that X is equivalent to PK for some K. Then K should be the kleisli object for (A,s). The following "recognition theorem" does not seem good enough. One would hope to be able to generalize the work of Mikkelsen on complete atomic boolean algebras in a topos to improve the result. For enriched categories the next theorem appears in [5] p189.

Theorem 36. *In a precosmos suppose* K *is legitimate and suppose* $z: K \longrightarrow X$ *is an admissible fully faithful, dense arrow such that* $\mathit{lex}\,z$ *exists and is preserved by* $\mathit{hom}_X(z,1)$. *Then* X *is equivalent to* PK.

Proof. We have the following left extensions.

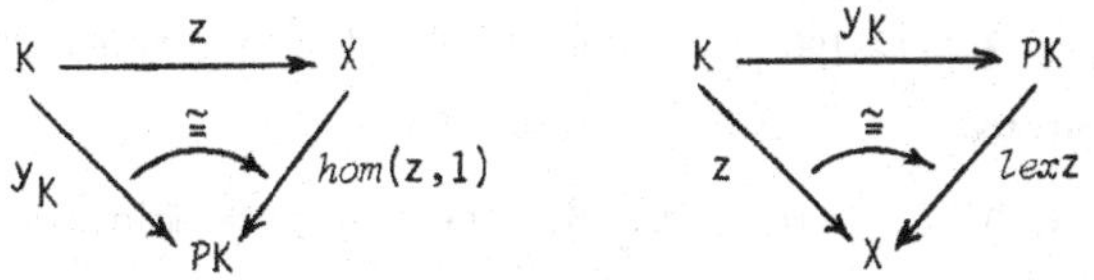

By Proposition 17, $\mathit{lex}\,z \dashv \mathit{hom}(z,1)$ and so $\mathit{lex}\,z$ preserves the first left extension. But $\mathit{lex}\,z.y_K \cong z$ and z is dense so the left extension of $\mathit{lex}\,z.y_K$ along z is 1. So $\mathit{lex}\,z.\mathit{hom}(z,1) \cong 1$. A similar argument proves $\mathit{hom}(z,1).\mathit{lex}\,z \cong 1.$//

§7. Examples

1) *Ordered objects in a topos*

A span (u,R,v) from A to B in a category $\mathcal{A}$ is called a *relation* when, given arrows $w,x:C \to R$, if $uw = ux$ and $vw = vx$ then $w = x$. Write $Rel(A,B)$ for the full subcategory of $SPN(A,B)$ consisting of the relations. There is at most one arrow of spans between any two relations from A to B; that is, $Rel(A,B)$ is an ordered set. An *ordered object* in $\mathcal{A}$ is a category object $\underset{\sim}{A}$ for which the span (d_0,A_1,d_1) is a relation; the span is then called an *order* on A_0. Objects A of $\mathcal{A}$ are regarded as ordered objects via the discrete order $(1,A,1)$. Write $\underset{\sim}{A}^{op}$ for A_0 with the reverse order (d_1,A_1,d_0). Given ordered objects $\underset{\sim}{A},\underset{\sim}{B}$, we say that an arrow $f:A_0 \to B_0$ is *order preserving* when there exists a functorial arrow $\underset{\sim}{f}: \underset{\sim}{A} \to \underset{\sim}{B}$ with $f_0 = f$. Since this $\underset{\sim}{f}$ is uniquely determined by f we often write $f: \underset{\sim}{A} \to \underset{\sim}{B}$.

The following definition is equivalent to that of Lawvere-Tierney [17]. An *(elementary) topos* is a category E which has finite limits and, for each object A, has an object PA and a relation $\in_A$ from A to PA satisfying the following "power-object" condition:

–given a relation R from A to B, there exists a unique arrow $h: B \to PA$ such that $R \cong h^{*}\circ\in_A$.

Given an arrow $f: A \to B$, define $f: PB \to PA$ by the condition $\in_B \circ f = (Pf)^{*}\circ\in_A$. In this way we obtain a functor $P: E^{op} \to E$. The composite span $\in_A \circ \in_A^{*}$ from PA to PA also comes equipped with a projection into A and so leads to a relation from A to $PA \times PA$. This relation corresponds under the power-object condition to an arrow $\wedge: PA \times PA \to PA$. The equalizer

$$C_A \xrightarrow{\binom{d_0}{d_1}} PA \times PA \underset{proj_1}{\overset{\wedge}{\rightrightarrows}} PA$$

defines an order (d_0,C_A,d_1) on PA. Henceforth we shall write PA for this ordered object.

Let K denote the full sub-2-category of $CAT(E)$ consisting of the ordered objects. Each of the categories $K(\underset{\sim}{A},\underset{\sim}{B})$ is an ordered set. The essential property of the order on PA is that

$$B \underset{s}{\overset{r}{\;\Lambda\!\!\backslash\;}} PA \quad \text{if and only if} \quad R \le S$$

where r,s correspond to the relations R,S from A to B under the power-object condition. Note that K is a representable 2-category with finite 2-limits. In particular, $\Phi\underset{\sim}{A}$ is A_1 with the order $(\hat{\partial}_1, A_1\circ A_1, \hat{\partial}_1)$.

There is a kind of lax 2-limit which we did not mention in §6 but which can be constructed in any representable 2-category with finite 2-limits. Given any ordered pair of arrows $f,g\colon \underset{\sim}{A} \longrightarrow \underset{\sim}{B}$, their *subequalizer* (Lambek [13]) is a universal diagram of the form

$$\begin{array}{ccc} \underset{\sim}{E} & \xrightarrow{\;k\;} & \underset{\sim}{A} \\ {\scriptstyle k}\downarrow & \overset{\kappa}{\nearrow} & \downarrow{\scriptstyle g} \\ \underset{\sim}{A} & \xrightarrow[\;f\;]{} & \underset{\sim}{B}\,. \end{array}$$

A construction for the subequalizer is the pullback

$$\begin{array}{ccc} \underset{\sim}{E} & \xrightarrow{\;\underline{\kappa}\;} & \Phi\underset{\sim}{B} \\ {\scriptstyle k}\downarrow & & \downarrow{\scriptstyle \binom{d_0}{d_1}} \\ \underset{\sim}{A} & \xrightarrow[\binom{f}{g}]{} & \underset{\sim}{B}\times\underset{\sim}{B}\,. \end{array}$$

We now wish to extend our functor $P\colon E^{op} \longrightarrow K$ to a 2-functor $P\colon K^{coop} \longrightarrow K$. For $\underset{\sim}{A}$ in K, let $P\underset{\sim}{A}$ be the subequalizer of $Pd_1, Pd_0 \colon PA_0 \longrightarrow PA_1$.

$$\begin{array}{ccc} P\underset{\sim}{A} & \xrightarrow{\;inc\;} & PA_0 \\ {\scriptstyle inc}\downarrow & \Leftarrow & \downarrow{\scriptstyle Pd_0} \\ PA_0 & \xrightarrow[\;Pd_1\;]{} & PA_1\,. \end{array}$$

The 2-functor structure of P is induced using the enriched "limit" property.

Suppose $\underset{\sim}{A},\underset{\sim}{B}$ are objects of K. An *ideal* from $\underset{\sim}{A}$ to $\underset{\sim}{B}$ is a split bifibration from $\underset{\sim}{A}$ to $\underset{\sim}{B}$ which is a relation. Let $Idl(\underset{\sim}{A},\underset{\sim}{B})$ denote the full subcategory of $SPN(\underset{\sim}{A},\underset{\sim}{B})$ consisting of the ideals. Given a relation (u, R_0, v) from A_0 to B_0 in E, there is a unique order (d_0, R_1, d_1) on R_0 such that $(u,\underset{\sim}{R},v)$ is a relation from $\underset{\sim}{A}$ to $\underset{\sim}{B}$ in K. So $Id\ell(\underset{\sim}{A},\underset{\sim}{B})$ is a subcategory of $Re\ell(A_0,B_0)$.

Proposition 37. The composite functor

$$K(\underset{\sim}{B},P\underset{\sim}{A}) \xrightarrow{K(1,\,inc)} K(\underset{\sim}{B},PA_0) \hookrightarrow E(B_0,PA_0) \approx Re\ell(A_0,B_0)$$

induces an equivalence of categories

$$K(\underset{\sim}{B},P\underset{\sim}{A}) \approx Id\ell(\underset{\sim}{A},\underset{\sim}{B}).$$

Proof. Each functor in the composite is clearly fully faithful so it remains to show that the composite is surjective up to isomorphism onto the ideals. By Proposition 12 of the last paper, a relation R from A_0 to B_0 is an ideal from $\underset{\sim}{A}$ to $\underset{\sim}{B}$ precisely when $B_1 \circ R \le R$ and $R \circ A_1 \le R$ (the extra conditions are diagrams in the ordered set $Re\ell(A_0,B_0)$ and hence automatically commute). Let $r: B_0 \longrightarrow PA_0$ correspond to R under $E(B_0,PA_0) \approx Re\ell(A_0,B_0)$. The conditions $B_1 \circ R \le R$, $R \circ A_1 \le R$ translate to the following conditons on r.

$$\begin{array}{ccc} B_1 & \xrightarrow{d_1} & B_0 \\ {\scriptstyle d_0}\downarrow & \le & \downarrow{\scriptstyle r} \\ B_0 & \xrightarrow[r]{} & PA_0 \end{array} \qquad\qquad \begin{array}{ccc} B_0 & \xrightarrow{r} & PA_0 \\ {\scriptstyle r}\downarrow & \le & \downarrow{\scriptstyle Pd_0} \\ PA_0 & \xrightarrow[Pd_1]{} & PA_1 \end{array}$$

The first of these says precisely that $r: \underset{\sim}{B} \longrightarrow PA_0$ is order preserving and the second says precisely that $r: \underset{\sim}{B} \longrightarrow PA_0$ factors uniquely through *inc* to yield $r: \underset{\sim}{B} \longrightarrow P\underset{\sim}{A}$.//

An arrow of spans between ideals from $\underset{\sim}{A}$ to $\underset{\sim}{B}$ is automatically a homomorphism of split bifibrations from $\underset{\sim}{A}$ to $\underset{\sim}{B}$ since the homomorphism axioms are diagrams in an ordered set. So $Id\ell(\underset{\sim}{A},\underset{\sim}{B})$ is a full subcategory of $SPL(\underset{\sim}{A},\underset{\sim}{B})$. Furthermore, $Id\ell$ is a sub-pseudo-functor of SPL (see early §1). With this pseudo functoriality of $Id\ell$, the inclusion $Id\ell(\underset{\sim}{A},\underset{\sim}{B}) \longrightarrow Re\ell(A_0,B_0)$ becomes pseudo natural in $\underset{\sim}{A},\underset{\sim}{B}$. The composite functor of Proposition 37 is clearly pseudo natural in $\underset{\sim}{A},\underset{\sim}{B}$. It follows that the equivalence $K(\underset{\sim}{B},P\underset{\sim}{A}) \approx Id\ell(\underset{\sim}{A},\underset{\sim}{B})$ is pseudo natural in $\underset{\sim}{A},\underset{\sim}{B}$.

Theorem 38. The 2-category K *of ordered objects in a topos* E *is a cosmos in which the attributes are precisely the ideals.*

Proof. We have already seen that the fully faithful functor

$$K(\underset{\sim}{B},P\underset{\sim}{A}) \approx \mathit{Idl}(\underset{\sim}{A},\underset{\sim}{B}) \subset \mathit{SPL}(\underset{\sim}{A},\underset{\sim}{B})$$

is pseudo natural in $\underset{\sim}{A},\underset{\sim}{B}$. Set $B = PA$ and evaluate at the identity to obtain an ideal $\in_{\underset{\sim}{A}}$ from $\underset{\sim}{A}$ to $P\underset{\sim}{A}$ which consequently endows K with attributes. We have immediately that the attributes are the ideals so the precosmos condition is satisfied. One readily verifies that P has a left 2-adjoint P^* given by $P^*\underset{\sim}{A} = (P\underset{\sim}{A}^{op})^{op}$.//

For an ordered object $\underset{\sim}{A}$, the split bifibration ΦA from $\underset{\sim}{A}$ to $\underset{\sim}{A}$ is a relation and hence an ideal. So K is a uniform cosmos.

Corollary 39. For each order-preserving arrow $f\colon \underset{\sim}{A} \longrightarrow \underset{\sim}{B}$, *the order-preserving arrow* $Pf\colon P\underset{\sim}{B} \longrightarrow P\underset{\sim}{A}$ *has both a left and a right adjoint.*//

2) *The pre-Spanier constuction as a* P.

The 2-category Simp and 2-functor P presented here are taken from unpublished work of Day-Kelly on categories like categories of topological spaces.

A function $f\colon X \longrightarrow Y$ between sets X,Y is called *constant* when it factors through the one-point set 1. A *simple category* is a category A together with a functor $|\ |\colon A \longrightarrow \mathrm{Set}$ satisfying:

SC1. $|\ |_{a,b}\colon A(a,b) \longrightarrow \mathrm{Set}(|a|,|b|)$ is an inclusion of sets;

SC2. the image of $|\ |_{a,b}$ contains all the constant functions;

SC3. there is an object a of A with $|a| \neq 0$.

For objects a, a' in A, we write $a \leq a'$ when $|a| = |a'|$ and $1_{|a|}\colon |a| \longrightarrow |a'|$ is in $A(a,a')$. A functor $f\colon A \longrightarrow B$ between simple categories is simple when the following diagram commutes.

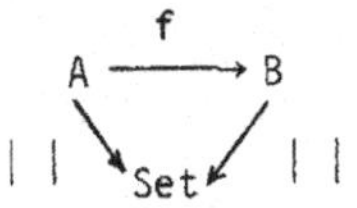

Let *Simp* denote the 2-category whose objects are simple categories, whose arrows are simple functors, and whose 2-cells are natural transformations. Then, for each A,B in *Simp*, the category *Simp*(A,B) is an ordered set; indeed, there is a natural transformation $A \overset{f}{\underset{g}{\Downarrow\sigma}} B$ precisely when $fa \le ga$ for all a in A.

Pullback in *Simp* is that of *CAT* and *Simp* is a representable 2-category. For simple functors $A \xrightarrow{r} D$, $B \xrightarrow{s} D$, the simple category r/s is defined as follows. The objects are pairs (a,b) with a in A, b in B and $ra \le sb$ in D. The arrows are pairs $(\alpha,\beta): (a,b) \longrightarrow (a',b')$ where $\alpha: a \longrightarrow a'$, $\beta: b \longrightarrow b'$ are in A,B respectively, and the following square commutes.

$$\begin{array}{ccc} ra & \xrightarrow{1_{|a|}} & sb \\ {\scriptstyle r\alpha}\downarrow & & \downarrow{\scriptstyle s\beta} \\ ra' & \xrightarrow[1_{|a'|}]{} & sb' \end{array}$$

Let A be a simple category. A new simple category PA is defined as follows. An object x of A is a set $|x|$ together with a subset $Ad(a,x)$ of $Set(|a|,|x|)$ for each a in A, satisfying:

AD1. every constant function $|a| \longrightarrow |x|$ is in $Ad(a,x)$;

AD2. if θ is in $Ad(b,x)$ and α is in $A(a,b)$ then $\theta|\alpha|$ is in $Ad(a,x)$.

An arrow $\kappa: x \longrightarrow y$ in PA is a function $\kappa: |x| \longrightarrow |y|$ such that $\kappa\theta$ is in $Ad(a,y)$ whenever θ is in $Ad(a,x)$.

Suppose $f: A \longrightarrow B$ is a simple functor. Define a simple functor $Pf: PB \longrightarrow PA$ by setting

$$Ad(a,(Pf)y) = Ad(fa,y)$$

This gives a 2-functor $P: Simp^{coop} \longrightarrow Simp$ which has a left 2-adjoint P^*; an object x of P^*A is a set $|x|$ together with a subset $Ad^*(x,a)$ of $Set(|x|,|a|)$ for each a in A satisfying the obvious axioms AD*1 and AD*2.

Define the simple functor $y_A: A \longrightarrow PA$ by

$$Ad(a,y_A a') = \{|\alpha|: |a| \longrightarrow |a'| \;\big|\; \alpha: a \longrightarrow a' \text{ in } A\}.$$

One readily proves that the functors

$$Simp(B,PA) \longrightarrow COV(A,B)$$
$$f \longmapsto y_A/f$$

are all equivalences and are the components of a pseudo natural transformation.

Theorem 40. Simp is a uniform cosmos. //

From Corollary 32 we know that $Pf: PB \longrightarrow PA$ has a left adjoint $\exists f: PA \longrightarrow PB$. It is given by the formula

$$Ad(b,(\exists f)x) = \{\theta|\beta| \,|\, a \text{ in } A;\ \beta{:}b \longrightarrow fa \text{ in } B;\ \theta \text{ in } Ad(a,x)\}$$

3) Categories

We regard the category *Cat* of small categories as an object of the 2-category CAT of all categories. For categories A,B, let $Sp\ell(A,B)$ denote the full subcategory of $SPL(A,B)$ consisting of the split bifibrations E from A to B such that the "fibre" E(a,b) over the pair $\mathbb{1} \xrightarrow{a} A$, $\mathbb{1} \xrightarrow{b} B$ is in *Cat*. There is a pseudo-natural equivalence of categories

$$Sp\ell(A,B) \approx CAT(B,[A^{op},Cat]).$$

For any full subcategory C of *Cat*, we consequently obtain a pseudo natural family of fully faithful functors

$$CAT(B,[A^{op},C]) \longrightarrow SPL(A,B).$$

Put $B = [A^{op},C]$ and evaluate at the identity to obtain $\in_A$, and put $PA = [A^{op},C]$. This endows CAT with attributes. Furthermore, P has a left 2-adjoint P^* given by $P^*A = [A,C]^{op}$. The attributes are split bifibrations with fibres in C.

In order that this structure should be a cosmos recall that the arrows of spans between attributes should automatically be homomorphisms of split bifibrations. This happens when the fibres of the attributes over pairs of objects are all discrete. Hence, with C any full subcategory of the category *Set* of small sets, CAT becomes a cosmos with $PA = [A^{op},C]$.

The case C = *Set* gives the paradigmatic example of a cosmos. The legitimate objects in this case are the categories with small hom-sets. The author conjectured in this case that a category A is equivalent to a category with a small set of objects if and only if both A and PA are legitimate. Peter Freyd has pointed out that this is a consequence of his paper [9]. So a category is small in the sense of §5 if and only if it is equivalent to a small category in the usual sense. A functor $j: A \longrightarrow B$ with A small and B legitimate is kan in the sense of §5.

If C is the category of finite sets, the legitimate objects are the categories with finite hom-sets.

Take C to consist of the empty set 0 and the one element set 1. The legitimate objects are the partially ordered sets. Attributes between legitimate objects are jointly monomorphic split bifibrations (= ideals).

If C consists just of the one-element set, the legitimate objects are the indiscrete categories (that is, those categories with precisely one arrow between any two objects).

Bibliography

[1] J. Benabou, *Catégories avec multiplication*. C.R. Acad. Sc. Paris 256 (1963) 1887-1890.

[2] J. Benabou, *Introduction to bicategories*. Lecture Notes in Mathematics 47 (1967) 1-77.

[3] J. Benabou and J. Roubaud, *Monades et descente*. C.R. Acad. Sc. Paris 270 (1970) 96-98.

[4] C. Chevalley, *Séminaire sur la descente 1964-65* (unpublished)

[5] B.J. Day and G.M. Kelly, *Enriched functor categories*. Lecture Notes in Math. 106 (1969) 178-191.

[6] E.J. Dubuc, *Kan extensions in enriched category theory*. Lecture Notes in Math. 145 (1970) 1-173.

[7] S. Eilenberg and G.M. Kelly, *Closed categories*. Proc. Conference on Categorical Algebra at La Jolla (Springer, 1966) 421-562.

[8] P. Freyd, *Abelian categories*. (Harper and Row, 1964)

[9] P. Freyd, *Several new concepts*. Lecture Notes in Math. 99 (1969) 196-241.

[10] J.W. Gray, *Report on the meeting of the Midwest Category Seminar in Zurich*. Lecture Notes in Math. 195 (1971) 248-255.

[11] J.W. Gray, *Formal category theory*. Springer Lecture Notes in Math. (to appear)

[12] G.M. Kelly and R.H. Street, *Review of the elements of 2-categories*. (this volume)

[13] J. Lambek, *Subequalizers*. Canad. Math. Bull. 13 (1970) 337-349.

[14] F.W. Lawvere, *The category of categories as a foundation for mathematics*. Proc. Conference on Categorical Algebra at La Jolla (Springer, 1966) 1-20.

[15] F.W. Lawvere, *Ordinal sums and equational doctrines*. Lecture Notes in Math. 80 (1969) 141-155.

[16] F.W. Lawvere, *Equality in hyperdoctrines and comprehension schema as an adjoint functor*. Proc. of Symposia in Pure Math. 17 (A.M.S. 1970) 1-14.

[17] F.W. Lawvere, *Introduction*. Lecture Notes in Mathematics 274 (1972) 1-12.

[18] F.W. Lawvere, *Metric spaces, generalized logic, and closed categories*. (Preprint from Istituto di Matematica, Università di Perugia, 1973)

[19] F.E.J. Linton, *The multilinear Yoneda lemmas*. Lecture Notes in Math. 195 (1971) 209-229.

[20] S. Mac Lane, *Categories for the working mathematician*. Graduate texts in Math. 5 (Springer 1971).

[21] R.H. Street, *The formal theory of monads*. Journal of Pure and Applied Algebra 2 (1972) 149-168.

[22] R.H. Street, *Fibrations and Yoneda's lemma in a 2-category*. (this volume)

ON CLUBS AND DOCTRINES

by

G.M. Kelly

<u>Doctrines</u> (= strict 2-monads) were introduced, under the name of <u>equational doctrines</u>, by Lawvere [16], and are discussed to some extent in [12] above, which is our general reference for notation, terminology, and basic concepts. <u>Clubs</u> were introduced, in connection with coherence problems, by the present author in [5], [6], and [7], where <u>covariant clubs</u> were distinguished from <u>mixed-variance clubs</u>; only the first concern us in this paper. Such a club K gives rise to a doctrine $K\circ-$ on the 2-category CAT of all categories. I suggested in [5] §4.1 and in [6] §4.1 that the simple concept of (covariant) club used there doubtless admitted of wide generalization, and indicated the way I expected this to go. Such a widening of the meaning of "club" would increase the number of doctrines writable in the form $K\circ-$ for a "club" K, and I was in effect toying there with the idea that perhaps every doctrine was so writable.

The purpose of the present paper (some aspects of which were announced in [11]) is to carry out this generalization of the "club" idea to what seem to be its natural limits. I must say at once that I withdraw from the position that every doctrine then comes from some club; in particular I withdraw my assertion in [6] §4.1 that we can get in this way the doctrine whose algebras are categories-with-equalizers. I, at any rate, can see no generalization wide enough to achieve this.

What we do construct is a monoidal 2-category $CAT\int{*}CAT$, an object of which is a category K together with an "augmentation functor" $\Gamma\colon K \to CAT$, and the "tensor product" for which is denoted by $\circ$. We embed CAT fully in $CAT\int{*}CAT$ by assigning to $A \in CAT$ the "trivial augmentation" $A \to CAT$ sending each object of A to the empty category; then $K\circ A \in CAT$ whenever $A \in CAT$. Thus $K \mapsto K\circ-$ gives a 2-functor Φ

from $CAT\int{}^{*}CAT$ to the 2-category $[CAT,CAT]$ of endo-2-functors of CAT. This 2-functor Φ is 2-fully-faithful, and carries the tensor product $\circ$ into composition of endo-2-functors. A club is now by definition a $\circ$-monoid K in $CAT\int{}^{*}CAT$, and is carried by Φ into the doctrine $K\circ -$.

The doctrines arising thus from a club have the advantage of a very concrete presentation: in the sense that a category K with its augmentation $\Gamma: K \to CAT$ is a more concrete object for study than a general 2-functor $D: CAT \to CAT$. Still more special are the doctrines $K\circ -$ where K is a $\circ$-monoid in certain sub-monoidal-2-categories of $CAT\int{}^{*}CAT$, such as CAT/Set. It is easy to recognize certain structures on a category as being actions of a doctrine of the latter kind, and to describe in theory the process of constructing K from the basic operations and axioms; finding K explicitly in practice is what we may call the <u>coherence problem</u> for the given structure. Again, it is to doctrines D of this latter kind that the work of Day [1] seems to extend, giving a D-structure on $[A^{op},Set]$ from one on A, and leading to the notions of a pro-D-structure and of a <u>closed</u> D-structure - one in which each structural functor has a right adjoint in every argument.

There are lots of things I still do not understand. I have no criterion <u>in terms of the doctrine</u> D for it to be $K\circ -$ where K is a club in $CAT\int{}^{*}CAT$, or a club in CAT/Set; although as I said I can recognize the latter case from certain presentations of the structure. I know that clubs in CAT/Set play an important role, but have no special feeling for those in $CAT\int{}^{*}CAT$, except that one is ineluctibly led to this generalization. In the CAT/Set case, I know what it means to ask the coherence question "which diagrams commute?"; for more general structures I do not. In the still more restricted $CAT/\underline{\underline{P}}$ case, where $\underline{\underline{P}}$ is the category of natural numbers and permutations, I know how to ask the coherence question even for <u>closed</u> D-categories (e.g. closed monoidal categories), in terms of mixed-variance functors and the generalized natural transformations of Eilenberg-Kelly [2]; but in the CAT/Set case

I do not. For instance, what kind of coherence result does one hope for in the case of cartesian closed categories?

Because of these uncertainties, which may attract the interest of others, I choose to devote §1 to a brief account of the original club idea as suggested by coherence problems, before embarking on the above generalizations; besides providing motivation, it may suggest avenues that I have not seen. The substance of the present paper begins in §2 where, changing gear completely, we begin with the most general before passing to the particular. Our path differs in two ways from the brief outline above: first, some things loosely stated there have to be made more precise; and secondly, to minimize the long verifications so common to 2-categorical studies, we take a round-about route.

1. THE ORIGINAL CLUB IDEA

1.1 It arose naturally, originally in the mixed-variance case, from an attempt to refine the formulation of certain coherence problems - such as that for a closed category [9], that for two closed categories connected by a closed functor [17], and that for a closed category $\mathcal{V}$ together with a $\mathcal{V}$-natural transformation between two $\mathcal{V}$-functors between two $\mathcal{V}$-categories [10].

In each of these cases the diagrams whose commutativity was in question could be conceived of, to a first approximation, as having functors (of many variables and of mixed variances) for vertices, and natural transformations (of the generalized kind introduced by Eilenberg-Kelly [2]) for edges. Precision however required, as was already recognized in [9], that the vertices be not actual functors in the model but their abstract descriptions in the theory; for otherwise one would have unwanted composites of $f: T \to S$ and $g: S' \to R$, where S and S' although formally different had identical realizations in a particular model. Yet in [9] the edges were still actual natural transformations.

This served well enough while only positive results - certain diagrams do commute - were sought. In Mac Lane's original coherence theorem [19] for a monoidal or for a symmetric monoidal category, the result had in fact been that all diagrams commute. For a closed category [9] it was no longer so; some diagrams failed to commute in general, although they might do so in a sufficiently degenerate model. My student Lewis [17] found such non-commutativity even in the purely covariant case of two monoidal categories connected by a monoidal functor. What was crucial was that Lewis found a perfectly correct necessary and sufficient condition for commutativity in this case - which made no sense at the level of models. We clearly had to go further and replace the edges too by their abstract descriptions; it is noteworthy that Isbell independently suggested doing this in his review [4] of [9]. (The preliminary account of Lewis' work in [17] is superseded by a far better one in his forthcoming thesis [18].)

1.2 We now had a category K whose vertices were abstract descriptions of those functors obtainable by iterated composition of the basic structural ones, and whose morphisms were abstract descriptions of those natural transformations obtainable from the basic structural ones by composing them in all possible ways with the basic functors and with one another. This category K had a good deal of structure, and it was the abstraction of this structure that I called a club. Each club was an invariant description of an extra structure that a category, or a family of categories, might carry; and the coherence problem became that of determining the club K from its generators (the basic structural functors and natural transformations) and its relations (the axioms to which these were subjected). The analogy with Lawvere's theories is clear; although in fact a closer analogy is with props or operads.

The structure on K was as follows. First, each formal functor T had a type ΓT, specifying its arity, the category (when there was more than one) from which the i-th argument was to be drawn, and the variance of this argument. Next, each formal natural transformation $f: T \to S$ also had a type Γf, specifying which arguments it was to pair off; this was called by Eilenberg-Kelly [2], and has since been called by me, not too happily I think, the graph of f; I shall try to avoid this over-used word. These types ΓT and Γf may be considered as objects and morphisms of a suitable category T, so that Γ appears as a functor $\Gamma: K \to T$, and K as an object of the category CAT/T.

The next piece of structure is the recognition that objects of K can be composed with one another - or rather substituted into one another, a better expression for functors of many variables. Thus from T of arity n and $S_1, \ldots, S_n$ of arities $m_1, \ldots, m_n$ we can form $T(S_1, \ldots, S_n)$ of arity $m_1 + \ldots + m_n$, also written $n(m_1, \ldots, m_n)$. Similarly for the morphisms of K; one can form $f(g_1, \ldots, g_n)$, with its special cases $T(g_1, \ldots, g_n)$ and $f(S_1, \ldots, S_n)$. It turns out that this can be expressed as a multiplication $\mu: K \circ K \to K$, where $\circ$ is the "tensor product" for a certain monoidal structure on CAT/T.

Finally one must distinguish among the objects of K the formal identity functor (or functors, when many categories are involved); it turns out that this comes to giving a unit $\eta: J \to K$, where J is the identity for $\circ$. The associativity of composition (or of substitution), and the identity property of η, are then given by subjecting μ and η to the monoid axioms. Thus we end with the definition of club as "$\circ$-monoid in the monoidal category CAT/T ".

We can embed CAT in CAT/T by giving to $A \in CAT$ the trivial augmentation $\Gamma: A \to T$ assigning arity zero to each object of A. Then to give to a category A the extra structure characterized by the club K is to give an action $K \circ A \to A$ of the monoid K on A. Thus the categories with this structure are the algebras for the monad $K \circ -$ on CAT.

Similarly when the structure is borne by a family of categories, indexed say by Λ; then CAT/Λ replaces CAT.

1.3 We must now point out that what we have just said is, in the mixed-variance cases from which we drew our examples, over-simplified for expository purposes: the phenomenon of incompatibility for the generalized natural transformations of [2] causes the "tensor product" $\circ$ to be not everywhere defined on CAT/T. This is discussed at length in [5], [6], and [7]; we pursue it no further here because we henceforth restrict to the purely covariant case. Then all we have said is exact, and in fact more is true. Namely, $\circ$ is a 2-functor and CAT/T is a monoidal 2-category; so that $K\circ-$ for a club K is not only a monad but a 2-monad or doctrine. Moreover the monoidal 2-category CAT/T is actually closed: its internal-hom $\{B,C\}$ is a very rich functor category, whose objects are functors $B^n \to C$ of all arities and whose morphisms are natural transformations of all types.

In the covariant case the type Γf of a natural transformation of the Eilenberg-Kelly kind is nothing but a permutation of the arguments in passing from domain to codomain, as in the symmetry $A \otimes B \to B \otimes A$ in a monoidal category, or in the associativity $(A \otimes B) \otimes C \to A \otimes (B \otimes C)$ where this permutation is the identity. If we restrict for simplicity to the one-category case the type ΓT of a functor is just its arity $n \in \underline{\underline{N}}$, so that in this case the category T is just the category $\underline{\underline{P}}$ with natural numbers as objects and permutations as morphisms. It is to clubs at this $CAT/\underline{\underline{P}}$ level that [5] and [6] are devoted, except that the many-category case is allowed for and the corresponding mixed--variance case is also considered.

1.4 The suggestion naturally occurs that many more coherence problems can be formulated in this way if we further widen the notion of "natural transformation f of type Γf ". For instance, a category with finite coproducts has a binary functor $+$ and a nullary functor 0, with "natural transformations" $A \to A + B$, $B \to A + B$, $A + A \to A$, $0 \to A$, subjected to equational axioms. The types of these natural transformations are not permutations but functions: the two functions $1 \to 2$, and the unique functions $2 \to 1$ and $0 \to 1$. Similarly for a category with two monoidal structures $(\otimes, I)$ and $(\oplus, N)$, connected by distributivity maps (not isomorphisms)

(1.1) $\quad d: (A\otimes B)\oplus(A\otimes C) \to A\otimes(B\oplus C), \quad e: N \to A\otimes N,$

for which the coherence problem has been studied in another formulation by Laplaza [14] and [15] (whose results we generalize in [8] below). In other words we expect a generalization with $CAT/\underline{\underline{P}}$ replaced by $CAT/\underline{\underline{S}}$, where $\underline{\underline{S}}$ is the skeletal category of finite sets, with natural numbers as objects and functions as morphisms.

Having gone this far it is natural to expect still more: why should the arities n be finite sets? We can still speak of a functor $B^n \to C$ when n is a <u>small category</u>: and if a category A with finite coproducts has functors $A^2 \to A$ and $A^0 \to A$, so a category A with coequalizers has a functor $A^{\rightrightarrows} \to A$. So we may expect a further generalization with $CAT/\underline{\underline{S}}$ replaced by $CAT/\mathcal{C}at$.

That such a generalization does exist was announced in [11]; in the sense that we still have a monoidal structure $\circ$ on $CAT/\mathcal{C}at$, the monoids K for which give doctrines on CAT. (I am ignoring for the moment the fact that we must, like Kock in [13], use here a rather special version of $\mathcal{C}at$.) Yet this is not altogether a generalization of the "coherence problem"; for K is not now so simply given by its generators and relations. It was my failure to anticipate this last point that led me falsely to predict that the doctrine for categories-with-coequalizers would be seen as coming from a club.

The attempt to carry out the details of this generalization automatically led to a further, and final, one: the 2-functor $\Phi: CAT/\mathcal{C}at \to [CAT, CAT]$ is not <u>full</u>; we can make it so by adding more 1-cells and 2-cells to $CAT/\mathcal{C}at$ to get a new 2-category $CAT\int{}^*\mathcal{C}at$; and $\circ$ can still be defined, indeed cries out to be so, on $CAT\int{}^*\mathcal{C}at$ or even

$CAT\int{}^*CAT$. This is the level of generalization we promised in the introduction, to the details of which we now pass, after first adding a final remark to this section.

1.5 If $D = K\circ -$ is the doctrine for categories-with-finite-coproducts, that for categories-with-finite-products is the opposite doctrine D^* (cf. [16]) given by $D^*A = (DA^{op})^{op}$. We could if we liked define a club of the second kind, with augmentation landing not in CAT but in CAT^{op} - or in $\underline{\underline{S}}^{op}$ in this particular case - and then D^* would be $K^{op}\circ -$, where this is a new "$\circ$". For the basic natural transformations are now $A\times B \to A$, $A\times B \to B$, $A \to A\times A$, $A \to 1$, which can clearly be said to have "types" in $\underline{\underline{S}}^{op}$. So we get a second distinguished subclass of doctrines, those coming from clubs of the second kind; but they are just the opposite doctrines of those coming from clubs, and we shall not complicate the present paper by making special reference to them.

This does, however, raise the question of how we approach the coherence problem for more complicated structures, such as a category with both finite coproducts and finite products, where we have natural transformations like

$$A+A \to A \to A\times A;$$

or for the distributivity problem of §1.4 when the maps (1.1) are isomorphisms, so that $d^{-1}d$ is a natural transformation

$$(A\otimes B)\oplus(A\otimes C) \to (A\otimes B)\oplus(A\otimes C).$$

It is tempting to say that here the type of the natural transformation is neither in $\underline{\underline{S}}$ nor in $\underline{\underline{S}}^{op}$, but is something of the form $n \to p \leftarrow m$, that is, an op-span. However I have so far been unable to extend the club idea to this case; I am quite puzzled about what a good "coherence result" would be, other than the explicit determination of the doctrine; and I make no further reference in this paper to the possibility of such further extensions. (P.s. Some later thoughts on what a "coherence result" should be are contained in my paper [8] below in this volume.)

2. LAX NATURAL TRANSFORMATIONS

2.1 We use a single arrow $\rightarrow$ for 1-cells and 3-cells, in particular for functors or 2-functors and for modifications; we use a double arrow $\Rightarrow$ for 2-cells, in particular for natural or 2-natural transformations; and we use a wiggly arrow $\rightsquigarrow$ for the lax natural transformations we are about to introduce.

We identify a set with the corresponding discrete category, and a category with the corresponding 2-category with only identity 2-cells. If $\mathcal{A}$ is a 2-category, $\mathcal{A}_0$ is the category obtained by discarding the 2-cells, and $|\mathcal{A}| = \mathrm{ob}\,\mathcal{A}$ is the set obtained by discarding also the 1-cells. For 2-categories $\mathcal{A}$ and $\mathcal{B}$, we denote by $[\mathcal{A},\mathcal{B}]$ the 2-category of 2-functors, 2-natural transformations, and modifications; recall that we use these terms in the sense of [12] above. It is necessary in this context to distinguish clearly $\mathcal{A}$ from $\mathcal{A}_0$, for $[\mathcal{A},\mathcal{B}]$ and $[\mathcal{A}_0,\mathcal{B}]$ are quite different when $\mathcal{B}$ is a 2-category. Therefore we shall always write $\mathcal{CAT}$ for the 2-category and $\mathcal{CAT}_0$ for the category, and so on.

2.2 For 2-categories $\mathcal{A}$ and $\mathcal{T}$ and 2-functors $F,G\colon \mathcal{A} \rightarrow \mathcal{T}$, a <u>lax natural transformation</u> $\alpha\colon F \rightsquigarrow G$ is what Gray [3] called a "2-natural" one: it assigns to each object A of $\mathcal{A}$ a morphism $\alpha_A\colon FA \rightarrow GA$ in $\mathcal{T}$, and to each morphism $a\colon A \rightarrow A'$ in $\mathcal{A}$ a 2-cell α_a in $\mathcal{T}$ of the form

(2.1)
$$\begin{array}{ccc} FA & \xrightarrow{\ \alpha_A\ } & GA \\ {\scriptstyle Fa}\downarrow & \Downarrow{\scriptstyle \alpha_a} & \downarrow{\scriptstyle Ga} \\ FA' & \xrightarrow[\ \alpha_{A'}\ ]{} & GA' \end{array} .$$

These data are to satisfy the axioms

(2.2) $\quad \alpha_{1_A} = 1_{\alpha_A}, \qquad \alpha_{a'a} = \alpha_{a'}\alpha_a$ (pasting composite), and the axiom

(2.3) 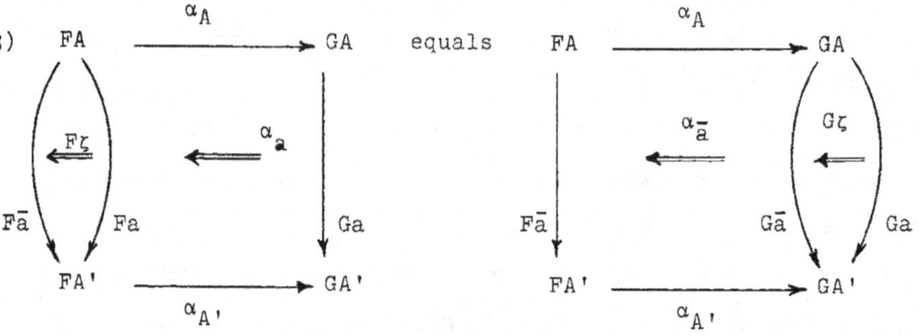

for each 2-cell $\zeta\colon a \Rightarrow \bar a$ in A. This last axiom (2.3) is vacuous when A, as it shall be in all our applications, is merely a category. The 2-natural transformations $\alpha\colon F \Rightarrow G$ (which are just the natural ones when A is merely a category) are those lax ones $\alpha\colon F \rightsquigarrow G$ for which each α_a is the identity 2-cell.

We compose lax n.t.'s by horizontal pasting of diagrams like (2.1), getting a category $[\![A,T]\!]_0$, of which $[A,T]_0$ is a subcategory. It becomes a 2-category $[\![A,T]\!]$ when we take as 2-cells the <u>modifications</u>

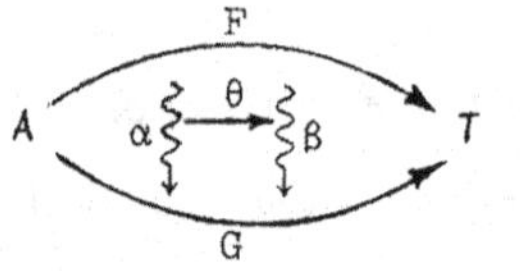 .

Such a modification $\theta\colon \alpha \to \beta$ assigns to each $A \in A$ a 2-cell $\theta_A\colon \alpha_A \Rightarrow \beta_A$ rendering equal the left two diagrams below, for whose common value we use the notation of the diagram on the right:

(2.4) 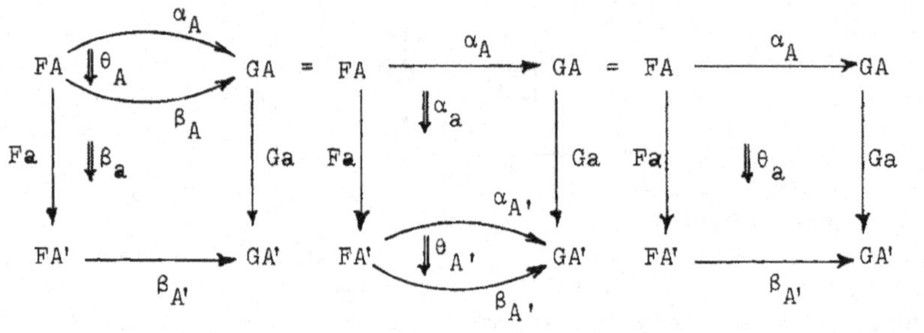 .

Horizontal and vertical composition of modifications is defined in the obvious way. When α and β are 2-natural, the above is the usual definition of modification as in [12], so that the inclusion $[A,T] \to [\![A,T]\!]$ is "full on 2-cells".

We write I for the unit category with one object $*$ and one morphism. There is an evident isomorphism

(2.5) $$\ulcorner\ \urcorner : T \overset{\cong}{\to} [I,T] = [\![I,T]\!] ,$$

where for example $\ulcorner B \urcorner$, called the <u>name</u> of B, for $B \in T$, is the 2-functor sending $*$ to B.

<u>2.3</u> Let A, B be merely categories and T a 2-category. From the situation

(2.6) 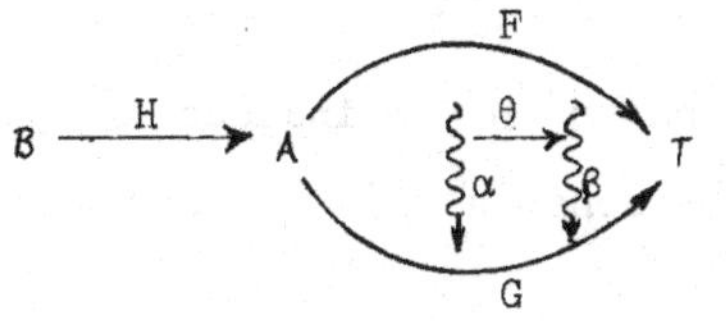 we get 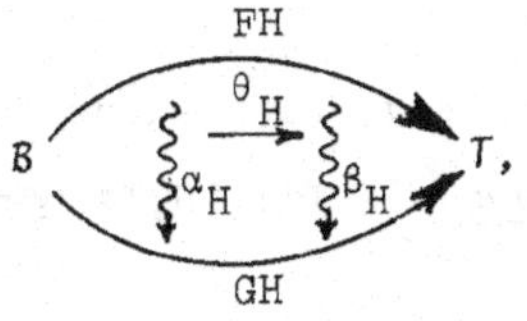

where

(2.7) $$(\alpha_H)_B = \alpha_{HB},\ (\alpha_H)_b = \alpha_{Hb},\ (\theta_H)_B = \theta_{HB}.$$

Thus the functor H induces a 2-functor $[\![H,T]\!]: [\![A,T]\!] \to [\![B,T]\!]$.

Given a second functor $K: B \to A$ and a natural transformation $\eta: H \Rightarrow K$, we get a modification

(2.8) $$\begin{array}{ccc} FH & \overset{\alpha_H}{\rightsquigarrow} & GH \\ {\scriptstyle F\eta}\Downarrow & \downarrow{\scriptstyle \alpha_\eta} & \Downarrow{\scriptstyle G\eta} \\ FK & \underset{\alpha_K}{\rightsquigarrow} & GK \end{array} ,$$

where

(2.9) $$(\alpha_\eta)_B = \alpha_{\eta_B} .$$

Moreover the two modifications on the left below coincide, and we denote their common value as in the diagram on the right:

(2.10) 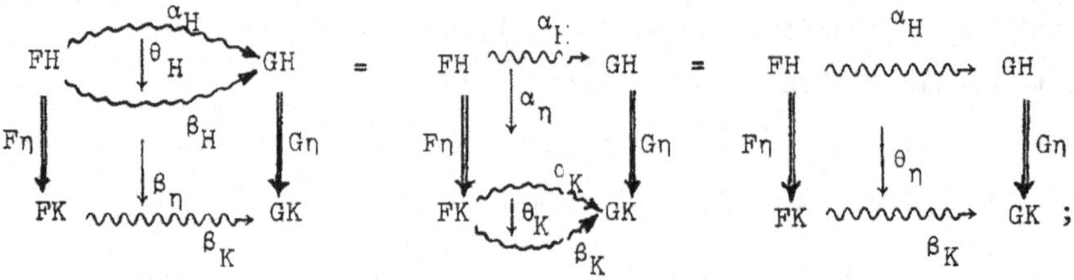

here we have, in the notation of (2.4),

(2.11) $$(\theta_\eta)_B = \theta_{\eta_B} .$$

We may say, with an obvious meaning, that η: H $\Rightarrow$ K induces an <u>op-lax-natural</u> $[\![\eta,T]\!]$: $[\![H,T]\!] \rightsquigarrow [\![K,T]\!]$ with components

(2.12) $$[\![\eta,T]\!]_F = F\eta, \quad [\![\eta,T]\!]_\alpha = \alpha_\eta .$$

3. THE 2-CATEGORIES CAT∫T AND CAT∫*T

3.1 Let T be a 2-category. An object of CAT∫T consists of a category A and a functor Γ: $A \to T$. A morphism from (A,Γ) to (B,Δ) consists of a functor T: $A \to B$ together with a lax natural transformation τ as in

(3.1) 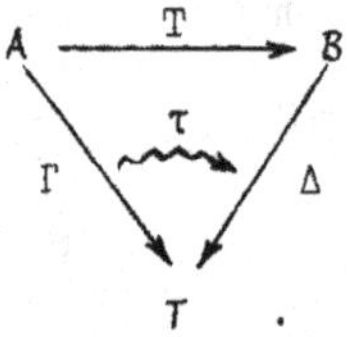

Composition of these is by the obvious pasting of diagrams:

(3.2) 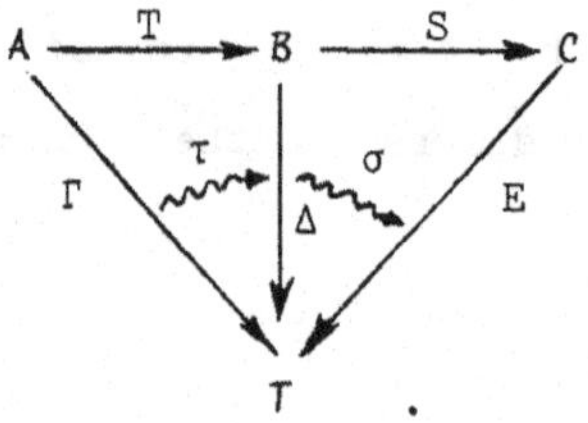

That is, the lax n.t. in the composite is the composite

(3.3)

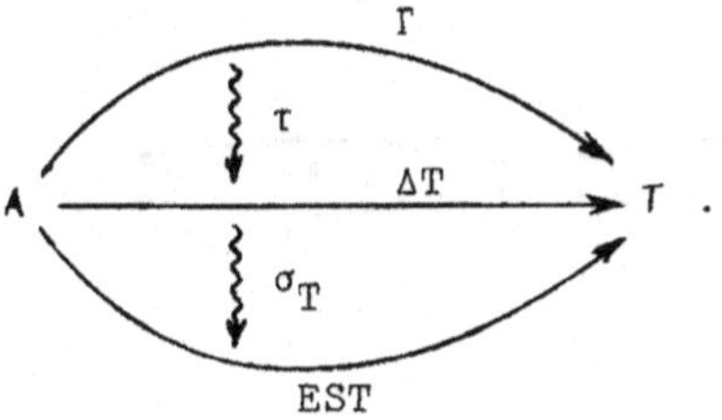

(The reader is asked for his cooperation in seeing "E" here as upper--case Greek epsilon.) A 2-cell $(T,\tau) \Rightarrow (T',\tau')$: $(A,\Gamma) \to (B,\Delta)$ consists of a natural transformation η: $T \Rightarrow T'$ together with a modification θ as in

(3.4)

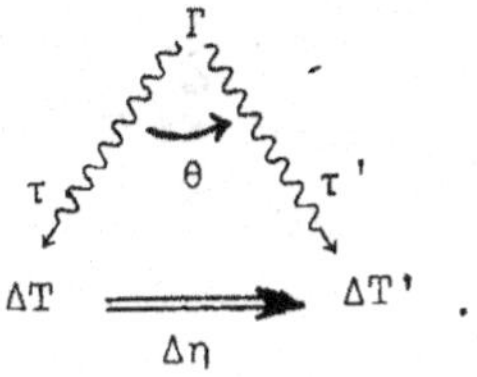

Vertical composition of these is given by

(3.5)

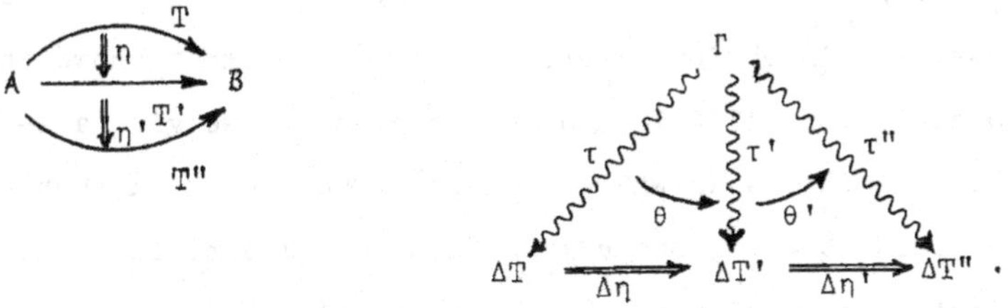

The more complicated horizontal composition of

(3.6)

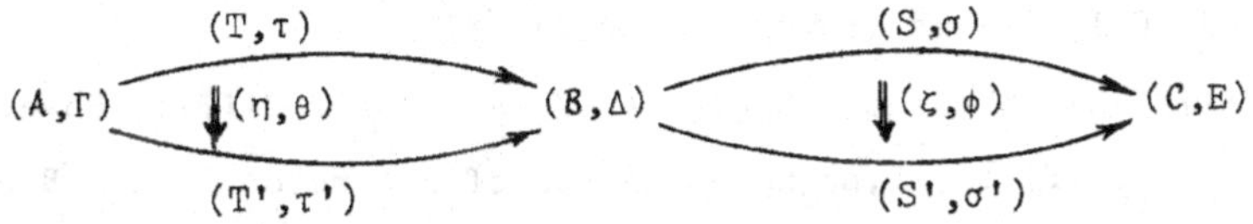

is given by

(3.7)

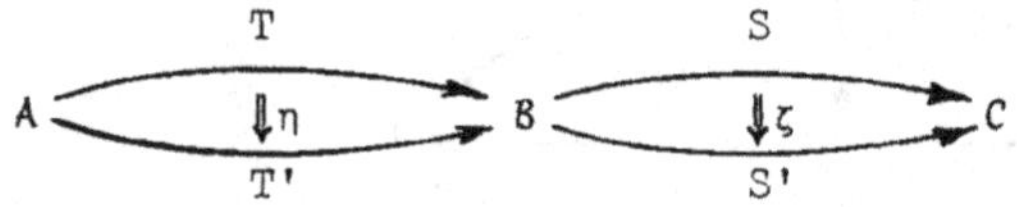

together with

(3.8)

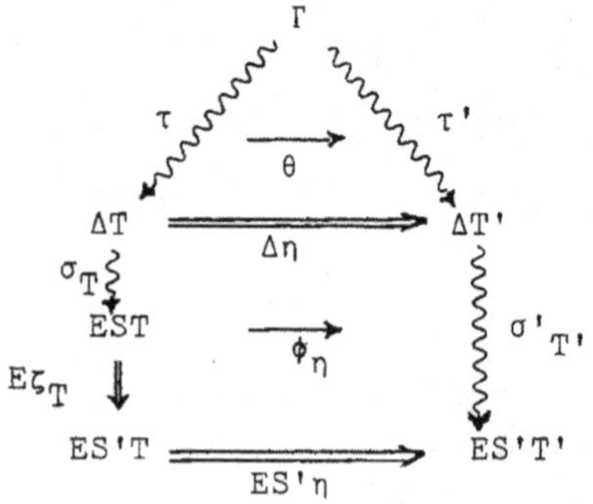

;

here ϕ_η is formed as in (2.10) from ϕ: $E\zeta.\sigma \to \sigma'$ and η: $T \Rightarrow T'$.

A direct verification that $CAT\int T$ is a 2-category is frightfully tedious, and to be manageable at all needs some lemmas about the behaviour under pasting of derived modifications like ϕ_η. Happily it is not necessary. An obvious representability argument shows that it suffices to do it for $CAT\int CAT$, and in §6.3 below we give a 2-full embedding of this into a known 2-category. Once we check that this embedding respects the various compositions and identities, the 2-category axioms are automatic. Alternatively, a different full embedding given in §5.2 below does the same for $CAT\int{*}CAT$ which we are about to introduce; this extends by representability arguments to $CAT\int{*}T$; and $(CAT\int T)^{co}$ is isomorphic to $CAT\int{*}T^{coop}$.

3.2 $CAT\int{*}T$ has the same objects $(A,\Gamma\colon A \to T)$ as $CAT\int T$. However a morphism from (A,Γ) to (B,Δ) now consists of a functor $T\colon A \to B$ together with a lax n.t. τ as in

(3.9)

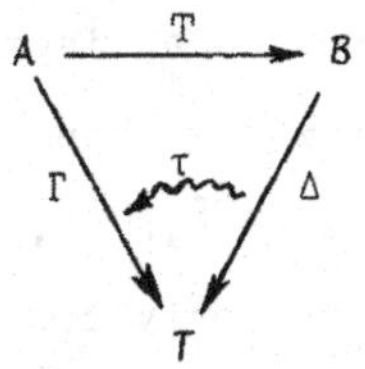

(so that T has the same sense as in (3.1), but τ the opposite sense). A 2-cell from (T,τ) to (T',τ') now consists of (η,θ) where $\eta: T \Rightarrow T'$ as before, but θ is a modification of the form

(3.10)

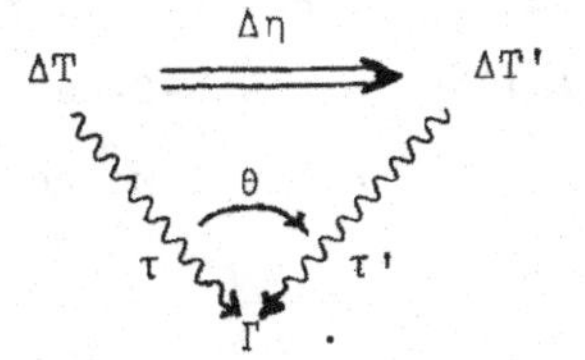

With τ and σ reversed in (3.2), the new analogue of (3.3) is

(3.11)

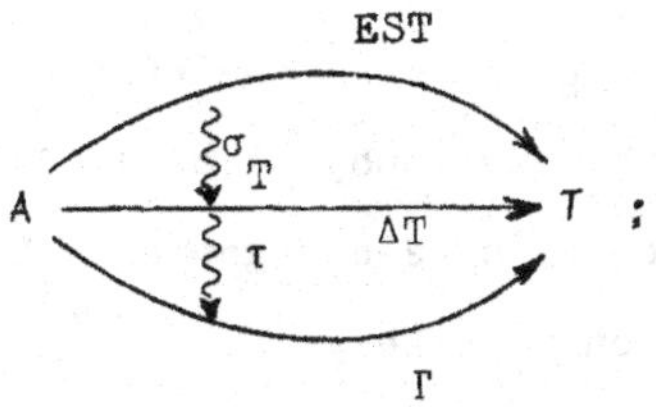

there is no change in (3.5) except the reversal of τ,τ',τ''; and (3.8) gets replaced by

(3.12)

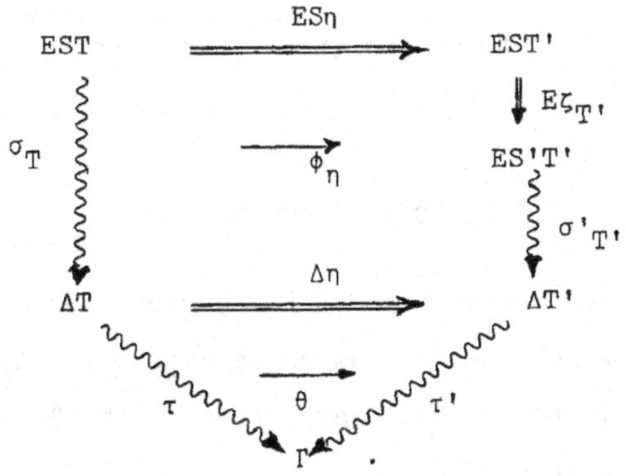

It is easy to see that there is an isomorphism $(CAT\int T)^{co} \to CAT\int{*T^{coop}}$ sending (in what I hope is, in the light of [12], an obvious notation) (A,Γ) to (A^{op},Γ^{op}), (T,τ) to (T^{op},τ^{coop}), and (η,θ) to (η^{op},θ^{co}). (In fact we apply coop to everything, but co is trivial on some elements and op on others.)

3.3 It is clear what is meant by $L\int T$, $L\int{*T}$ where L is a sub-2-category of CAT.

There are various sub-2-categories of $CAT\int T$ that stand out. By fixing A and requiring that $T = 1$ and $\eta = 1$, we get $[\![A,T]\!]$ as a sub-2-category; similarly $[\![A,T]\!]^{op}$ as a sub-2-category of $CAT\int{*T}$. In particular taking A as the unit category I we get the composite

(3.13) $\ulcorner\ \urcorner : T \cong [\![I,T]\!] \to CAT\int T$,

where the first isomorphism is what we called $\ulcorner\ \urcorner$ in (2.5).

If instead we allow all T and η, but require $\tau = 1$ and $\theta = 1$, we get CAT/T_0 (cf. [12] §1.3) as a sub-2-category both of $CAT\int T$ and of $CAT\int{*T}$. (We should note I suppose that the strictly accurate name for CAT/T_0 is $CAT/\ulcorner T_0\urcorner$: it is the comma-object of $1: CAT \to CAT$ and $\ulcorner T_0\urcorner : I \to CAT$ in the 2-category of 2-categories. However we may omit the corners when no confusion is likely.)

Finally we observe that we have evident forgetful 2-functors

(3.14) $CAT\int T \to CAT \leftarrow CAT\int{*T}$,

sending (A,Γ) to A, (T,τ) to T, and (η,θ) to η.

4. LAX COMMA CATEGORIES

4.1 Let T be a 2-category, A and B categories, and $\Gamma: A \to T$, $\Delta: B \to T$ functors. We define the lax comma category $\Gamma/\!\!/\Delta$, or more precisely $(A,\Gamma)/\!\!/(B,\Delta)$, to be what Gray [3] calls the "2-comma-category". An object of $\Gamma/\!\!/\Delta$ is a triple (A,k,B),

(4.1) $$A, \qquad \Gamma A \xrightarrow{k} \Delta B, \qquad B,$$

where $A \in \mathcal{A}$, $B \in \mathcal{B}$, and k is a morphism in $\mathcal{T}$. A morphism from (A,k,B) to (A',k',B') is a triple (a,κ,b),

(4.2) $$\begin{array}{ccccccc} A & \quad & \Gamma A & \xrightarrow{k} & \Delta B & \quad & B \\ {\scriptstyle a}\downarrow & & {\scriptstyle \Gamma a}\downarrow & \Downarrow{\scriptstyle \kappa} & \downarrow{\scriptstyle \Delta b} & & \downarrow{\scriptstyle b} \\ A' & & \Gamma A' & \xrightarrow[k']{} & \Delta B' & & B' \end{array},$$

where $a \in \mathcal{A}(A,A')$, $b \in \mathcal{B}(B,B')$, and κ is a 2-cell in $\mathcal{T}$. Composition is the evident pasting operation.

There are the obvious projection functors $d_0: \Gamma /\!/ \Delta \to \mathcal{A}$, $d_1: \Gamma /\!/ \Delta \to \mathcal{B}$, sending (a,κ,b) to a and to b respectively. There is also a lax natural transformation $\lambda: \Gamma d_0 \rightsquigarrow \Delta d_1$ with components $\lambda_{(A,k,B)} = k$, $\lambda_{(a,\kappa,b)} = \kappa$. The easy proofs of the following two propositions are left to the reader.

Proposition 4.1 $\Gamma /\!/ \Delta$ *is determined by the following universal property. For a category* $\mathcal{C}$, *the equations*

(4.3) $$P = d_0 G, \quad R = d_1 G, \quad \alpha = \lambda_G,$$

or in brief

(4.4) 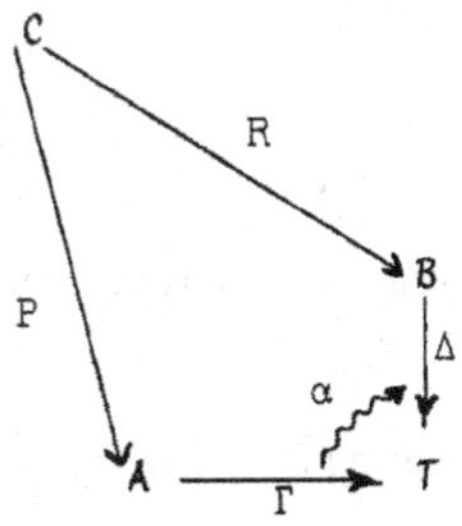

equals

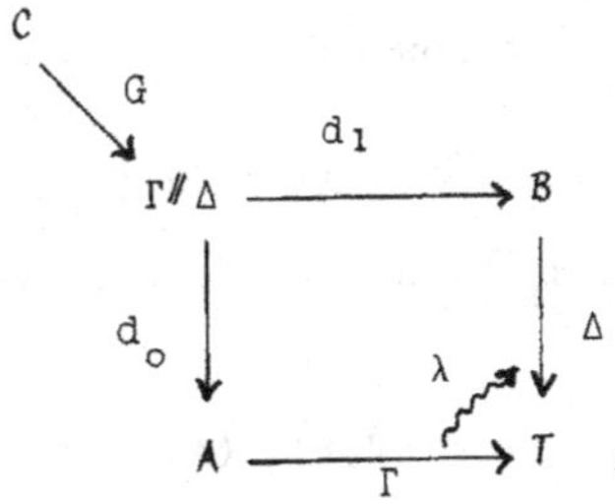

set up a bijection between functors $G: \mathcal{C} \to \Gamma /\!/ \Delta$ *and triples* $(P: \mathcal{C} \to \mathcal{A},\ R: \mathcal{C} \to \mathcal{B},\ \alpha: \Gamma P \rightsquigarrow \Delta R)$. *To wit*, G *must carry*

(4.5) $$\begin{array}{ccc} C \\ {\scriptstyle c}\downarrow \\ C' \end{array} \quad \underline{\text{to}} \quad \begin{array}{ccc} PC \\ {\scriptstyle Pc}\downarrow \\ PC' \end{array} \qquad \begin{array}{ccc} \Gamma PC & \xrightarrow{\alpha_C} & \Delta RC \\ {\scriptstyle \Gamma Pc}\downarrow & \Downarrow{\scriptstyle \alpha_c} & \downarrow{\scriptstyle \Delta Rc} \\ \Gamma PC' & \xrightarrow[\alpha_{C'}]{} & \Delta RC' \end{array} \qquad \begin{array}{c} RC \\ \downarrow{\scriptstyle Rc} \\ RC'. \end{array} \quad \square$$

Proposition 4.2 _If in the above situation we also have_ G' _corresponding to_ (P',R',α'), _there is a bijection between natural transformations_ $\gamma: G \Rightarrow G'$ _and triples_ (π,ρ,μ) _where_ $\pi: P \Rightarrow P'$, $\rho: R \Rightarrow R'$, _and_ μ _is a modification_

(4.6) $$\begin{array}{ccc} \Gamma P & \overset{\alpha}{\rightsquigarrow} & \Delta R \\ {\scriptstyle \Gamma\pi}\Downarrow & \downarrow{\scriptstyle \mu} & \Downarrow{\scriptstyle \Delta\rho} \\ \Gamma P' & \underset{\alpha'}{\rightsquigarrow} & \Delta R' \end{array} .$$

In terms of γ , _we have in the notation of_ (2.8)

(4.7) $$\pi = d_0\gamma, \qquad \rho = d_1\gamma, \qquad \mu = \lambda_\gamma;$$

while the C-_component of_ γ _is_

(4.8) $$\begin{array}{c} PC \\ {\scriptstyle \pi_C}\downarrow \\ P'C \end{array} \qquad \begin{array}{ccc} \Gamma PC & \xrightarrow{\alpha_C} & \Delta RC \\ {\scriptstyle \Gamma\pi_C}\downarrow & \Downarrow{\scriptstyle \mu_C} & \downarrow{\scriptstyle \Delta\rho_C} \\ \Gamma P'C & \xrightarrow[\alpha'_C]{} & \Delta R'C \end{array} \qquad \begin{array}{c} RC \\ \downarrow{\scriptstyle \rho_C} \\ R'C. \end{array} \quad \square$$

4.2 We now exhibit $/\!/$ as a 2-functor

(4.9) $$/\!/ : \mathcal{CAT}\!\int^{*}\!\mathcal{T} \times \mathcal{CAT}\!\int\!\mathcal{T} \to \mathcal{CAT};$$

in fact the definitions of $\mathcal{CAT}\!\int\!\mathcal{T}$ and $\mathcal{CAT}\!\int^{*}\!\mathcal{T}$ were chosen to make this so.

From $(T,\tau): (\mathcal{A},\Gamma) \to (\mathcal{A}',\Gamma')$ in $\mathcal{CAT}\!\int^{*}\!\mathcal{T}$ and $(S,\sigma): (\mathcal{B},\Delta) \to (\mathcal{B}',\Delta')$ in $\mathcal{CAT}\!\int\!\mathcal{T}$ we get

(4.10)

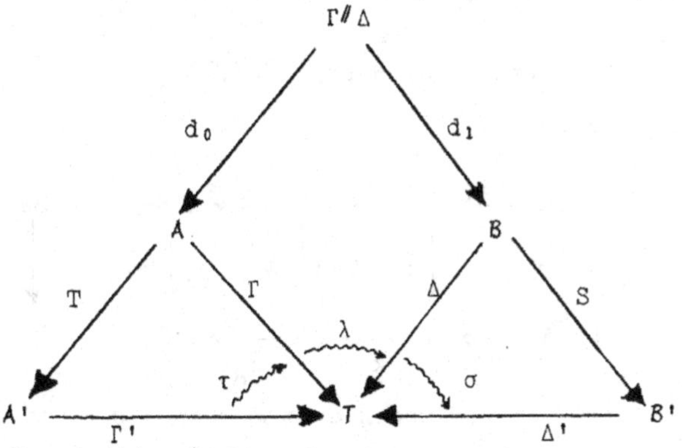

and hence, by Proposition 4.1, a functor

(4.11) $(T,\tau)/\!\!/(S,\sigma)\colon (A,\Gamma)/\!\!/(B,\Delta) \to (A',\Gamma')/\!\!/(B',\Delta')$

which sends the typical morphism (4.2) to

(4.12)
$$\begin{array}{ccccccccc}
TA & & \Gamma'TA & \xrightarrow{\tau_A} & \Gamma A & \xrightarrow{k} & \Delta B & \xrightarrow{\sigma_B} & \Delta'SB & & SB \\
{\scriptstyle Ta}\downarrow & & {\scriptstyle \Gamma'Ta}\downarrow & \Downarrow\tau_a & \downarrow{\scriptstyle \Gamma a} & \Downarrow\kappa & \downarrow{\scriptstyle \Delta b} & \Downarrow\sigma_b & \downarrow{\scriptstyle \Delta'Sb} & & \downarrow{\scriptstyle Sb} \\
TA' & & \Gamma'TA' & \xrightarrow[\tau_{A'}]{} & \Gamma A' & \xrightarrow[k']{} & \Delta B' & \xrightarrow[\sigma_{B'}]{} & \Delta'SB' & & SB'.
\end{array}$$

That the functor (4.11) depends functorially on (T,τ) and (S,σ) is perhaps best seen by pasting onto (4.10) a $(T',\tau')\colon (A',\Gamma') \to (A'',\Gamma'')$ and an $(S',\sigma')\colon(B',\Delta') \to (B'',\Delta'')$; alternatively by direct use of (4.12) in the light of (3.3) and (3.11).

Next consider $(\eta,\theta)\colon (T,\tau) \Rightarrow (\bar{T},\bar{\tau})$ in $CAT\int^{*}T$ and $(\zeta,\phi)\colon (S,\sigma) \Rightarrow (\bar{S},\bar{\sigma})$ in $CAT\int T$. From the natural transformations η_{d_0} and ζ_{d_1} together with the modification

(4.13)
$$\begin{array}{ccccccc}
\Gamma'Td_0 & \overset{\tau_{d_0}}{\rightsquigarrow} & \Gamma d_0 & \overset{\lambda}{\rightsquigarrow} & \Delta d_1 & \overset{\sigma_{d_1}}{\rightsquigarrow} & \Delta'Sd_1 \\
{\scriptstyle \Gamma'\eta_{d_0}}\Downarrow & \downarrow{\scriptstyle \theta_{d_0}} & \Downarrow{\scriptstyle 1} & & {\scriptstyle 1}\Downarrow & \downarrow{\scriptstyle \phi_{d_1}} & \Downarrow{\scriptstyle \Delta'\zeta_{d_1}} \\
\Gamma'\bar{T}d_0 & \underset{\bar{\tau}_{d_0}}{\rightsquigarrow} & \Gamma d_0 & \underset{\lambda}{\rightsquigarrow} & \Delta d_1 & \underset{\bar{\sigma}_{d_1}}{\rightsquigarrow} & \Delta'\bar{S}d_1
\end{array}$$

we get by Proposition 4.2 a natural transformation

$$(4.14) \qquad (\eta,\theta)/\!\!/(\zeta,\phi): (T,\tau)/\!\!/(S,\sigma) \Rightarrow (\bar{T},\bar{\tau})/\!\!/(\bar{S},\bar{\sigma})$$

whose (A,k,B)-component is

$$(4.15) \qquad \begin{array}{ccccccccccc} TA & & \Gamma'TA & \xrightarrow{\tau_A} & \Gamma A & \xrightarrow{k} & \Delta B & \xrightarrow{\sigma_B} & \Delta'SB & & SB \\ {\scriptstyle \eta_A}\downarrow & & {\scriptstyle \Gamma'\eta_A}\downarrow & \Downarrow\theta_A & \downarrow{\scriptstyle 1} & & {\scriptstyle 1}\downarrow & \Downarrow\phi_B & \downarrow{\scriptstyle \Delta'\zeta_B} & & \downarrow{\scriptstyle \zeta_B} \\ \bar{T}A & & \Gamma'\bar{T}A & \xrightarrow[\bar{\tau}_A]{} & \Gamma A & \xrightarrow[k]{} & \Delta B & \xrightarrow[\bar{\sigma}_B]{} & \Delta'\bar{S}B & & \bar{S}B \end{array} .$$

It is very easy from (4.15) to check that $/\!\!/$ respects vertical and horizontal composition of 2-cells in $CAT\int^*T$ and $CAT\int T$; thus $/\!\!/$ is indeed a 2-functor of the form (4.9).

<u>4.3</u> The projections of $CAT\int^*T \times CAT\int T$ onto its factors, composed with the forgetful 2-functors (3.14), give new forgetful 2-functors

$$(4.6) \qquad forg_0,\ forg_1: CAT\int^*T \times CAT\int T \to CAT ,$$

sending $((A,\Gamma), (B,\Delta))$ to A and B respectively. The following is clear from (4.12) and (4.15):

<u>Proposition 4.3</u> <u>The functors</u> $d_0: (A,\Gamma)/\!\!/(B,\Delta) \to A$ <u>and</u> $d_1: (A,\Gamma)/\!\!/(B,\Delta) \to B$ <u>are the components of</u> 2-<u>natural transformations</u> $d_0: /\!\!/ \Rightarrow forg_0$ <u>and</u> $d_1: /\!\!/ \Rightarrow forg_1$. □

5. THE 2-FUNCTOR ∘ AND THE EMBEDDING $CAT\int^*CAT \to [CAT,CAT]$

<u>5.1</u> Form the composite 2-functor

$$CAT\int^*T \times T \xrightarrow[1\times\ulcorner\ \urcorner]{} CAT\int^*T \times CAT\int T \xrightarrow[/\!\!/]{} CAT,$$

where $\ulcorner\ \urcorner: T \to CAT\int T$ is the 2-functor (3.13) sending $B \in T$ to $(I,\ulcorner B\urcorner)$, etc. Our interest is in the special case $T = CAT$; we then write the above 2-functor as

$$(5.1) \qquad \circ: CAT\int^*CAT \times CAT \to CAT.$$

We usually abbreviate its value $(A,\Gamma)\circ B$ on objects to $A\circ B$; for its value on morphisms or 2-cells we write $(T,\tau)\circ S$ and $(\eta,\theta)\circ\zeta$ in full; except that we are sometimes interested in the restriction of $\circ$ to $CAT/CAT_0 \times CAT$, where τ and θ are identities, and then we write simply $T\circ S$ and $\eta\circ\zeta$.

An object of $A\circ B$ is a pair (A,X) where $A \in A$ and X is a functor $\Gamma A \to B$; we write this object as $A[X]$. A morphism $A[X] \to A'[X']$ is a pair (a,x) of the form

(5.2)

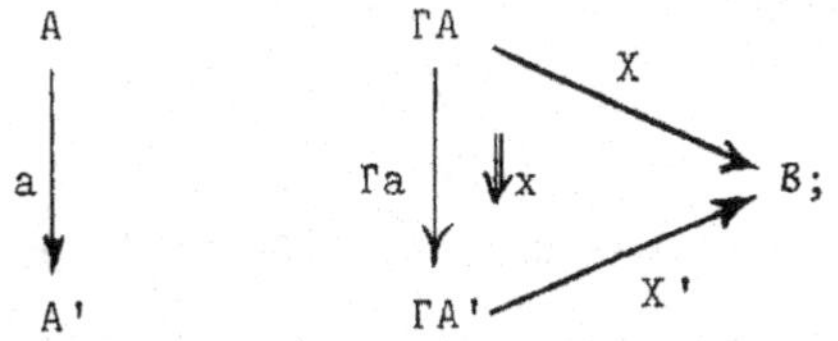

here a is a morphism in A and x is a natural transformation. We write the morphism (5.2) as

(5.3)

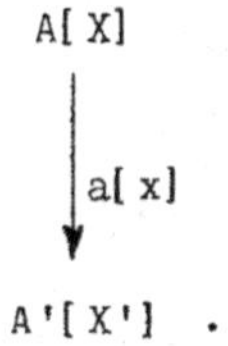

If $(T,\tau)\colon (A,\Gamma) \to (A',\Gamma')$ and $S\colon B \to B'$, the functor $(T,\tau)\circ S\colon A\circ B \to A'\circ B'$ sends, as a special case of (4.12), the morphism (5.2) to

(5.4)

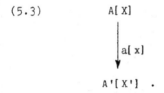

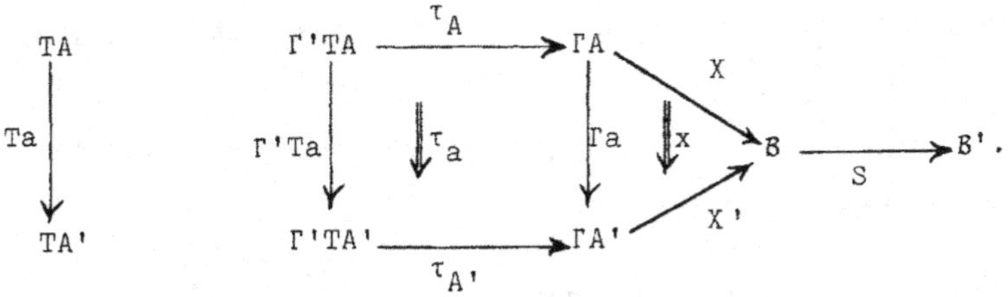

If $(\eta,\theta)\colon (T,\tau) \Rightarrow (\bar{T},\bar{\tau})$ and $\zeta\colon S \Rightarrow \bar{S}$, the natural transformation $(\eta,\theta)\circ\zeta\colon (T,\tau)\circ S \Rightarrow (\bar{T},\bar{\tau})\circ\bar{S}$ has, as a special case of (4.15), the $A[X]$-
-component

(5.5)

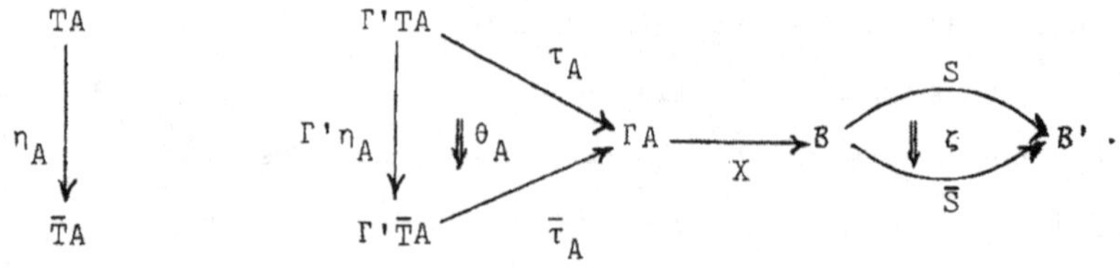

We have as in §4.1 the projection functor

(5.6) $$d_0 \colon A \circ B \to A$$

sending (5.2) to $a \colon A \to A'$. As in Proposition 4.3 this is the component of a 2-natural transformation

(5.7) $$d_0 \colon \circ \Rightarrow \mathrm{forg}_0,$$

where $\mathrm{forg}_0 \colon CAT\int{}^{*}CAT \times CAT \to CAT$ sends $((A,\Gamma),B)$ to A, etc.

5.2 The 3-category of 2-categories is of course cartesian closed; in particular there is a bijection between 2-functors $K \times L \to M$ and 2-functors $K \to [L,M]$. Under this bijection the 2-functor $\circ$ of (5.2) gives a 2-functor

(5.8) $$\Phi \colon CAT\int{}^{*}CAT \to [CAT,CAT],$$

to wit

(5.9) $$\Phi(A,\Gamma) = A\circ -, \quad \Phi(T,\tau) = (T,\tau)\circ -, \quad \Phi(\eta,\theta) = (\eta,\theta)\circ -.$$

Our first main result is:

Theorem 5.1 *The 2-functor Φ is 2-fully-faithful.*

Proof. Take a 2-natural transformation $G \colon (A,\Gamma)\circ - \Rightarrow (A',\Gamma')\circ -$, with components $G_B \colon A\circ B \to A'\circ B$. We are to show that $G = (T,\tau)\circ -$ for a unique (T,τ).

The uniqueness of (T,τ) is immediate; for if G is $(T,\tau)\circ -$ it follows from (5.4) that

(5.10) $$G_{\Gamma A}(A[1_{\Gamma A}]) = TA[\tau_A],$$

and that $G_{\Gamma A'}$ sends the morphism

(5.11)

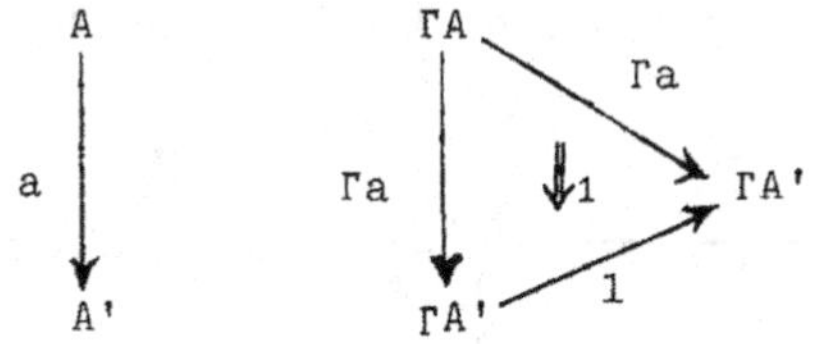

of $A \circ \Gamma A'$ to the morphism

(5.12)

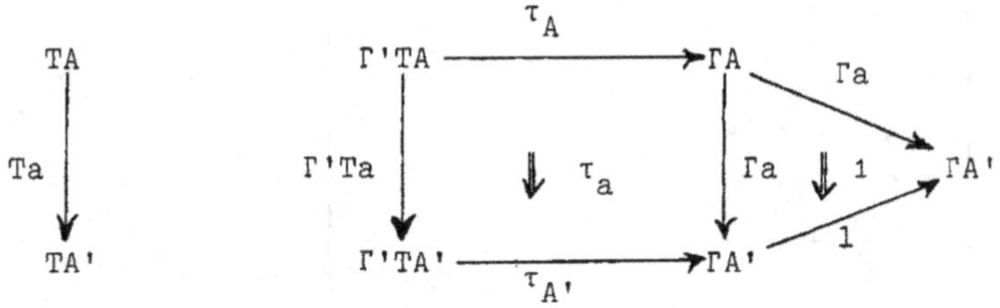

of $A' \circ \Gamma A'$; that is to say,

(5.13) $\qquad G_{\Gamma A'}(a[1_{\Gamma a}]) = Ta[\tau_a]$.

So by (5.10) and (5.13) G uniquely determines T and τ.

Now let any 2-natural G be given, let $A[X] \in A \circ \mathcal{B}$, and observe that by the naturality of G we have commutativity in

(5.14)

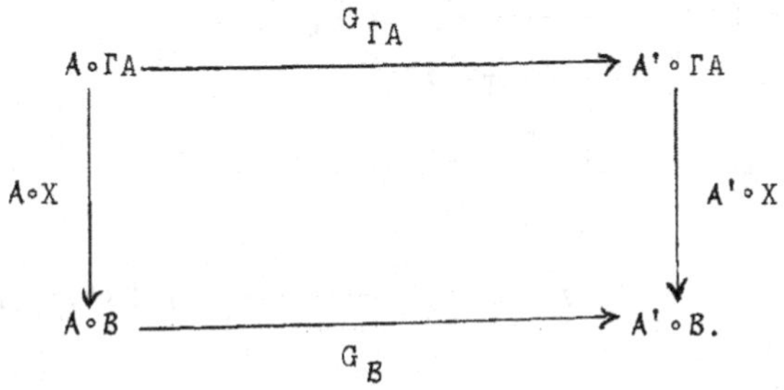

If we <u>define</u> TA, τ_A by (5.10) and evaluate both legs of (5.14) at the object $A[1_{\Gamma A}]$ we get

(5.15) $\qquad G_{\mathcal{B}}(A[X]) = TA[X.\tau_A]$.

In particular, $G_{\Gamma A'}(A[\Gamma a]) = TA[\Gamma a.\tau_A]$, the domain of (5.12). With this said, it now <u>makes sense</u> to <u>define</u> Ta, τ_a by (5.13).

Next, replace A,X by A',X' in (5.14) and evaluate both legs at the morphism $a[1_{\Gamma a}]$ of (5.11). By (5.13) this gives

(5.16) $\qquad G_{\mathcal{B}}(a[1_{X'.\Gamma a}]) = Ta[X'.\tau_a]$.

Now let a[x] be the morphism (5.2) of A∘B. By the 2-naturality of G we have commutativity in

(5.17)

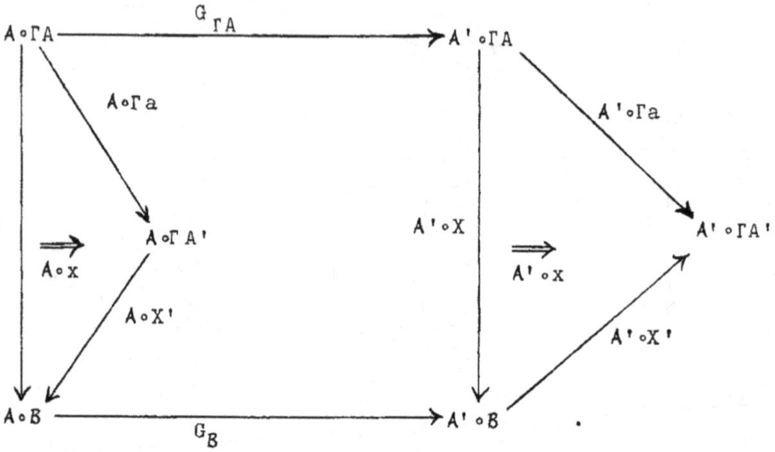

Taking the $A[1_{\Gamma A}]$-component of each leg we get, using (5.10) and (5.5),

(5.18) $$G_B(1_A[x]) = 1_{TA}[x.\tau_A].$$

But the a[x] of (5.2) is the composite

$$A[X] \xrightarrow[1_A[x]]{} A[X'.\Gamma a] \xrightarrow[a[1_{X'.\Gamma a}]]{} A'[X'],$$

and G_B is a functor; so $G_B(a[x])$ is, by (5.16) and (5.18), the composite

$$TA[X.\tau_A] \xrightarrow[1_{TA}[x.\tau_A]]{} TA[X'.\Gamma a.\tau_A] \xrightarrow[Ta[X'.\tau_a]]{} TA'[X'.\tau_{A'}] ;$$

that is,

(5.19) $G_B(a[x])$ is

$$\begin{array}{ccc}
TA & \Gamma'TA \xrightarrow{\tau_A} \Gamma A & \\
\downarrow Ta & \Big\downarrow \Gamma'Ta \quad \Downarrow\tau_a \quad \Big\downarrow \Gamma a \quad \Downarrow x & X \searrow B. \\
TA' & \Gamma'TA' \xrightarrow{\tau_{A'}} \Gamma A' & X' \nearrow
\end{array}$$

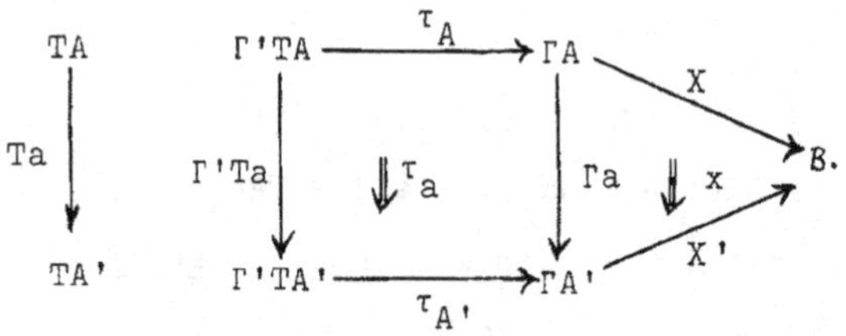

Using the fact that G_B is a functor, we now easily get that T is a functor and τ a lax n.t. by applying $G_{\Gamma A}$ to $1_A[1_{1_{\Gamma A}}]$ and by applying

$G_{\Gamma A''}$ to the composite

$$A[\Gamma a'.\Gamma a] \xrightarrow[a[1]]{} A'[\Gamma a'] \xrightarrow[a'[1]]{} A''[1_{\Gamma A''}] .$$

Then by comparing (5.19) with (5.4) we see that G_B is indeed $(T,\tau)\circ B$.

It remains to prove that Φ is fully faithful on 2-cells. So let (T,τ), $(\bar{T},\bar{\tau})$: $(A,\Gamma) \to (A',\Gamma')$ and let γ: $(T,\tau)\circ - \to (\bar{T},\bar{\tau})\circ -$ be a modification with components γ_B: $(T,\tau)\circ B \Rightarrow (\bar{T},\bar{\tau})\circ B$. We are to show that $\gamma = (\eta,\theta)\circ -$ for a unique (η,θ).

If γ is indeed of this form, it follows from (5.5) that the $A[1_{\Gamma A}]$ component of $\gamma_{\Gamma A}$ is

$$(5.20) \qquad (\gamma_{\Gamma A})_{A[1_{\Gamma A}]} = \eta_A[\theta_A],$$

which proves the uniqueness of η and θ. If now for any γ we <u>define</u> η_A and θ_A by (5.20), we have because γ is a modification commutativity in

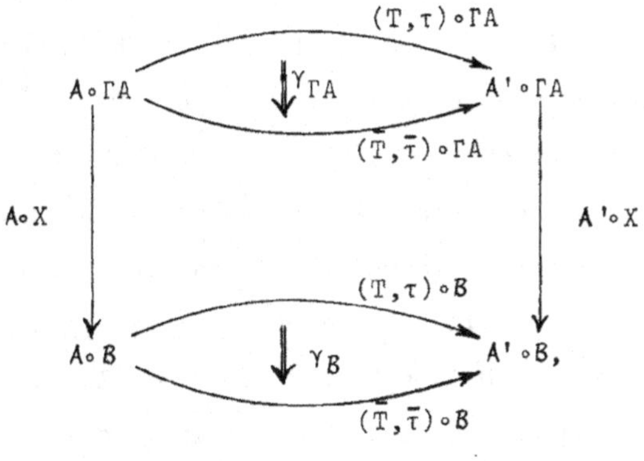

and calculating the $A[1_{\Gamma A}]$-component of both legs gives

$$(5.21) \qquad (\gamma_B)_{A[X]} = \eta_A[X\theta_A],$$

in agreement with (5.5). That η is indeed natural, and that θ is indeed a modification, follow at once when we express the <u>naturality</u> of $\gamma_{\Gamma A'}$ for the particular morphism (5.11) of $A\cdot\Gamma A'$. □

<u>Corollary 5.2</u> <u>CAT∫T and CAT∫*T are indeed 2-categories.</u>

<u>Proof</u> By the remarks in §3.1 and §4.2, with the above theorem. □

5.3 We devote this section to some comments, largely informal, on this full embedding Φ: $CAT\int{}^*CAT \to [CAT,CAT]$ of 2-categories.

First, its image is not the whole of $[CAT\ CAT]$. To see this, note that we have clearly

(5.22) $(A,\Gamma)\circ I \cong A,$

where I is the unit category. So for a 2-functor D: $CAT \to CAT$ we have:

(5.23) If $D \cong (A,\Gamma)\circ -$ then $A \cong DI$.

It follows that if $DI \cong I$ and if $D \cong (A,\Gamma)\circ-$ then $A \cong I$, so that $\Gamma = \ulcorner n \urcorner$ for some category n. But it is also clear that

(5.24) $(I,\ulcorner n \urcorner)\circ B \cong [n,B]$;

so any D in the image of Φ with $DI \cong I$ must be $[n,-]$ for some n. Now the 2-functor D: $CAT \to CAT$ sending B to $[[B,2],B]$, where 2 is the discrete category with 2 objects, is certainly not of the form $[n,-]$. (For $[n,B] = [m,B]$ when B is discrete, where m is the number of path--components of n; and no m is consistent with both $D2 = 2^4$ and $D3 = 3^8$.)

Even the 2-functor part of a doctrine D need not be in the image of Φ. Consider the doctrine, which can be given explicitly, for a category-with-coequalizers. There it is clear that DI is at any rate equivalent to I; this is enough to make DB equivalent to $[n,B]$ if D were in the image of Φ; but D is not of this form. I think however that this D is a quotient of something in the image of Φ, to wit of $[n,-]$ where n is $\rightrightarrows$.

I have no real idea how big the image of Φ in fact is; chiefly because, while I can pick out the putative A by (5.23), I don't see how to fix Γ in terms of D, and hence have no test for D to be in the image of Φ.

We are going to show in §7 that the image of Φ is closed under composition of 2-functors. What else is it closed under? Certainly products and coproducts.

It suffices to illustrate by binary ones. The product in $CAT\int^{*}CAT$ of (A,Γ) and (A',Γ') is $(A \times A',\Delta)$ where $\Delta(A,A') = \Gamma A + \Gamma'A'$; and $(A \times A',\Delta)\circ B \cong (A,\Gamma)\circ B \times (A',\Gamma')\circ B$. The coproduct in $CAT\int^{*}CAT$ of (A,Γ) and (A',Γ') is $(A + A',\Delta)$ where $\Delta|A = \Gamma$ and $\Delta|A' = \Gamma'$; and $(A + A',\Delta)\circ B \cong (A,\Gamma)\circ B + (A',\Gamma')\circ B$.

I haven't checked whether $CAT\int^{*}CAT$ has equalizers: chiefly because it would be a long business not relevant to our main purpose. But if it does have equalizers, my rough calculations indicate that Φ would not preserve them. (If Φ preserved them it would mean that equalizers of lax n.t.'s were constructed pointwise, which seems false.) If it has coequalizers, on the other hand, they are probably preserved.

For then Φ would seem to have a left adjoint, modulo size considerations. Consider a 2-natural $G\colon D \Rightarrow \Phi(A,\Gamma) = (A,\Gamma)\circ -$, with components $G_B\colon DB \to A\circ B$. By Proposition 4.1 and the definition of $A\circ B$ as $(A,\Gamma)/\!\!/(I, \ulcorner B \urcorner)$, such a G_B would correspond to $P\colon DB \to A$ together with $\alpha\colon \Gamma P \rightsquigarrow B!$, where $B!$ is the composite

$$DB \to I \underset{\ulcorner B \urcorner}{\to} CAT,$$

that is, the constant functor at B. In other words the G_B are in bijection with morphisms $(DB,B!) \to (A,\Gamma)$ in $CAT\int^{*}CAT$. The G's themselves would then be in bijection with the morphisms $\int^{B}(DB,B!) \to (A,\Gamma)$ in $CAT\int^{*}CAT$ if the indicated coend exists.

On the other hand it seems that Φ is as far as it can be from having a right adjoint; in the sense that any $D \in [CAT,CAT]$ with a 2-reflexion in $CAT\int^{*}CAT$ already lies in the latter. For if

$$(5.25)\quad [CAT,CAT]((A,\Gamma)\circ -,D) \cong CAT\int^{*}CAT((A,\Gamma),(B,\Delta))$$

2-naturally in (A,Γ), take $(A,\Gamma) = (I, \ulcorner n \urcorner)$. Then the left side of (5.25) is, by (5.24) and Yoneda, Dn; while the right side is easily seen to be $(B,\Delta)\circ n$. So $D \cong (B,\Delta)\circ -$.

5.4 We shall always henceforth use T to denote a sub-2-category of CAT, not necessarily full on 1-cells, but always taken to be full on 2-cells.

We have an obvious inclusion

(5.26) $\mathcal{CAT}\int^{*}\mathcal{T} \to \mathcal{CAT}\int^{*}\mathcal{CAT}$,

which composed with Φ gives

(5.27) $\Phi_{\mathcal{T}}: \mathcal{CAT}\int^{*}\mathcal{T} \to [\mathcal{CAT},\mathcal{CAT}]$.

It is clear that (5.26) is full on 2-cells, and is full on 1-cells precisely when $\mathcal{T} \to \mathcal{CAT}$ is; so that the same is true of (5.27).

6. THE GROTHENDIECK CONSTRUCTION

6.1 Our next main goal is the proof in §7 below that $\mathcal{CAT}\int^{*}\mathcal{CAT}$ is closed in $[\mathcal{CAT},\mathcal{CAT}]$ under composition of objects (that is, of 2-functors $\mathcal{CAT} \to \mathcal{CAT}$). We shall do this by showing that $\mathcal{A}\circ(\mathcal{B}\circ -) \cong (\mathcal{A}\circ\mathcal{B})\circ -$; here $\mathcal{A}\circ\mathcal{B}$ is the category so denoted in §5, but now provided with an augmentation $\mathcal{A}\circ\mathcal{B} \to \mathcal{CAT}$ derived from the augmentations of $\mathcal{A}$ and $\mathcal{B}$.

The considerations of §1.2 above suggest that the augmentation $n = \Gamma A$ of $A \in \mathcal{A}$ is a kind of "arity" of A. An object A[X] of $\mathcal{A}\circ\mathcal{B}$ consists of A and X: $\Gamma A = n \to \mathcal{B}$, so that X is a kind of n-ad of objects of $\mathcal{B}$. If these objects of $\mathcal{B}$ also have arities, provided by an augmentation $\Delta: \mathcal{B} \to \mathcal{CAT}$, then A[X] should have an arity n(m), where m is the composite $\Delta X: n \to \mathcal{B} \to \mathcal{CAT}$, and where n(m) generalizes the $n(m_1,\dots, m_n) = m_1 +\dots+ m_n$ of §1.2. In fact the appropriate generalization is precisely the <u>Grothendieck construction</u> assigning to a category n and a functor m: $n \to \mathcal{CAT}$ the corresponding fibred category over n, which we may call n(m).

6.2 We therefore define the <u>Grothendieck</u> 2-functor

(6.1) $\Theta: \mathcal{CAT}\int\mathcal{CAT} \to \mathcal{CAT}$

by setting

(6.2) $\Theta = {}^{\ulcorner}I^{\urcorner}/\!\!/-$,

where of course $/\!\!/$ is the 2-functor of (4.9) with T put equal to CAT. We also use, where confusion is unlikely, the "parenthesis" notation suggested above: namely

(6.3) $\Theta(A,\Gamma) = A(\Gamma)$, $\Theta(T,\tau) = T(\tau)$, $\Theta(\eta,\theta) = \eta(\theta)$.

An object of $A(\Gamma)$ is a pair A,Y where $A \in A$ and $Y\colon I \to \Gamma A$ (that is, $Y \in \Gamma A$; but it is useful to write Y as a functor with domain I). We write this object as $A\langle Y\rangle$. A morphism $A\langle Y\rangle \to A'\langle Y'\rangle$ is a pair a,y of the form

(6.4)

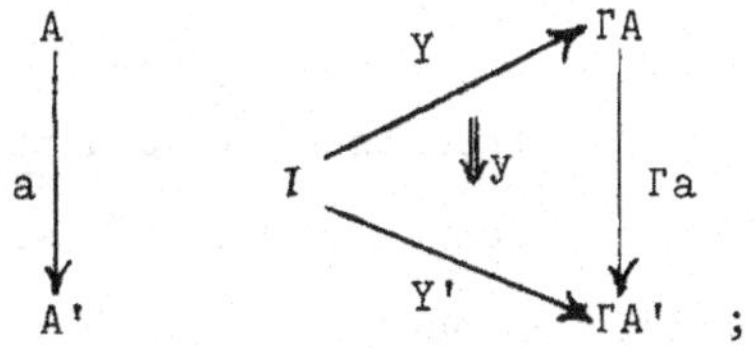

;

which is a fancy way of saying that y is a morphism $\Gamma a.Y \to Y'$ in $\Gamma A'$. We write this morphism as $a\langle y\rangle$.

If $(T,\tau)\colon (A,\Gamma) \to (A',\Gamma')$ in $CAT\!\int\! CAT$, then by (4.12) the functor $T(\tau)$ sends (6.4) to

(6.5)

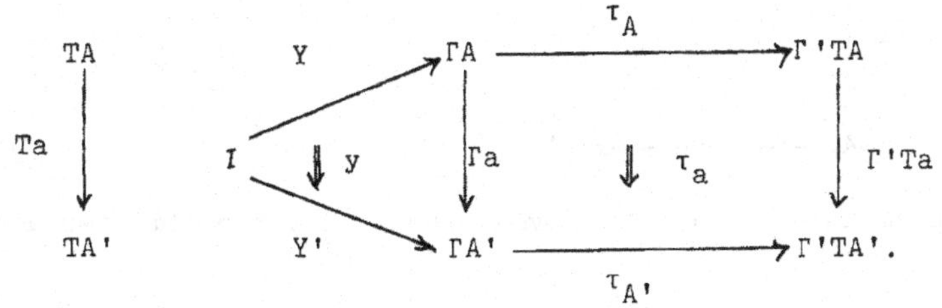

If $(\eta,\theta)\colon (T,\tau) \Rightarrow (\bar{T},\bar{\tau})$ in $CAT\!\int\! CAT$, the natural transformation $\eta(\theta)$ has by (4.15) its $A\langle Y\rangle$ component given by

(6.6)

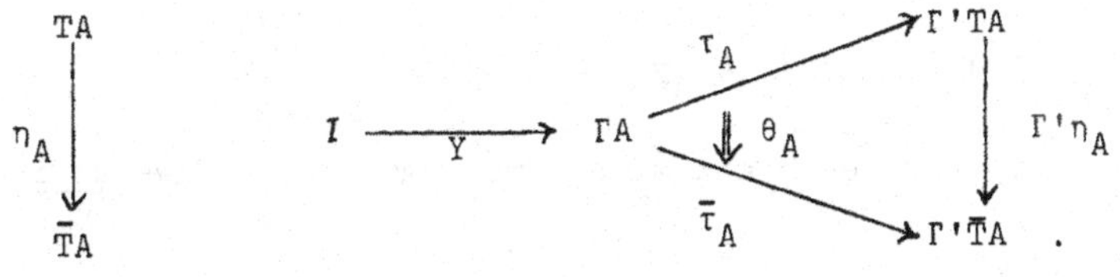

.

We have as in §4.1 the projection functor

(6.7) $\quad d_1 \colon A(\Gamma) \to A$

sending (6.4) to a: $A \to A'$. As in Proposition 4.3 this is the component of a 2-natural transformation

(6.8) $\quad d_1 \colon \Theta \Rightarrow \text{forg},$

where forg: $CAT\int CAT \to CAT$ is the forgetful 2-functor sending (A,Γ) to A, etc. Writing $\underline{\underline{2}}$ for the arrow category $0 \to 1$, we can regard Θ, forg, and d_1 in (6.8) as constituting a 2-functor

(6.9) $\quad \bar{\Theta} \colon CAT\int CAT \to [\underline{\underline{2}}, CAT]$

sending (A,Γ) to $d_1 \colon A(\Gamma) \to A$, etc. We can call $\bar{\Theta}$ the <u>augmented</u> Grothendieck 2-functor.

<u>6.3</u> The results we need about the Grothendieck construction are most easily derived from the following theorem, which is a slight extension of Gray's "Yoneda-Like Lemma" on page 290 of [3].

<u>Theorem 6.1</u> <u>The</u> 2-<u>functor</u> $\bar{\Theta} \colon CAT\int CAT \to [\underline{\underline{2}}, CAT]$ <u>is</u> 2-<u>fully-faithful</u>.

<u>Proof</u>. Let

(6.10)
$$\begin{array}{ccc} A(\Gamma) & \xrightarrow{\;M\;} & A'(\Gamma') \\ {\scriptstyle d_1}\downarrow & & \downarrow{\scriptstyle d_1} \\ A & \xrightarrow[\;T\;]{} & A' \end{array}$$

be a morphism in $[\underline{\underline{2}}, CAT]$. We have to show that $M = T(\tau)$ for a unique τ.

If M is indeed $T(\tau)$ we conclude from (6.5) with $a = 1_A$ that

(6.11) $\quad M(A\langle Y\rangle) = TA\langle \tau_A Y\rangle,$

(6.12) $\quad M(1_A\langle y\rangle) = 1_{TA}\langle \tau_A y\rangle,$

which fixes τ_A on objects and morphisms. Moreover if we <u>define</u> τ_A by

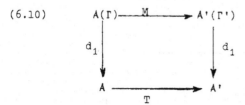

(6.11) and (6.12), the fact that M is a functor and the fact that the composite of $1_A\langle y'\rangle$ and $1_A\langle y\rangle$ is $1_A\langle y'y\rangle$ show that τ_A is a functor.

Again, if M is $T(\tau)$ we conclude from (6.5) with $Y' = \Gamma a.Y$ and $y = 1$ that

(6.13) $$M(a\langle 1_{\Gamma a.Y}\rangle) = Ta\langle(\tau_a)_Y\rangle ,$$

which fixes the components of τ_a. If we <u>define</u> these components by (6.13), the naturality of τ_a follows from the functoriality of M together with the following equality, where $y\colon Y \to Y'$ in ΓA:

$$1_{A'}\langle \Gamma a.y\rangle \;.\; a\langle 1_{\Gamma a.Y}\rangle = a\langle 1_{\Gamma a.Y'}\rangle \;.\; 1_A\langle y\rangle .$$

The fact that the typical morphism $a\langle y\rangle$ of (6.4) is the composite $1_{A'}\langle y\rangle \;.\; a\langle 1_{\Gamma a.Y}\rangle$, along with the fact that M is a functor and the definitions (6.12) and (6.13), now easily gives that $M(a\langle y\rangle)$ is (6.5). The lax naturality of τ then follows by applying M to the composite $a'\langle 1_{\Gamma(a'a).Y}\rangle \;.\; a\langle 1_{\Gamma a.Y}\rangle$.

Now let $\mu\colon T(\tau) \Rightarrow \bar{T}(\bar{\tau})$ be a natural transformation rendering commutative

(6.14)

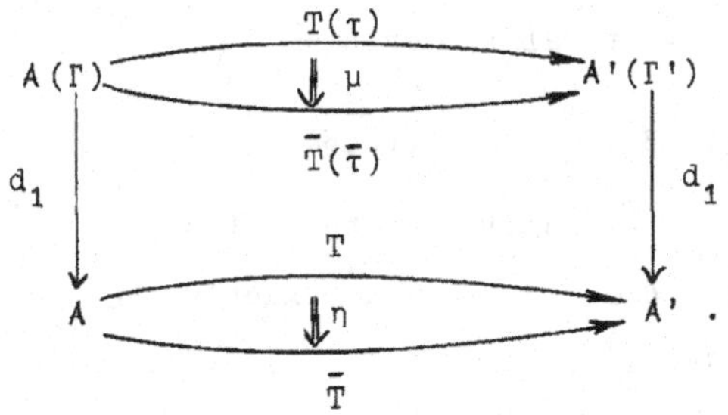

We have to show that $\mu = \eta(\theta)$ for a unique θ.

If μ is indeed $\eta(\theta)$ we have by (6.6) that

(6.15) $$\mu_{A\langle Y\rangle} = \eta_A\langle(\theta_A)_Y\rangle ,$$

which fixes the Y-component of θ. If we now <u>define</u> $(\theta_A)_Y$ by (6.15), the naturality of θ_A follows from the naturality of μ with respect to a morphism $1_A\langle y\rangle$; and the fact that θ is a modification follows from the naturality of μ with respect to a morphism $a\langle 1_{\Gamma a.Y}\rangle$. □

Remark 6.2 The above theorem could serve as well as Theorem 5.1 to complete the verification, in Corollary 5.2, that $CAT\int T$ and $CAT\int^{*} T$ are 2-categories.

6.4 As a result of the above theorem we have an isomorphism of categories

$$(6.16)\quad CAT\int CAT((A,\Gamma),(A',\Gamma')) \cong [\underline{2},CAT]\left(\begin{array}{c} A(\Gamma) \\ d_1\downarrow \\ A \end{array}, \begin{array}{c} A'(\Gamma') \\ \downarrow d_1 \\ A' \end{array}\right),$$

which is moreover 2-natural because it is given by the 2-functor $\bar{\Theta}$.

There is an evident 2-functor (not full)

$$(6.17)\quad CAT \times CAT \to CAT\int CAT$$

sending A,B to $(A,B!)$ etc., where $B!\colon A \to CAT$ is the constant functor at B, composite of $A \to I$ and $\ulcorner B \urcorner\colon I \to CAT$. It is also clear that

$$(6.18)\quad \bar{\Theta}(A,B!) = \begin{array}{c} A \times B \\ \downarrow pr_1 \\ A \end{array},$$

and similarly on 1-cells and 2-cells. Putting $(A',\Gamma') = (I,\ulcorner B \urcorner)$ in (6.16) therefore gives a 2-natural isomorphism

$$(6.19)\quad CAT\int CAT((A,\Gamma), (I,\ulcorner B \urcorner)) \cong CAT(A(\Gamma),B),$$

which exhibits Θ as the left-2-adjoint of the 2-functor $\ulcorner\ \urcorner$ of (3.13). (Note that, for categories C and B, we mean exactly the same by $CAT(C,B)$ and by $[C,B]$.) The left side of (6.19) can equally be written as the left side of

$$(6.20)\quad [\![A,CAT]\!](\Gamma,B!) \cong CAT(A(\Gamma),B).$$

An object $\alpha\colon \Gamma \rightsquigarrow B!$ of the left side of (6.20) is called a lax cocone over Γ with vertex B; its components look like

$$(6.21)\quad \begin{array}{ccc} \Gamma A & \xrightarrow{\alpha_A} & \\ \Gamma a \downarrow & \Downarrow \alpha_a & B \\ \Gamma A' & \xrightarrow{\alpha_{A'}} & \end{array}.$$

So in view of (6.20) it is reasonable to call $A(\Gamma)$ the <u>lax colimit</u> of $\Gamma: A \to CAT$; this is in effect Gray's theorem on page 289 of [3].

We give the explicit form of (6.20). By (6.5), the image of $\alpha: \Gamma \rightsquigarrow B!$ is the functor $P: A(\Gamma) \to B$ sending $a\langle y\rangle$ to

(6.22)

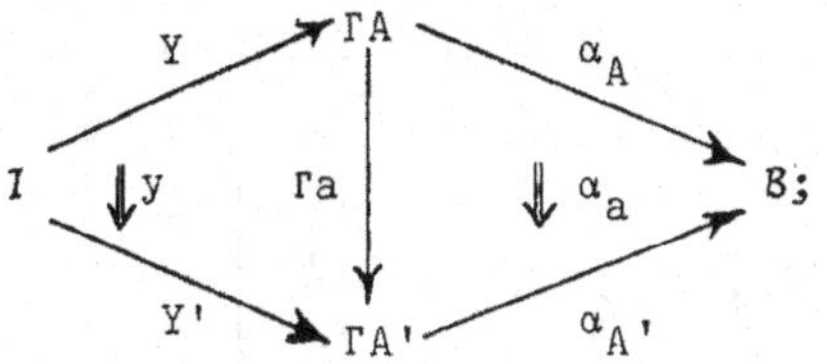

and by (6.6) the image of a modification $\rho: \alpha \to \bar{\alpha}$ is the natural transformation $\pi: P \Rightarrow \bar{P}$ whose $A\langle Y\rangle$ component is

(6.23)

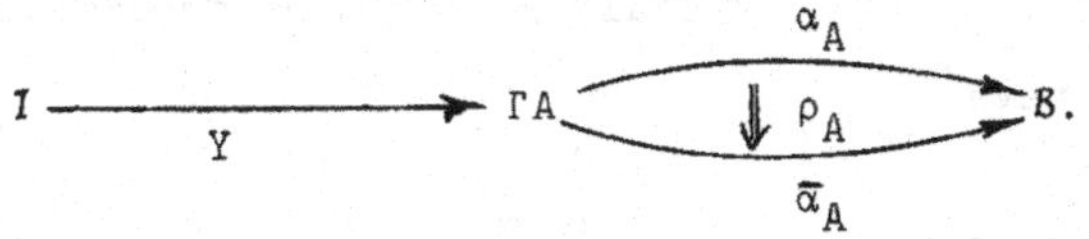

By (6.11) - (6.13), or directly from the above, α is refound in terms of P by

(6.24) $\alpha_A Y = P(A\langle Y\rangle)$, $\alpha_A y = P(1_A\langle y\rangle)$, $(\alpha_a)_Y = P(a\langle 1_{\Gamma a.Y}\rangle)$;

and ρ is refound in terms of π by

(6.25) $(\rho_A)_Y = \pi_{A\langle Y\rangle}$.

<u>6.5</u> In (6.20), we describe functors with <u>domain</u> $A(\Gamma)$. To get information about functors with <u>codomain</u> $A(\Gamma)$, we can either go back to the definition of $A(\Gamma)$ as $\ulcorner I \urcorner / \Gamma$ and use Propositions 4.1 and 4.2 directly; or use (6.16) again, this time replacing (A,Γ) by $(C,I!)$ and (A',Γ') by (A,Γ). In either case we get

(6.26) $CAT\int CAT((C,I!),(A,\Gamma)) \cong CAT(C,A(\Gamma))$.

An object of the left side of (6.26) consists of a functor $T: C \to A$ together with a lax n.t. $\tau: I! \rightsquigarrow \Gamma T$ (that is, a <u>lax cone</u> over ΓT with vertex I). A nicer-looking result is given by looking at the subcategories of both sides of (6.16) obtained by <u>fixing</u> the T of

(6.10).

In the general case this gives, for $T\colon A \to A'$,

$$(6.27) \quad [\![A,CAT]\!](\Gamma,\Gamma'T) \cong CAT/A' \left(\begin{array}{c} A(\Gamma) \\ {\scriptstyle Td_1}\big\downarrow \\ A' \end{array} \;,\; \begin{array}{c} A'(\Gamma') \\ \big\downarrow{\scriptstyle d_1} \\ A' \end{array} \right) .$$

Making again the above substitutions of $(C,I!)$ for (A,Γ) and (A,Γ) for (A',Γ') this gives

$$(6.28) \quad [\![C,CAT]\!](I!,\Gamma T) \cong CAT/A \left(\begin{array}{c} C \\ {\scriptstyle T}\big\downarrow \\ A \end{array} \;,\; \begin{array}{c} A(\Gamma) \\ \big\downarrow{\scriptstyle d_1} \\ A \end{array} \right) .$$

We remark, without making use of the fact, that another expression for the right side of (6.28) is classical; by e.g. the proposition on page 255 of Gray [3] we have

$$(6.29) \quad [A,CAT](\Omega T,\Gamma) \cong CAT/A \left(\begin{array}{c} C \\ {\scriptstyle T}\big\downarrow \\ A \end{array} \;,\; \begin{array}{c} A(\Gamma) \\ \big\downarrow{\scriptstyle d_1} \\ A \end{array} \right) ,$$

where the functor Ω sends T to the functor $\Omega T\colon A \to CAT$ given by $(\Omega T)A = T/\ulcorner A \urcorner$, the <u>ordinary</u> comma category of $T\colon C \to A$ and $\ulcorner A \urcorner\colon I \to A$.

<u>6.6</u> Taking $A = A'$ and $T = 1$ in (6.27) gives

$$(6.30) \quad [\![A,CAT]\!](\Gamma,\Gamma') \cong CAT/A \left(\begin{array}{c} A(\Gamma) \\ {\scriptstyle d_1}\big\downarrow \\ A \end{array} \;,\; \begin{array}{c} A(\Gamma') \\ \big\downarrow{\scriptstyle d_1} \\ A \end{array} \right) .$$

If we define the 2-functor Θ_A as the restriction of Θ given by

$$(6.31) \quad [\![A,CAT]\!] \to CAT\!\int\! CAT \xrightarrow[\Theta]{} CAT ,$$

and define $\bar{\Theta}_A\colon [\![A,CAT]\!] \to CAT/A$ as the 2-functor defined by $\bar{\Theta}$ and sending Γ to $d_1\colon A(\Gamma) \to A$, then (6.30) is the special case of Theorem 6.1 asserting

<u>Proposition 6.3</u> $\bar{\Theta}_A$ <u>is 2-fully-faithful</u>. □

We can look upon (6.20) as asserting that Θ_A has the right adjoint $B \mapsto B!$; and on (6.29) as asserting that the composite

(6.32) $$[A, CAT] \to [\![A, CAT]\!] \xrightarrow[\bar{\Theta}_A]{} CAT/A$$

has left adjoint Ω.

7. THE MONOIDAL STRUCTURE ON $CAT\int^* CAT$

7.1 Consider the category $CAT_0 \circ CAT_0$, where the first CAT_0 has as augmentation the inclusion $CAT_0 \to CAT$, and the second CAT_0 is merely a category. Consider a typical morphism $T[\gamma] : A[\Gamma] \to A'[\Gamma']$ in $CAT_0 \circ CAT_0$; it has the form

(7.1)

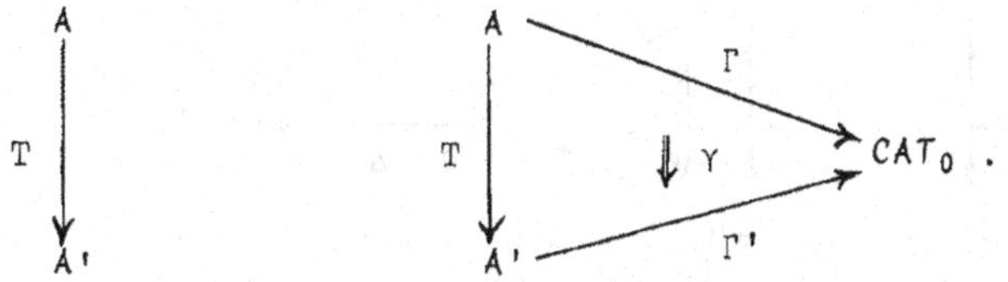

Here A is a category, and Γ a functor from A to CAT_0, or equally to CAT; so we may identify the objects $A[\Gamma]$ with the objects (A,Γ) of $CAT\int CAT$. We can then identify the morphism $T[\gamma]$ in (7.1) with the morphism $(T,\gamma)\colon (A,\Gamma) \to (A',\Gamma')$ of $CAT\int CAT$. Thus $CAT_0 \circ CAT_0$ may be identified with a subcategory of $(CAT\int CAT)_0$: to wit, it has all the objects, but only those morphisms (T,τ) for which the lax n.t. τ is in fact an honest natural transformation γ. This being so, there should be no confusion if we write

(7.2) $$\Theta\colon CAT_0 \circ CAT_0 \to CAT$$

for the functor obtained by restricting the 2-functor Θ of (6.1).

Given functors $\Gamma\colon A \to CAT$ and $\Delta\colon B \to CAT$, we write as usual $A \circ B$ for the more precise $(A,\Gamma)\circ B$, and we define an augmentation functor $E\colon A\circ B \to CAT$ as the composite

(7.3) $$A\circ B \xrightarrow[\Gamma\circ\Delta]{} CAT_0 \circ CAT_0 \xrightarrow[\Theta]{} CAT\,.$$

Here Γ stands for the morphism $(\Gamma, 1)$ of $CAT\int{*CAT}$ shown in

(7.4)

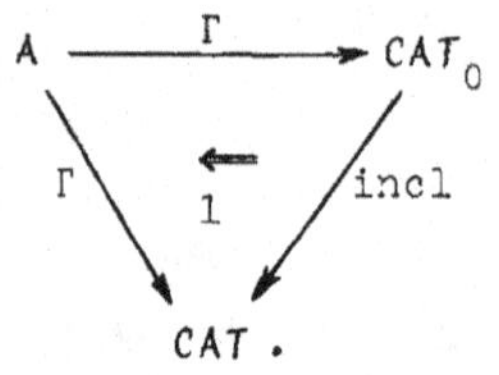

Thus $\Gamma \circ \Delta$ in (7.3) sends the typical morphism $a[x]$ of $A \circ B$ exhibited in (5.2) to the morphism

(7.5)

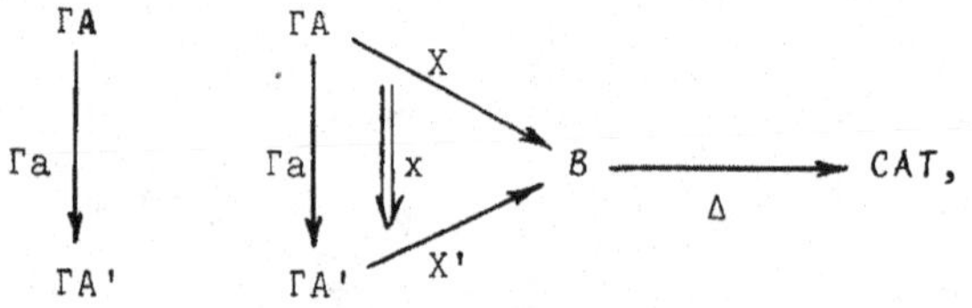

so that E sends $a[x]$ to

(7.6) $\Gamma a(\Delta x)\colon \Gamma A(\Delta X) \to \Gamma A'(\Delta X')$.

The basic result asserting that the image of the embedding Φ of (5.8) is closed under composition of endo-2-functors of CAT is now the following:

<u>Proposition 7.1</u> <u>Given categories</u> A <u>and</u> B <u>with augmentations</u> Γ <u>and</u> Δ, <u>assign to</u> $A \circ B$ <u>the augmentation</u> E <u>above</u>. <u>Then for</u> $C \in CAT$ <u>there is</u>, 2-<u>naturally in</u> C, <u>an isomorphism</u>

(7.7) $a\colon (A \circ B) \circ C \to A \circ (B \circ C)$.

<u>Proof</u> A typical morphism of $(A \circ B) \circ C$ is an $a[x][p]\colon A[X][P] \to A'[X'][P']$ of the form

(7.8)
$$\begin{array}{ccc} A[X] & \qquad & \Gamma A(\Delta X) \xrightarrow{P} C \\ \Big\downarrow a[x] & & \Gamma a(\Delta x)\Big\downarrow \quad \Downarrow p \qquad \\ A'[X'] & & \Gamma A'(\Delta X') \xrightarrow{P'} C \end{array}$$

where a[x] itself is the typical morphism (5.2) of $A \circ B$.

By (6.19) or (6.20), to give P is equivalent to giving a lax n.t.

(7.9) $\alpha: \Delta X \rightsquigarrow C!$

where $\Delta X, C!: \Gamma A \to CAT$. If similarly P' corresponds to α', the 2-naturality of (6.19) shows that $P'.\Gamma a(\Delta x)$ corresponds to the composite

(7.10)

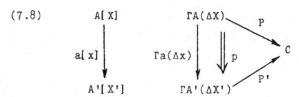

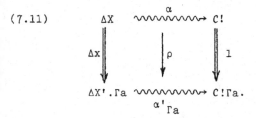

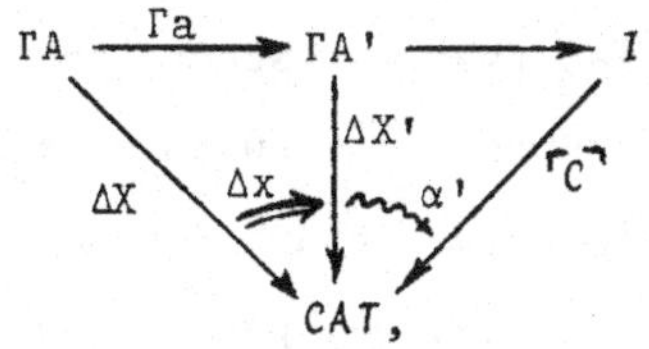

that is, to $\alpha'_{\Gamma a}.\Delta x$. By (6.19) or (6.20) again, to give p is then equivalent to giving a modification

(7.11)
$$\begin{array}{ccc} \Delta X & \overset{\alpha}{\rightsquigarrow} & C! \\ \Delta x \Big\Downarrow & \Big\downarrow \rho & \Big\Downarrow 1 \\ \Delta X'.\Gamma a & \underset{\alpha'_{\Gamma a}}{\rightsquigarrow} & C!\Gamma a. \end{array}$$

Since however $B \circ C$ is $\Delta /\!/ \ulcorner C \urcorner$, to give the pair $X: \Gamma A \to B$ and α is equivalent by Proposition 4.1 to giving a functor

(7.12) $Z: \Gamma A \to B \circ C.$

If similarly the pair X', α' corresponds to $Z': \Gamma A' \to B \circ C$, then by (4.4) the pair $X'.\Gamma a$, $\alpha'_{\Gamma a}$ corresponds to the composite

(7.13) $\Gamma A \xrightarrow[\Gamma a]{} \Gamma A' \xrightarrow[Z']{} B \circ C.$

Now by Proposition 4.2 to give the pair x,ρ as in (7.11) is equivalent to giving a natural transformation z of the form

(7.14)

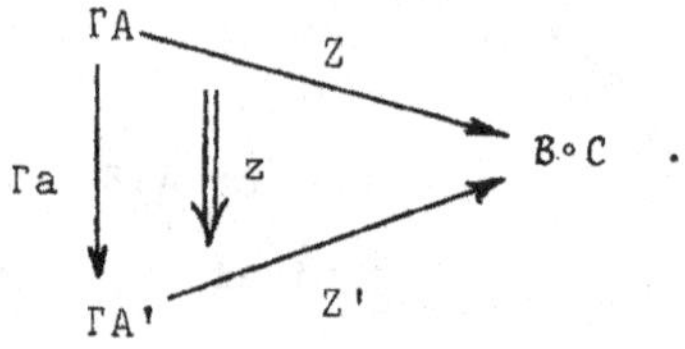

We henceforth write Z,z as

(7.15) $$Z = X[P], \quad z = x[p];$$

and we define the isomorphism α by

(7.16) $$\alpha(a[x][p]) = a[x[p]].$$

It has of course to be verified that α respects identies - this is obvious - and composites. If we compose (7.8) with $a'[x'][p']$: $A'[X'][P'] \to A''[X''][P'']$ to get $a''[x''][p'']$, we actually have $p'' = p'_{\Gamma A(\Delta X)}.p$. By the 2-naturality of (6.19), $p'_{\Gamma A(\Delta X)}$ corresponds to $\rho'_{\Gamma a}.\Delta x$, when we take account of what horizontal composition of 2-cells means in $CAT\int CAT$. Since (6.19) is functorial, it follows that the ρ'' corresponding to p'' is the composite

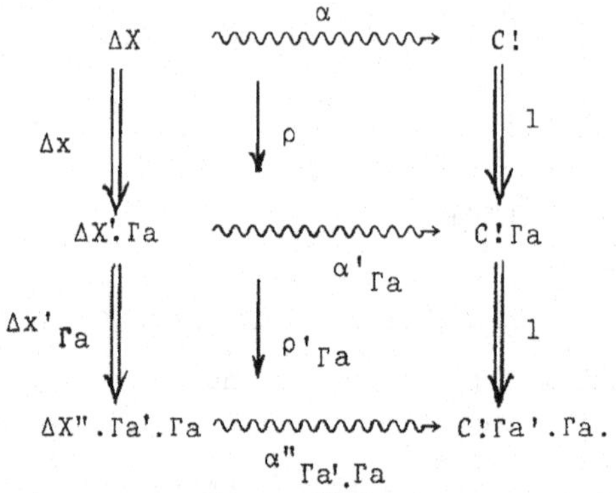

It now follows easily from (4.8) that $z'' = z'_{\Gamma a}.z$, as required.

Finally we have to verify the 2-naturality of α in C. First, if we compose (7.8) with $R: C \to C'$ then α,ρ are replaced by their composites with $R!: C! \Rightarrow C'!$ by the 2-naturality of (6.19), and it follows by §4.2 that (7.14) is replaced by its composite with $\mathcal{B}\circ R$. This is the naturality of a. For its 2-naturality we just take the domain $A[X][P]$ of (7.8) and compose it with

$$C \underset{\bar{R}}{\overset{R}{\rightrightarrows}} C' \quad (\Downarrow \zeta),$$

to get the morphism $1_A[1_X][\zeta_P]$, and apply a to this. The corresponding "ρ" is, by the 2-naturality of (6.19), given by

$$\Delta X \overset{}{\underset{\alpha}{\rightsquigarrow}} C! \underset{\bar{R}!}{\overset{R!}{\rightrightarrows}} C'! \quad (\Downarrow \zeta!)$$

so that by §4.2 it easily follows that a sends $1_A[1_X][\zeta_P]$ to

$$\Gamma A \xrightarrow[Z]{} \mathcal{B}\circ C \underset{\mathcal{B}\circ\bar{R}}{\overset{\mathcal{B}\circ R}{\rightrightarrows}} \mathcal{B}\circ C', \quad (\Downarrow \mathcal{B}\circ\zeta)$$

as required. □

We record the detailed form of a more explicitly than by (7.16) by using Propositions 4.1 and 4.2, along with (6.24) and (6.25), to describe explicitly $Z = X[P]$ and $z = x[p]$. Namely Z sends the morphism $\phi\colon n \to n'$ of ΓA to

(7.17)
$$\begin{array}{ccc} Xn & \qquad & \Delta Xn \xrightarrow{\alpha_n} C \\ \downarrow{\scriptstyle X\phi} & & \downarrow{\scriptstyle \Delta X\phi} \;\; \Downarrow{\scriptstyle \alpha_\phi} \;\; \nearrow{\scriptstyle \alpha_{n'}} \\ Xn' & & \Delta Xn' \end{array}$$

and z has n-component

(7.18)
$$\begin{array}{ccc} Xn & \qquad & \Delta Xn \xrightarrow{\alpha_n} C \\ \downarrow{\scriptstyle x_n} & & \downarrow{\scriptstyle \Delta x_n} \;\; \Downarrow{\scriptstyle \rho_n} \;\; \nearrow{\scriptstyle \alpha'_{\Gamma a.n}} \\ X'\Gamma a.n & & \Delta X'\Gamma a.n \end{array}$$

where, for $y\colon Y \to Y'$ in ΔXn,

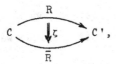

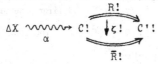

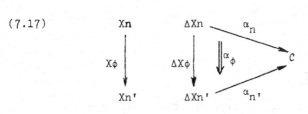

(7.19) $\alpha_n Y = P(n\langle Y\rangle)$, $\alpha_n y = P(1_n\langle y\rangle)$, $(\alpha_\phi)_Y = P(\phi\langle 1_{\Delta X\phi.Y}\rangle)$,

(7.20) $(\rho_n)_Y = p_{n\langle Y\rangle}$.

7.2 We now get automatically the desired monoidal structure on $CAT\int{*}CAT$. First, by Theorem 5.1, there is a unique way of making

(7.21) $\circ\colon CAT\int{*}CAT \times CAT\int{*}CAT \to CAT\int{*}CAT$,

which is so far only defined on objects, into a 2-functor such that the a of (7.7) is 2-natural in (A,Γ) and (B,Δ) as well as in C. Using the explicit forms (7.17) - (7.20) above, and (5.10), (5.13), and (5.20), it is a purely mechanical process to calculate

(7.22) $(T,\tau)\circ(S,\sigma)\colon A\circ B \to A'\circ B'$

and

(7.23) $(\eta,\theta)\circ(\zeta,\phi)\colon (T,\tau)\circ(S,\sigma) \Rightarrow (\bar{T},\bar{\tau})\circ(\bar{S},\bar{\sigma})$.

One part of the calculation is easy: if

(7.24) $(T,\tau)\circ(S,\sigma) = (N,\nu)$ and $(\eta,\theta)\circ(\zeta,\phi) = (\xi,\psi)$

then

(7.25) $N = (T,\tau)\circ S$ and $\xi = (\eta,\theta)\circ\zeta$;

this follows on observing that the I-component of the a of (7.7) is clearly

(7.26) $(A\circ B)\circ I \xrightarrow[r]{} A\circ B \xrightarrow[A\circ r^{-1}]{} A\circ(B\circ I)$

where $r\colon A\circ I \cong A$ is the 2-natural isomorphism of (5.22). For the calculation of ν and ψ, however, we have to follow through the above process, and I merely give the results.

ν is to be a lax n.t.

(7.27)
$$\begin{array}{ccc} A\circ B & \xrightarrow{(T,\tau)\circ S} & A'\circ B' \\ & {\scriptstyle E}\searrow \quad \overset{\nu}{\Leftarrow} \quad \swarrow{\scriptstyle E'} & \\ & CAT & \end{array}$$

where E is given by (7.6) and E' analogously. Take $a[x] : A[X] \to A'[X']$ in $\mathcal{A}\circ\mathcal{B}$ as in (5.2), and write $Ta[w] : TA[W] \to TA'[W']$ for its image under $(T,\tau)\circ\mathcal{B}$, as displayed on the right side of (5.19). Then its image under $(T,\tau)\circ S$ is $Ta[Sw] : TA[SW] \to TA'[SW']$ as in (5.4). In view of (7.6), the ν we want is to have components

(7.28)
$$\begin{array}{ccc} \Gamma'TA(\Delta'SW) & \xrightarrow{\quad\nu_{A[X]}\quad} & \Gamma A(\Delta X) \\ {\scriptstyle \Gamma'Ta(\Delta'Sw)}\Big\downarrow & \Downarrow \nu_{a[x]} & \Big\downarrow{\scriptstyle \Gamma a(\Delta x)} \\ \Gamma'TA'(\Delta'SW') & \xrightarrow[\quad\nu_{A'[X']}\quad]{} & \Gamma A'(\Delta X') \end{array} \quad .$$

Now we have

(7.29)
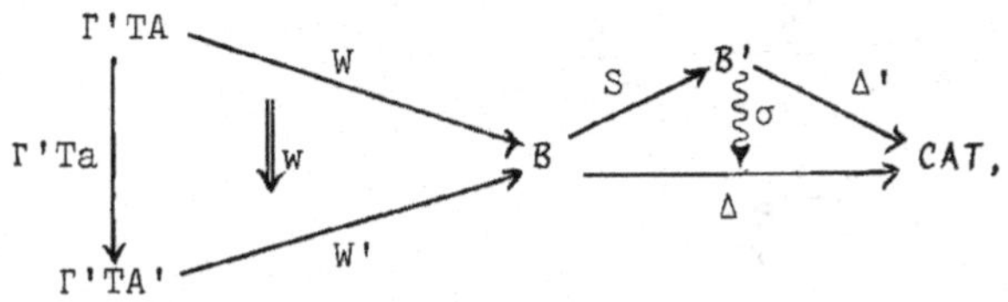

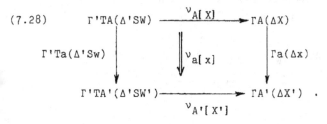

whence as in (2.8) we get

(7.30)
$$\begin{array}{ccc} \Delta'SW & \overset{\sigma_W}{\rightsquigarrow} & \Delta W \\ {\scriptstyle \Delta'Sw}\Big\Downarrow & \Big\downarrow \sigma_w & \Big\Downarrow{\scriptstyle \Delta w} \\ \Delta'SW'.\Gamma'Ta & \underset{\sigma_{W'}.\Gamma'Ta}{\rightsquigarrow} & \Delta W'.\Gamma'Ta \end{array} \quad .$$

Since $W = X\tau_A$ by (5.19), the pair (τ_A, σ_W) is a morphism $(\Gamma'TA, \Delta'SW) \to (\Gamma A, \Delta X)$ in $\mathcal{CAT}\!\int\!\mathcal{CAT}$, and it turns out that we have

(7.31) $\quad \nu_{A[X]} = \tau_A(\sigma_W).$

The top leg of (7.28) is therefore (recalling the meaning of composition in $\mathcal{CAT}\!\int\!\mathcal{CAT}$) $\Gamma a.\tau_A(\Delta x_{\tau_A}.\sigma_W)$, while the bottom leg is $\tau_{A'}.\Gamma'Ta(\sigma_{W'}.\Gamma'Ta.\Delta'Sw)$. Since by (5.19) $w = X'\tau_a.x_{\tau_A}$, we can write (7.30) as

(7.32)

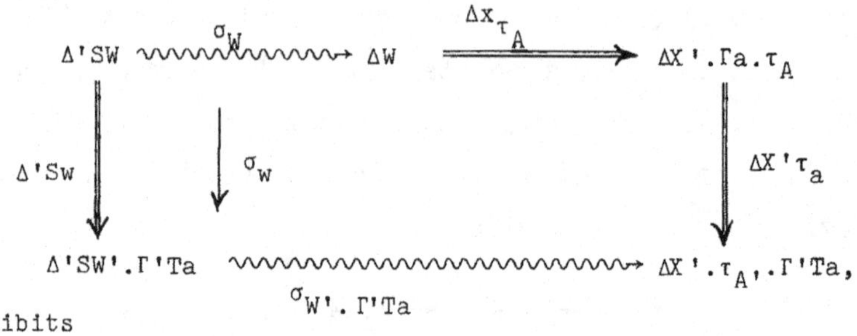

which exhibits

$$(\tau_a, \sigma_W) : (\Gamma a.\tau_A, \Delta x_{\tau_A}.\sigma_W) \to (\tau_{A'}.\Gamma' Ta, \sigma_{W'.\Gamma' Ta}.\Delta' Sw)$$

as a 2-cell in *CAT*∫*CAT*. It turns out that we have

(7.33) $$\nu_{a[x]} = \tau_a(\sigma_W).$$

It remains to give the modification ψ in

(7.34)

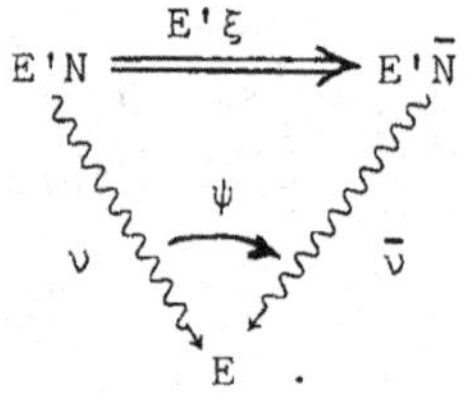

The A[X]-component of ν is $\tau_A(\sigma_W)$ by (7.31); that of $\bar{\nu}$ is $\bar{\tau}_A(\bar{\sigma}_{\bar{W}})$; here $W = X\tau_A$, $\bar{W} = X\bar{\tau}_A$. The A[X]-component of ξ, that is of $(\eta,\theta)\circ\zeta$ by (7.25), is $\eta_A[\zeta_{X\bar{\tau}_A.\Gamma'\eta_A}.SX\theta_A]$ by (5.5); hence that of $E'\xi$ is $\Gamma'\eta_A(\Delta'\zeta_{X\bar{\tau}_A.\Gamma'\eta_A}.\Delta' SX\theta_A)$ by (7.6). Thus the A[X]-component of $\bar{\nu}.E'\xi$ is

(7.35) $$\bar{\tau}_A.\Gamma'\eta_A(\bar{\sigma}_{X\bar{\tau}_A.\Gamma'\eta_A}.\Delta'\zeta_{X\bar{\tau}_A.\Gamma'\eta_A}.\Delta' SX\theta_A).$$

The A[X]-component of ψ is to be a 2-cell from $\tau_A(\sigma_W)$ to (7.35); it is in fact given by

(7.36) $$\psi_{A[X]} = \theta_A(\phi_{X\theta_A}).$$

Here $\phi_{X\theta_A}$ is the modification

(7.37)
$$\begin{array}{ccc}
\Delta' SX\tau_A & \xRightarrow{\quad \Delta' SX\theta_A \quad} & \Delta' SX\bar{\tau}_A . \Gamma'\eta_A \\
\sigma_{X\tau_A} \Big\downarrow & \xrightarrow{} \phi_{X\theta_A} & \Big\downarrow (\bar{\sigma}.\Delta'\zeta)_{X\tau_A . \Gamma'\eta_A} \\
\Delta X\tau_A & \xRightarrow[\quad \Delta X\theta_A \quad]{} & \Delta X\bar{\tau}_A . \Gamma'\eta_A
\end{array}$$

obtained as in (2.10) from $\phi\colon \sigma \to \bar{\sigma}.\Delta\zeta$ and $X\theta_A\colon X\tau_A \Rightarrow X\bar{\tau}_A.\Gamma'\eta_A$; it is clear that the pair $(\theta_A, \phi_{X\theta_A})$ is indeed a 2-cell in $CAT\int CAT$ with the right domain and codomain.

<u>7.3</u> Some comments now about the 2-functor (7.21) whose explicit form we have just given. Let us embed CAT 2-fully in $CAT\int^* CAT$ by identifying A with $(A, 0!)$ where $0!$ is the constant functor at the empty category 0. Then if $B \in CAT$, so that $\Delta = 0!$, we have for all $A[X]$ that $\Gamma A(\Delta X) = \Gamma A(0!) = 0$, so that $E\colon A\circ B \to CAT$ is also $0!$, and $A\circ B \in CAT$. Hence the 2-functor $\circ$ of (5.1) is just the restriction of the 2-functor $\circ$ of (7.21).

The expressions we have given for $(T,\tau)\circ(S,\sigma)$ and $(\eta,\theta)\circ(\zeta,\phi)$ above simplify when, as in many important examples, τ and σ are honest natural transformations, not lax ones, and θ and ϕ are identities; this is necessarily so for instance when the augmentations Γ and Δ of A and B factorize through $SET \subset CAT$. For then, using Γ in the two senses of (7.4), it is the case that $\tau = (\tau, 1)$ is a 2-cell

(7.38)
$$\begin{array}{ccccc}
 & & (A', \Gamma') & & \\
 & \overset{(T,\tau)}{\nearrow} & \Downarrow \tau & \overset{\Gamma'}{\searrow} & \\
(A, \Gamma) & & \xrightarrow[\Gamma]{\qquad\qquad} & & (CAT_0,\ \mathrm{incl})
\end{array}$$

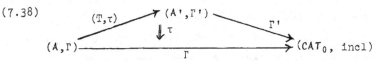

in $CAT\int^* CAT$, and the ν of (7.27) is the composite

(7.39)
$$\begin{array}{ccccccc}
 & & A'\circ B' & & & & \\
 & \overset{(T,\tau)\circ S}{\nearrow} & \Downarrow \tau\circ\sigma & \overset{\Gamma'\circ\Delta'}{\searrow} & & & \\
A\circ B & & \xrightarrow[\Gamma\circ\Delta]{\qquad\qquad} & & CAT_0\circ CAT_0 & \xrightarrow[\Theta]{\qquad} & CAT,
\end{array}$$

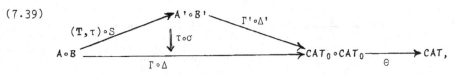

while the ψ of (7.34) is, by (7.36), the identity again.

There is a further simplification when we suppose that τ and

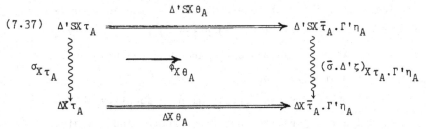

σ also are identities, so that we are working in the sub-2-category CAT/CAT of $CAT\int{*}CAT$; then ν too is an identity by (7.31) and (7.33). In other words CAT/CAT is closed under $\circ$; and $T\circ S$, $\eta\circ\zeta$ unambiguously denote the composites in (7.24) or (7.25).

7.4 That $\circ$ is coherently associative with an identity, making $CAT\int{*}CAT$ a monoidal 2-category, follows from Proposition 7.1 since $[CAT,CAT]$ is a strict monoidal 2-category with composition as its "tensor product". Turning first to the identity for $\circ$, we see at once that it may be taken to be the object

$$(7.40)\qquad J = (I, \ulcorner I \urcorner)$$

of $CAT\int{*}CAT$; for (5.24) gives a 2-natural isomorphism $\ell: J\circ C \cong C$ for $C \in CAT$, so that $J\circ-$ is isomorphic to $1 \in [CAT,CAT]$. (Note that, while in general we write n for the discrete category with n objects, and in particular 0 for the empty category, we write I rather than 1 to avoid confusion with identity maps. Note also that, since we identify $A \in CAT$ with $(A,0!) \in CAT\int{*}CAT$, and in particular I with $(I,\ulcorner 0 \urcorner)$, we need to distinguish I from $J = (I,\ulcorner I \urcorner)$.)

From Proposition 7.1 we now get well-determined 2-natural isomorphisms

$$(7.41)\qquad \ell: J\circ A \cong A, \qquad r : A\circ J \cong A$$

for $A \in CAT\int{*}CAT$. It is clear that the functor-part of ℓ is the ℓ above, that of (5.24), sending $*[\ulcorner a \urcorner]$ to a; and that the functor-part of r is what we called r in (7.26), that of (5.22), sending $a[1_{*!}]$ to a. However, as morphisms in $CAT\int{*}CAT$, ℓ and r involve also lax n.t.'s λ and ρ as in

(7.42)

where Δ, E are the appropriate augmentations. As we have defined things λ and ρ are not identities, only isomorphisms; for e.g. by (7.6) $\Delta(*[\ulcorner A \urcorner]) = I(\ulcorner \Gamma A \urcorner)$ which, while isomorphic to ΓA, is not the <u>same</u> as it.

Again Proposition 7.1 gives a well-determined 2-natural isomorphism

$$(7.43) \qquad a: (A \circ B) \circ C \to A \circ (B \circ C)$$

for $A, B, C \in CAT\int{}^* CAT$; the coherence of a, ℓ, r is automatic; and clearly the functor-part of a is just the a of (7.7). Once again, however, its other part α,

(7.44)

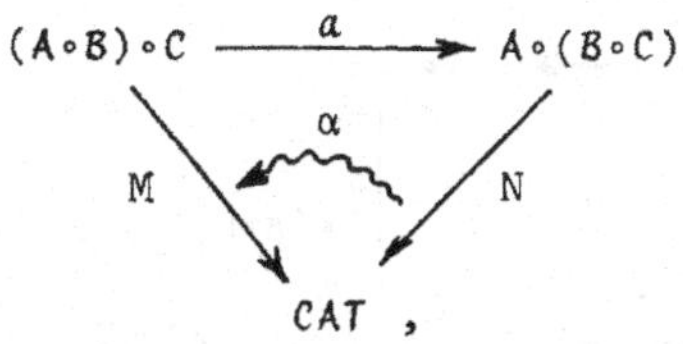

is only an isomorphism, not an identity. This is true even if the augmentations Γ, Δ, E of A, B, C factorize through $SET \subset CAT$; $M(A[X][P])$ and $N(A[X[P]])$ are different sets, in the same sense that, for sets U, V, W, the sets $(U \times V) \times W$ and $U \times (V \times W)$ are different sets.

<u>7.5</u> This failure of λ, ρ, α to be identities is of importance, and we look at it more closely.

The whole failure arises, as it were, from the special case $A = B = C = CAT_0$, where (7.42) and (7.44) take the forms

(7.45)

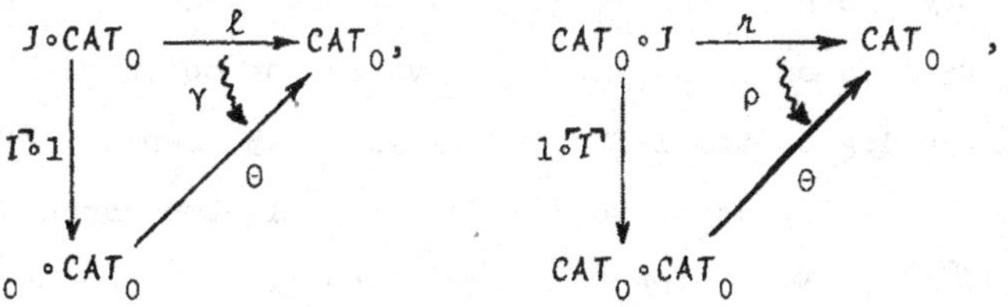

(7.46)

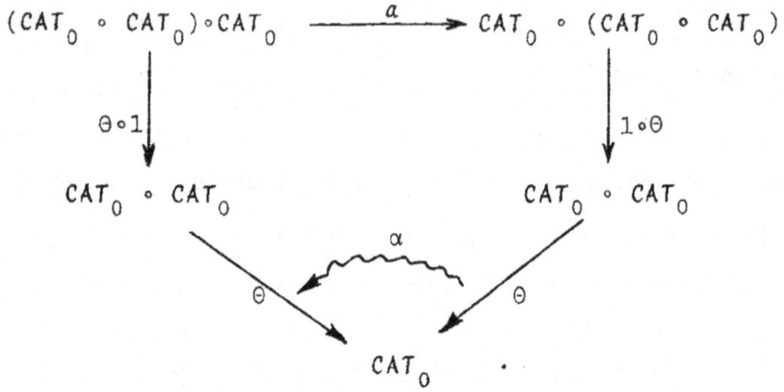

For example, the diagram (7.44) is just the result of pasting on to the top of (7.46) the commutative diagram

(7.47)
$$\begin{array}{ccc} (A\circ B)\circ C & \xrightarrow{\ a\ } & A\circ(B\circ C) \\ {\scriptstyle (\Gamma\circ\Delta)\circ E}\downarrow & & \downarrow{\scriptstyle \Gamma\circ(\Delta\circ E)} \\ (CAT_0\circ CAT_0)\circ CAT_0 & \xrightarrow[\ a\ ]{} & CAT_0\circ(CAT_0\circ CAT_0). \end{array}$$

To see this, recall that M is $\Theta.\Theta\circ 1.(\Gamma\circ\Delta)\circ E$ by (7.3), with a corresponding value for N; recall that the a of (7.43) is natural; express this naturality with respect to Γ,Δ,E regarded as morphisms as in (7.4); and use the observation at the end of §7.3 that $(\Gamma,1)\circ(\Delta,1) = (\Gamma\circ\Delta,1)$, etc. Similar observations can be made for λ and ρ.

It is easy to calculate precisely the λ,ρ,α of (7.45) and (7.46). First, they are not merely lax natural, but honest natural isomorphisms; so we need only look at their values on objects. First, for $A \in CAT$, the top leg of the left diagram in (7.45) sends $*[A]$ to A, and the bottom leg sends it to $I(\ulcorner A \urcorner)$; $\lambda_{*[A]}$ is the canonical isomorphism $A \to I(\ulcorner A \urcorner)$. Next, the top leg of the right diagram in (7.45) sends $A[!]$ to A, while the bottom leg sends it to $A(I!)$; $\rho_{A[!]}$ is the canonical isomorphism $A \to A(I!)$. Finally, consider the effect of the two legs of (7.46) on $A[\Gamma][\Delta]$ where $\Gamma\colon A \to CAT$ and

$\Delta: A(\Gamma) \to CAT$. The top leg sends it to $A(E)$, where $E = \Theta(\Gamma[\Delta])$; the bottom leg to $A(\Gamma)(\Delta)$. If we write Z for $\Gamma[\Delta]: A \to CAT_0 \circ CAT_0$, it follows from (7.17) that $E = \Theta Z$ sends $a: A \to A'$ to $\Gamma a(\beta_a): \Gamma A(\beta_A) \to \Gamma A'(\beta_{A'})$ where, for $y: Y \to Y'$ in ΓA,

$$(7.48) \quad \beta_A(Y) = \Delta(A\langle Y\rangle), \ \beta_A(y) = \Delta(1_A\langle y\rangle), \quad (\beta_a)_Y = \Delta(a\langle 1_{\Gamma a.Y}\rangle).$$

The isomorphism $\alpha = \alpha_{A[\Gamma][\Delta]}: A(E) \to A(\Gamma)(\Delta)$ is now easily written down; an object of $A(E)$ is $A\langle K\rangle$ where $A \in A$, $K: I \to EA = \Gamma A(\beta_A)$; K corresponds to $Y\langle L\rangle$ where $Y \in \Gamma A$, $L: I \to \beta_A Y$; an object of $A(\Gamma)(\Delta)$ is $A\langle Y\rangle\langle L\rangle$ where $A\langle Y\rangle \in A(\Gamma)$ and $L: I \to \Delta(A\langle Y\rangle) = \beta_A Y$; similarly for morphisms.

So λ, ρ, α fail to be identities in the most trivial way; if we take A to be discrete, Γ to be constant at the discrete category B, and Δ to be constant at the discrete category C, the above component of α is in effect the isomorphism $(A \times B) \times C \to A \times (B \times C)$. However they do fail to be identities, and as a result, as we see by (7.45) and (7.46), CAT_0 fails to be a monoid in $CAT\int{}^*CAT$; so that $CAT_0\circ$- fails to be a doctrine; it is only a pseudo-doctrine (cf. [12], §1.5 and §3.5). The same is true of the more realistic "doctrine" $Cat_0\circ$-; and hence of the "doctrine" whose algebras are categories with small colimits, which Kock [13] shows to be a quotient doctrine of $Cat_0\circ$-.

The pseudoness here is a fact of life; Zöberlein in his thesis [20] accepts it and gets involved in the whole complexity of pseudo-doctrines. Perhaps this is the right thing to do; but it daunts me totally, and I shall avoid it as long as I am able. To do this I follow Kock's thesis [13], where he considers categories with small colimits indexed only by specially chosen isomorphs of small categories. I am the more inclined to do so since, when we specialize the small categories to be finite sets, we get what is obviously the "right" setting for the corresponding coherence problems.

We therefore set about to modify the above in such a way that

α,λ,ρ become identities; we do so by replacing $CAT\int{*}CAT$ by a sub-2-category $CAT\int{*}cat$. We could define a club to be a $\circ$-monoid in $CAT\int{*}CAT$, but it will be simpler if we use the name only for a $\circ$-monoid in $CAT\int{*}cat$.

8. THE MONOIDAL STRUCTURES ON $CAT\int{*}cat$ AND $CAT\int{*}T$

8.1 Up until now size considerations have clearly been of no importance, and we have not hesitated for instance to write CAT/CAT (meaning of course $CAT/\ulcorner CAT \urcorner$). As we now wish to move back step-by--step towards more practical situations, we should in any case want at this stage to consider $CAT\int{*}Cat$ rather than $CAT\int{*}CAT$; that is, to deal with those (A,Γ) for which each ΓA is a small category. This however would not be enough to make identities of the isomorphisms λ,ρ,α of §7.5, so we go further.

We write *set* for the full sub-category of *Set* determined by those small sets that are in fact ordinal numbers. We write *cat* for the full sub-2-category of *Cat* determined by those small categories A such that $|A| = \text{ob}A$ is an ordinal number and moreover each hom-set $A(A,B)$ is an ordinal number. We are led to use this 2-category *cat* because Kock's thesis [13] showed its success in overcoming our difficulty about λ,ρ,α; but it is in fact an entirely reasonable thing to do in analogy with §1.4 above, where the "types" ΓA were not arbitrary finite sets but elements of $\underline{\underline{N}}$, i.e. finite ordinals.

More generally we are going to let T denote a sub-2-category of *cat*, not necessarily full on morphisms, but at any rate full on 2-cells. As in §5.4, $CAT\int{*}T$ is then a sub-2-category of $CAT\int{*}CAT$, full on 2-cells, and full on 1-cells when $T \to cat$ is so. Similarly $CAT\int T$ is a sub-2-category of $CAT\int CAT$. Recall too that, if L is a full sub-2--category of CAT, $L\int T$ denotes the obvious full sub-2-category of $CAT\int T$, and similarly for $L\int{*}T$.

Of course we get the restriction

(8.1) $\circ\colon CAT\!\int\!{}^{*}T \times CAT \to CAT$

of the 2-functor $\circ$ of (5.1); and as pointed out in §5.4 nothing in §5 needs changing when T is full in cat. The non-trivial modifications we are going to make concern the Grothendieck construction of §6.

8.2 In place of the Θ of (6.1) we define a new Grothendieck 2-functor

(8.2) $\Theta'\colon cat\!\int\! cat \to cat$

which is an isomorph of the restriction of Θ to $cat\!\int\! cat$. To do this it suffices, when $A \in cat$ and $\Gamma\colon A \to cat \subset CAT$, to give an object $\Theta'(A,\Gamma)$ in cat and an isomorphism $\omega\colon \Theta(A,\Gamma) \to \Theta'(A,\Gamma)$. We then use this isomorphism to transfer to Θ' the 2-functor structure of Θ.

We begin by giving the restriction of Θ' to $set\!\int\! set$. For $A \in set$ and $\Gamma\colon A \to set$, $\Theta(A,\Gamma)$ is the discrete category with objects $A\langle Y\rangle$ where $A \in A$, $Y\colon I \to \Gamma A$. These objects can be ordered lexicographically: $A\langle Y\rangle < A'\langle Y'\rangle$ if $A < A'$ or if $A = A'$ and $Y < Y'$ (regarding Y, Y' as elements of ΓA). Then $\Theta(A,\Gamma)$ is a well-ordered set. We define $\Theta'(A,\Gamma)$ to be the corresponding ordinal, and $\omega\colon \Theta(A,\Gamma) \to \Theta'(A,\Gamma)$ to be the unique order-isomorphism.

Now let $A \in cat$, $\Gamma\colon A \to cat$. Define $|\Gamma|\colon |A| \to set$ by $|\Gamma|A = |\Gamma A|$. We define $\Theta'(A,\Gamma)$ to be B where

(8.3) $|B| = \Theta'(|A|,|\Gamma|)$,

(8.4) $B(\omega A\langle Y\rangle, \omega A'\langle Y'\rangle) = \Theta'(A(A,A'),Z)$,
where $Za = \Gamma A'(\Gamma a.Y, Y')$.

There is an obvious isomorphism of categories $\omega\colon \Theta(A,\Gamma) \to \Theta'(A,\Gamma)$, as desired. Clearly the Θ' we have defined on $set\!\int\! set$ is the restriction of this Θ' on $cat\!\int\! cat$.

We now simplify by changing our notation. When there is no

danger of confusion, we henceforth write Θ for Θ', $A\langle Y\rangle$ for $\omega A\langle Y\rangle$, and $a\langle y\rangle$ for $\omega a\langle y\rangle$. We also use the parenthesis notation again, with $A(\Gamma)$ for $\Theta'(A,\Gamma)$, etc.

8.3 Returning now to §6, we see that there is little to change when the Θ of (6.1) is replaced by the Θ', now itself called Θ, of (8.2). Of course $a\langle y\rangle$ is now not (6.4) but its image under ω, but (6.4) continues to represent it well. The $\bar{\Theta}$ of (6.9) becomes $\bar{\Theta}$: $\mathit{cat}\int\mathit{cat} \to [\underline{2},\mathit{cat}] \subset [\underline{2},\mathrm{CAT}]$; and Theorem 6.1 continues to hold for this new $\bar{\Theta}$. Similarly we still have (6.16) and its consequences (6.19), (6.20), and (6.26) - (6.31), as well as Proposition 6.3.

We can go further. Let T be a full sub-2-category of *cat* that is <u>closed under the Grothendieck construction</u>. By this we mean that $A(\Gamma) \in T$ if $A \in T$ and if $\Gamma: A \to \mathit{cat}$ factorizes through $T \subset \mathit{cat}$ (we then write $\Gamma: A \to T$). Then for the same reasons all the above results of §6 remain true with CAT replaced not only by *cat* but by T. Examples of such T are *set*; the full sub-2-category *ord* of *cat* whose objects are categories $A \in \mathit{cat}$ which are ordered sets (i.e. $A(A,A') = 0$ or 1); and the full subcategory $\underline{S}$ of *set* determined by the <u>finite</u> ordinals (the case of §1.4).

Those results of §6 depend of course on the fullness of T in *cat*. But we can at least <u>define</u> $\Theta: T\int T \to T$ and $\bar{\Theta}: T\int T \to [\underline{2},T]$ for certain non-full sub-2-categories T of *cat*: although we always require fullness on 2-cells. For this we need not only that $A(\Gamma) \in T$ when $A \in T$ and $\Gamma: A \to T$, but also that $T(\tau)$ is a morphism in T when T is a morphism in T and each τ_A is a morphism in T. The only cases that will concern us are that where $T = \underline{P}$, the subcategory of $\underline{S}$ with the same object-set $\underline{N}$ but with permutations as its only morphisms - this is the case of §1.3 - and the case where T is the discrete category $\underline{N}$ itself.

8.4 Coming now to §7, and taking a $T \subset \mathit{cat}$ as above, not necessarily full, we can again regard $T_0 \circ T_0$ as a subcategory of $(T\int^* T)_0$, and as in (7.2) define the restriction

(8.5) $\Theta: T_0 \circ T_0 \to T_0$

of Θ; of course when T is just a category, as in the cases *set*, $\underline{S}$, $\underline{P}$, this is no restriction of Θ but is Θ itself. Note in particular that (8.5) is the restriction of Θ: $cat_0 \circ cat_0 \to cat_0$.

Then for (A,Γ), $(B,\Delta) \in CAT\int{*T}$, the analogue of (7.3) becomes $E: A \circ B \to T$ given by

(8.6) $$A \circ B \xrightarrow[\Gamma \circ \Delta]{} T_0 \circ T_0 \xrightarrow[\Theta]{} T,$$

exhibiting $A \circ B$ as an object of $CAT\int{*T}$, where $\circ$ is the restriction (8.1) of (5.1). Since we have done no more than replace $A \circ B$ by an isomorph in $CAT\int{*CAT}$, we still have Proposition 7.1; even the explicit description in (7.17) - (7.20) of the a of (7.16) does not change, modulo of course the new meaning of $n\langle Y \rangle$ etc. in (7.19) and (7.20).

As a result we get in analogy to (7.21) an extension

(8.7) $$\circ: CAT\int{*T} \times CAT\int{*T} \to CAT\int{*T}$$

of (8.1) when T is full in *cat*; and moreover the $\circ$ of (8.7) is just the restriction of its value

(8.8) $$\circ: CAT\int{*cat} \times CAT\int{*cat} \to CAT\int{*cat}$$

for $T = cat$. Of course the detailed descriptions of this 2-functor in §7.2 also hold unchanged, again modulo the new meanings of $\tau_A(\sigma_W)$, etc.

When T is not full in *cat* we cannot conclude the same directly since we then lack the analogue of Theorem 5.1. Nevertheless we still do get (8.7); we have only to check that, for the ν of (7.24), each $\nu_{A[X]}$ lies in T; but by (7.31) this is $\tau_A(\sigma_W)$; and τ_A is a morphism in T by hypothesis; each $(\sigma_W)_n = \sigma_{X\tau_A n}$ for $n \in \Gamma' TA$ is also a morphism in T by hypothesis; whence the same is true of $\tau_A(\sigma_W)$ since T is closed under the Grothendieck construction.

The expressions for a, ℓ, r are also unchanged, as far as their functor parts go. However the α, λ, ρ of §§7.4 and 7.5 are now identities, since before replacing Θ by Θ' they are order-preserving isomorphisms of categories. Hence we have what we want. In particular a, ℓ, r actually live in CAT/cat_0, or in CAT/T_0; and we recall from §7.3 that these latter 2-categories are also closed under the 2-functor $\circ$. We sum all this up in the following theorem:

Theorem 8.1 $CAT\int^* cat$ is a monoidal 2-category with tensor product $\circ$ given by (8.8) and described explicitly in §7.2; with identity $J = (I, \ulcorner I \urcorner)$; and with a, ℓ, r as described in §7.4. It contains CAT as a full sub-2-category when we identify A with $(A, 0!)$; and $\circ$ restricts to $\circ$: $CAT\int^* cat \times CAT \to CAT$. The corresponding

$$\Phi : CAT\int^* cat \to [CAT, CAT] \tag{8.9}$$

is 2-fully-faithful, and is a strong monoidal functor (sending $\circ$ to composition, to within isomorphism). The monoidal structure on $CAT\int^* cat$ restricts to the sub-2-category $CAT\int^* T$, and also to the sub-2-categories CAT/cat_0 and CAT/T_0, when T is closed under the Grothendieck construction. Moreover the restriction of (8.9) given by

$$\Phi : CAT\int^* T \to [CAT, CAT] \tag{8.10}$$

is still 2-fully-faithful when T is 2-full in cat. $\square$

Note that the $\circ$ of (8.8) is not the restriction of that of (7.21); it is only isomorphic to this.

9. THE CLOSED STRUCTURE ON CAT/T_0

9.1 We have said that we shall use the word club in general to denote a $\circ$-monoid in $CAT\int^* cat$, which may happen to lie in some smaller $CAT\int^* T$, and which by Theorem 8.1 is essentially the same thing as a doctrine whose functor-part lies in the image of the Φ of (8.9) (or of (8.10)). The original clubs however, described in §1 above, and arising from coherence problems, were $\circ$-monoids in the non-full sub-2-category CAT/cat_0, often in a smaller CAT/T_0, and in particular in $CAT/\underline{\underline{S}}$ or $CAT/\underline{\underline{P}}$. Because these clubs are of especial practical importance, we now look more closely at the monoidal structure on CAT/T_0, and show that it is in fact <u>closed</u> (but not biclosed).

Consider first the relation between several such 2-categories. Let $T \subset T' \subset cat$ where T, T' are closed under the Grothendieck construction. Then, whether T and T' are full in cat or not, the inclusion 2-functors

$$(9.1) \qquad CAT/T_0 \to CAT/T'_0 \to CAT/cat_0$$

are full (which is not the case when / is replaced by $\int^*$). Moreover each has a right 2-adjoint. If we write $i\colon T \to T'$ for the inclusion, and write the first 2-functor in (9.1) as i_*, then it has the right 2-adjoint i^* sending (C,E) to its pullback (B,Δ) along i as in

$$(9.2) \qquad \begin{array}{ccc} B & \xrightarrow{\;\bar{i}\;} & C \\ {\scriptstyle\Delta}\downarrow & & \downarrow{\scriptstyle E} \\ T & \xrightarrow[\;i\;]{} & T'. \end{array}$$

Next, the restriction

$$(9.3) \qquad \Phi'\colon CAT/cat_0 \to [CAT,CAT]$$

of the Φ of (8.9) is certainly not full; nor of course is its further restriction

$$(9.4) \qquad \Phi'\colon CAT/T_0 \to [CAT,CAT].$$

What is the case, however, is that (9.3), and hence also (9.4), have right 2-adjoints — recall from the end of §5.3 that this was not the case for Φ itself; rather there it was "as far from true as it could be".

We define a 2-functor

(9.5) $\qquad \Psi\colon [CAT, CAT] \to CAT/T_0$

which shall be the right adjoint of the Φ' of (9.4). Let $k\colon T_0 \to CAT$ be the inclusion, and define Ψ as the composite of the 2-functors

(9.6) $$[CAT,CAT] \xrightarrow[{[k,1]}]{} [T_0, CAT] \xrightarrow[q]{} [\![T_0, CAT]\!] \xrightarrow[\bar{\Theta}]{} CAT/T_0,$$

where the middle 2-functor q is just the inclusion, and $\bar{\Theta}$ is $\bar{\Theta}_{T_0}$ in the sense of (6.31) (the original, unmodified, $\bar{\Theta}$). We observe that, in view of (6.32), we are asserting that Φ' is the composite of $\Omega\colon CAT/T_0 \to [T_0, CAT]$ and the Kan right 2-adjoint of $[k,1]$; but it seems to be just as easy to establish the present result without reference to Ω. We state it formally:

Theorem 9.1 $\qquad$ *Ψ is right 2-adjoint to Φ'.*

Proof $\quad$ We have to give a 2-natural isomorphism of categories

(9.7) $$\pi\colon CAT/T_0 \begin{pmatrix} A & & T_0(Dk) \\ \downarrow \Gamma & , & \downarrow d_1 \\ T_0 & & T_0 \end{pmatrix} \to [CAT, CAT](A_0-, D).$$

In view of (6.28), we need therefore an isomorphism

(9.8) $$\pi_1\colon [\![A, CAT]\!](I!, Dk\Gamma) \cong [CAT, CAT](A\circ -, D).$$

We might as well, as we have constantly been doing, write Γ for $k\Gamma$. So now $\Gamma\colon A \to CAT$, and we seek an isomorphism

(9.9) $$\pi_1\colon [\![A, CAT]\!](I!, D\Gamma) \cong [CAT, CAT](A\circ -, D).$$

For $\alpha\colon I! \rightsquigarrow D\Gamma$ we define $\pi_1\alpha$ to be $G\colon A\circ - \Rightarrow D$ whose B-component $G_B\colon A\circ B \to DB$ is the functor sending the typical morphism $a[x]$ of (5.2) to

(9.10)

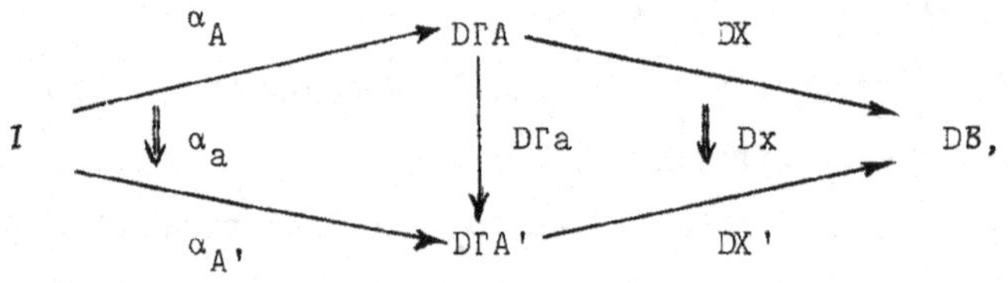

conceived of as a morphism in DB. For a modification

$\rho: \alpha \to \alpha': I! \rightsquigarrow D\Gamma$ we define $\pi_1\rho$ to be the modification $\sigma: G \to G'$ whose $\mathcal{B}$-component $\sigma_{\mathcal{B}}: G_{\mathcal{B}} \Rightarrow G'_{\mathcal{B}}$ is the natural transformation with $A[X]$-component given by

$$(9.11) \qquad I \underset{\alpha'_A}{\overset{\alpha_A}{\Downarrow \rho_A}} D\Gamma A \xrightarrow[DX]{} D\mathcal{B}.$$

It is clear enough that $G_{\mathcal{B}}$ is indeed a functor, and scarcely harder to see that $\sigma_{\mathcal{B}}$ is indeed a natural transformation. We leave to the reader the easy proof that G is 2-natural, that σ is a modification, and that π_1 is a functor.

One next shows that π_1 is an isomorphism; given $G: A\circ\!- \Rightarrow D$, one argues pretty much as in the proof of Theorem 5.1 to show that G arises as above from a unique α; for example (9.10) shows we must have

$$(9.12) \qquad \alpha_A = G_{\Gamma A}(A[1_{\Gamma A}]), \ \alpha_a = G_{\Gamma A'}(a[1_{\Gamma a}]),$$

and we use the 2-naturality of G to show that α is lax natural and that G is $\pi_1(\alpha)$. Similarly (9.11) shows that we must have

$$(9.13) \qquad \rho_A = (\sigma_{\Gamma A})_{A[1_{\Gamma A}]};$$

and we verify that ρ is a modification with $\pi_1(\rho) = \sigma$.

I have no really short way of verifying the 2-naturality of the isomorphism π thus established; conceptual proofs would take us into extensions for which we have no other use; I myself have found it easiest to verify the 2-naturality of the inverse π^{-1}, by direct calculation using (9.12), (9.13), and our detailed knowledge of (6.28).□

I have given the above proof in outline only, because I make no really central use of the theorem. I note here that the above theorem provides a new proof of Theorem 5.1, by specialization to the case where $D = \mathcal{B}\circ-$; it is a kind of generalization of Theorem 5.1. I note further that, just as Gray on page 290 of [3] calls our Theorem 6.1 "Yoneda-like", so too is this theorem "Yoneda-like"; it purports to give 2-natural transformations $A\circ - \Rightarrow D$ in terms of special "universal" components as in (9.12), just as the V-Yoneda lemma allows us to determine a V-natural transformation $A\otimes - \Rightarrow D$ in terms of a universal element $A \to DI$.

<u>9.2</u> We have said that CAT/T_0 is a <u>closed</u> category; $-\circ B\colon CAT/T_0 \to CAT/T_0$ has a right adjoint $\{B,-\}\colon (CAT/T_0)^{op} \to CAT/T_0$. We have also said that it is not <u>biclosed</u>; $A\circ -$ has no right adjoint. Let us dispose of this latter point first.

If $A\circ -$ had a right adjoint, it would preserve the initial object $0 \to T_0$ of CAT/T_0. Yet $A\circ 0$ has as objects those pairs $A[1_0]$ where $\Gamma A = 0$; it is therefore isomorphic to the fibre of $\Gamma\colon A \to T_0$ above 0; which need by no means be 0.

What we are claiming is that there is a 2-functor

(9.14) $$\{\ ,\ \}\colon (CAT/T_0)^{op} \times CAT/T_0 \to CAT/T_0$$

such that we have 2-naturally

(9.15) $$CAT/T_0(A\circ B,C) \cong CAT/T_0(A,\{B,C\}).$$

The case of greatest importance for us is that where $B,C \in CAT \subset CAT/T_0$; and here it is particularly simple to give $\{B,C\}$. So for the moment we are looking at

(9.16) $$CAT(A\circ B,C) \cong CAT/T_0(A,\{B,C\})$$

for $A \in CAT/T_0$, $B,C \in CAT$.

The left side of (9.16) is the same as $[CAT,CAT](A\circ -,D)$, where D is the right 2-Kan-extension of $\ulcorner C\urcorner\colon I \to CAT$ along $\ulcorner B\urcorner\colon I \to CAT$, that is to say,

(9.17) $$D = [[-,B],C]\colon CAT \to CAT.$$

This is obvious without explicit reference to Kan extensions; we have using 2-Yoneda, and writing E for $A\circ -$,

(9.18) $$[CAT,CAT](E,[[-,B],C]) \cong \text{2-nat}(En,[[n,B],C]) \cong \text{2-nat}([n,B],[En,C]) \cong [EB,C] = CAT(A \circ B,C).$$

It follows from Theorem 9.1 that we have (9.16) where

(9.19) $$\{B,C\} = \Psi D = \Psi[[-,B],C].$$

In view of the definition (9.6) of Ψ, we have the following description of $\{B,C\}$. An object consists of $n \in T_0$ together with a functor $T: [n,B] \to C$; which latter we may equally write as $T: B^n \to C$. A morphism consists of $\phi: n \to n'$ in T_0 together with a natural transformation

(9.20)

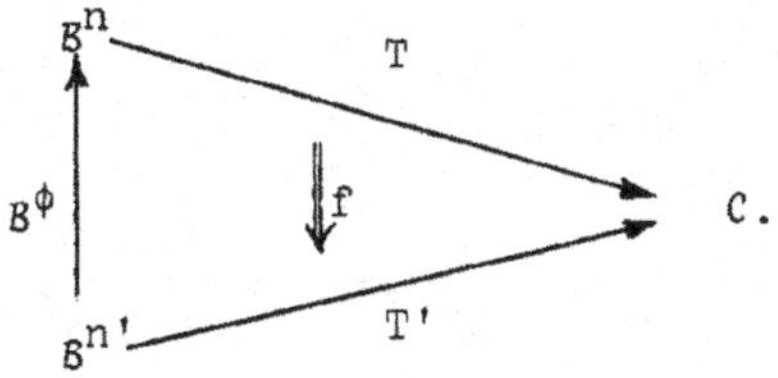

This last is precisely what we called, in §1.3 in the case $T = \underline{\underline{P}}$, or in §1.4 in the case $T = \underline{\underline{S}}$, a <u>natural transformation</u> $T \to T'$ <u>of type</u> $\phi : n \to n'$. So $\{B,C\}$ is a "rich functor category", with objects $(n, T: B^n \to C)$ and morphisms (ϕ, f) as in (9.20). The augmentation $\{B,C\} \to T_0$ of course sends (n,T) to n and (ϕ,f) to ϕ.

In fact one scarcely needs to appeal to Theorem 9.1, whose proof after all we have only sketched, in this case. With $\{B,C\}$ defined by (9.19), we know from (6.28) that a functor $H: A \to \{B,C\}$ in CAT/T_0 corresponds to an $\alpha: I! \rightsquigarrow [[\Gamma -,B],C]$; the components

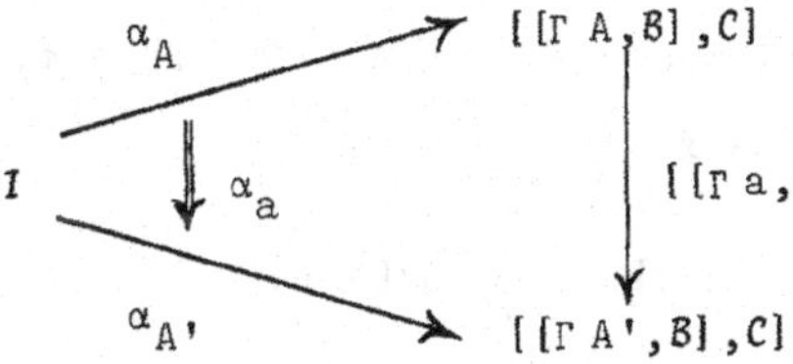

clearly may be written as

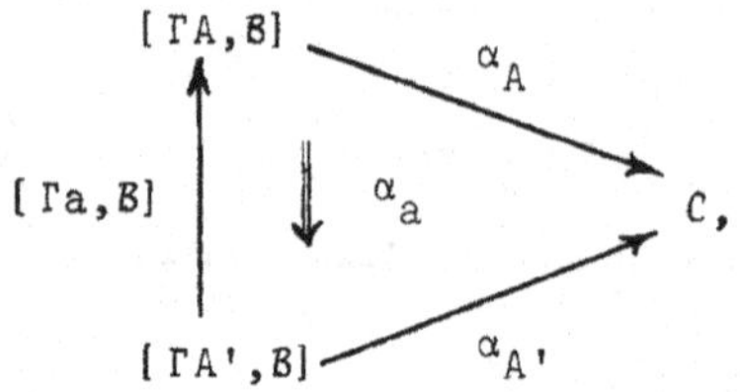

and evidently correspond to a functor $F\colon A \circ B \to C$, sending $A[X]$ to $\alpha_A X$, etc. We leave the reader to verify the following explicit description of the isomorphism (9.16), which can be used to provide a third, totally direct and elementary, proof of the isomorphism.

<u>9.3</u> The $F\colon A \circ B \to C$ and the corresponding $H\colon A \to \{B, C\}$ are connected as follows. H sends $a\colon A \to A'$, where $\Gamma A = n$, $\Gamma A' = n'$, $\Gamma a = \phi$, to (9.20) where

$$TX = F(A[X]), \quad Tx = F(1_A[x]), \quad f_{X'} = F(a[1_{X'\phi}]). \tag{9.21}$$

To give F in terms of H, let H send $a\colon A \to A'$ to (9.20). Then (9.21) gives F on objects; it also gives F on morphisms since the generic morphism $a[x]$ of (5.2) is the composite

$$A[X] \xrightarrow[1_A[x]]{} A[X'\phi] \xrightarrow[a[1_{X'\phi}]]{} A'[X']. \tag{9.22}$$

This suggests making an abbreviation which goes against the usual convention that the name of an object can be used as the name of its identity morphism. We do this because, especially in §10 below, the notational advantages outweigh the disadvantages.

We elect then to abbreviate the morphism $1_A[x]$ of $A \circ B$ to $A[x]$, but the morphism $a[1_{X'\phi}]$, where $a\colon A \to A'$ with $\Gamma a = \phi$, to $a[X']$. Then (9.22) gives the generic $a[x]$ as

$$A[X] \xrightarrow[A[x]]{} A[X'\phi] \xrightarrow[a[X']]{} A'[X'], \tag{9.23}$$

and (9.21) gets re-written as

(9.24) $\quad TX = F(A[X])$, $Tx = F(A[x])$, $f_{X'} = F(a[X'])$.

Next, we consider a natural transformation $\rho\colon F \Rightarrow \bar{F}$, and the corresponding $\sigma\colon H \Rightarrow \bar{H}$ in CAT/T_0. Writing again T for HA and $\bar{T}$ for $\bar{H}A$, we have

(9.25)

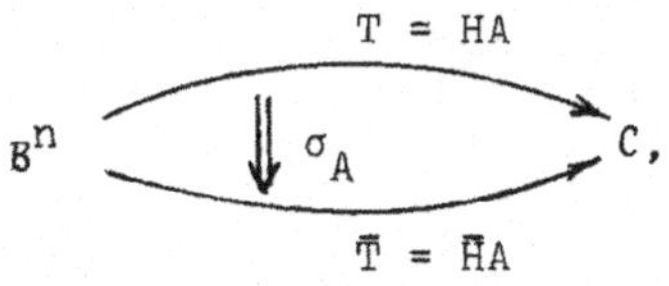

and it is given by

(9.26) $\quad (\sigma_A)_X = \rho_{A[X]}$.

9.4 We now give $\{B,C\}$, or more exactly $\{(B,\Delta),(C,E)\}$, in the general case where (B,Δ), $(C,E) \in CAT/T_0$. Just for the moment write $\overline{\{B,C\}}$ for the original $\{B,C\}$ of §9.2 when the augmentations of B and C are discarded (or replaced by the trivial ones), so that $B,C \in CAT$. Then $\{B,C\} = \{(B,\Delta),(C,E)\}$ is to be a subcategory of $\overline{\{B,C\}}$.

Recall that we have $\Theta\colon T\int T \to T$. For $n \in T_0$ define a functor $Y_n\colon {T_0}^n \to T_0$ by

(9.27) $\quad Y_n(m) = n(m)$, $\quad Y_n(\mu) = n(\mu) = 1_n(\mu)$,

where $\mu\colon m \Rightarrow m'$ in ${T_0}^n$. For $\nu\colon n \to n'$ in T_0 define a natural transformation

(9.28)

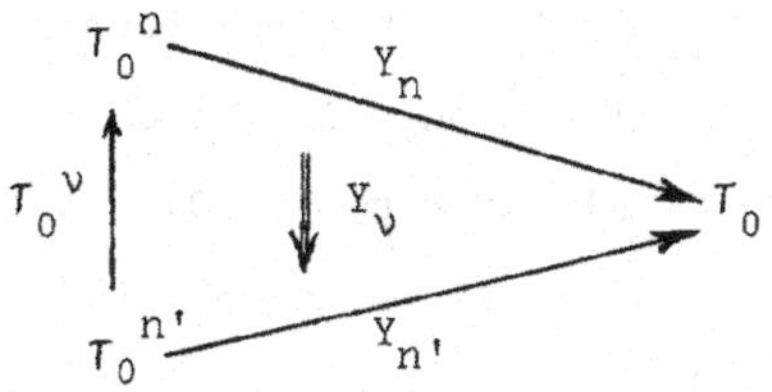

whose m'-component $(Y_\nu)_{m'}\colon n(m'\nu) \to n'(m')$, where $m'\colon n' \to T_0$, is given by

(9.29) $\quad (Y_\nu)_{m'} = \nu(m') = \nu(1_{m'\nu})$.

Note that in (9.27) and (9.29) we have used the same conventions with round brackets as we introduced in §9.3 with square ones: $1_n(\mu)$ is written $n(\mu)$ but $\nu(1_{m'\nu})$ as $\nu(m')$.

Now an object of $\{B,C\}$ is a pair n,T where $n \in T_0$ and where $T: B^n \to C$ renders commutative

(9.30)

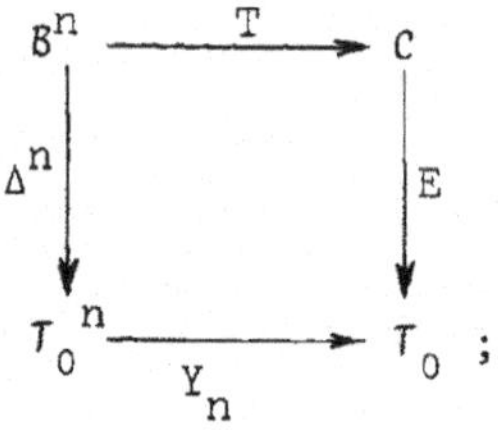

a morphism from n,T to n',T' consists of $\phi: n \to n'$ in T_0 and a natural transformation f as in (9.20), satisfying

(9.31)

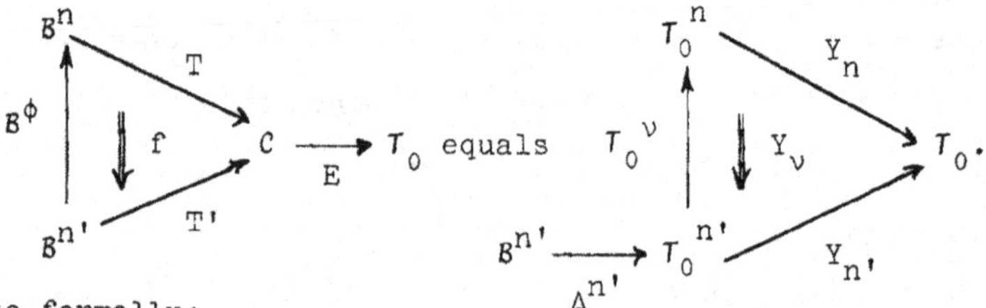

We now state formally:

<u>Theorem 9.2</u> <u>There is a 2-adjunction</u>

(9.24) $\qquad CAT/T_0(A \circ B, C) \cong CAT/T_0(A, \{B,C\})$. □

We have not given the proof in the general case; it is a very easy matter to get it once we have the special case where $B,C \in CAT$; one has only to check that a functor $A \circ B \to C$ (or a natural transformation) is consistent with the augmentations if and only if the corresponding $A \to \overline{\{B,C\}}$ lands in the subcategory $\{B,C\}$ given by (9.30) and (9.31). A detailed treatment in the case $T_0 = \underline{P}$ is given in [5], to which we refer any reader wanting more details.

Note of course that, while the $\circ$ of CAT/T_0 does not depend on T_0, being just the restriction of that of $CAT\int{}^*cat$, the internal-hom $\{B,C\}$ depends heavily on T_0. Of course if i is the inclusion $T_0 \subset T_0'$, then $\{B,C\}$ is just $i^*\{B,C\}'$ in the sense of the second paragraph of §9.1.

9.5 We end this section with some examples, or rather counter--examples, before passing in the next section to the still-more-special properties of $CAT/\underline{\underline{S}}$.

The terminal object in any monoidal category is a monoid; hence T_0, with its identity augmentation, is a club in CAT/T_0, with multiplication the θ of (8.5), and with identity the inclusion $J \to T$. Thus we have as examples of clubs cat, set, ord, $\underline{\underline{S}}, \underline{\underline{P}}, \underline{\underline{N}}$. Other examples of clubs in $CAT/\underline{\underline{P}}$ were given in [6]. If K is any club in CAT/T_0, then $\Gamma\colon K \to T_0$ is a map of monoids in CAT/T_0, inducing a doctrine-map $\Gamma\circ -\colon K\circ - \Rightarrow T_0\circ -$.

For an example of a club not in CAT/cat_0 we can take any $n \in cat$ and set $K = (I, \ulcorner n \urcorner)$. Then $K\circ A \cong A^n$, and $K\circ -$ is the 2-functor $A \mapsto A^n$. We know that this becomes a doctrine if we define its multiplication $(A^n)^n \to A^n$ and its unit $A \to A^n$ via the diagonal $n \to n \times n$ and the unique map $n \to 1$. The corresponding multiplication $\mu\colon K\circ K \to K$ and identity $\eta\colon J \to K$ for K certainly do not lie in CAT/cat_0.

Now we give an example of two clubs (K,Γ), (L,Δ) in $CAT/\underline{\underline{S}}$, indeed in $CAT/\underline{\underline{N}}$, and a doctrine-map $K\circ - \Rightarrow L\circ -$, where the $(T,\tau)\colon (K,\Gamma) \to (L,\Delta)$ inducing this does not lie in $CAT/\underline{\underline{S}}$, but only in $CAT\!\int\!{}^*\underline{\underline{S}}$ (and not in $CAT\!\int\!{}^*\underline{\underline{N}}$ or $CAT\!\int\!{}^*\underline{\underline{P}}$ either). We take $K = L = \underline{\underline{N}}$, with $\Gamma\colon \underline{\underline{N}} \to \underline{\underline{S}}$ as the constant functor at 1, and with $\Delta\colon \underline{\underline{N}} \to \underline{\underline{S}}$ as the inclusion. The 2-functor $K\circ -$ is $\underline{\underline{N}}\times -$; it becomes a doctrine via the sum $+\colon \underline{\underline{N}} \times \underline{\underline{N}} \to \underline{\underline{N}}$ and $\ulcorner 0 \urcorner\colon I \to \underline{\underline{N}}$; a K-structure on A consists in giving a single endofunctor $P\colon A \to A$. On the other hand L is the club (see [6]) whose algebras are strict monoidal categories. From an L-structure on A we get a K-structure on A by setting $PA = A\otimes A$. This algebraic 2-functor must arise from a doctrine-map $K\circ - \Rightarrow L\circ -$, and hence from a club-map $(T,\tau)\colon K \to L$. The latter is easily seen to be

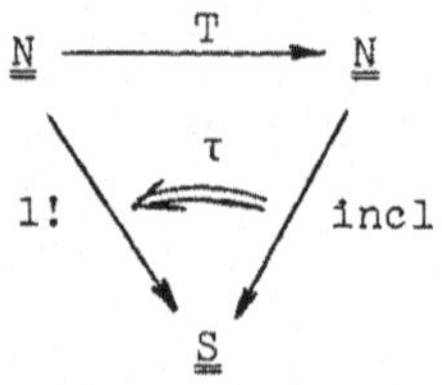

where $Tn = 2^n$ and τ_n is the unique map $2^n \to 1$; it lies in $CAT\int{}^*\underline{\underline{S}}$ but neither in $CAT/\underline{\underline{S}}$ nor in $CAT\int{}^*\underline{\underline{P}}$.

In general, by a club-map we mean a map of $\circ$-monoids in $CAT\int{}^*cat$. The above shows that, even if the clubs in $CAT/\underline{\underline{S}}$, to which we now turn, are of especial importance, we cannot handle all club maps without going to $CAT\int{}^*\underline{\underline{S}}$. However in some cases a club map too can be seen from the basic data to lie in $CAT/\underline{\underline{S}}$.

10. CLUBS IN CAT/S

10.1 We now return to this most important case, referred to in §1.4 above as arising in coherence problems, along with its subcases $CAT/\underline{\underline{P}}$ and $CAT/\underline{\underline{N}}$ (and, for that matter, $CAT/I = CAT$ itself; here $\circ$ reduces to cartesian product $\times$). As a matter of notation, we write $A[X] \in A\circ B$ in the form $A[X_1, \ldots, X_n]$ when $\Gamma A = n$; and we write the $a[x]$ of (5.2), when $\Gamma a = \phi\colon n \to n'$, as

$$a[x_1, \ldots, x_n]\colon A[X_1, \ldots, X_n] \to A'[X'_1, \ldots, X'_{n'}]; \tag{10.1}$$

here

$$x_i\colon X_i \to X'_{\phi i}. \tag{10.2}$$

We replace the square brackets by round ones to denote the image of $A[X]$ under an action $K\circ A \to A$, and in particular under a multiplication $K\circ K \to K$; so the image of $A[X]$ becomes $A(X)$, or in more detail that of $A[X_1, \ldots, X_n]$ becomes $A(X_1, \ldots, X_n)$, and similarly for morphisms. This agrees with our "parenthesis notation" for Θ; since n(m) or

$n(m_1, \ldots, m_n) = m_1 + \ldots + m_n$ is indeed the image of $n[m]$ under the multiplication θ: $\underline{\underline{S}} \circ \underline{\underline{S}} \to \underline{\underline{S}}$. When we have a club K, we also write $\underline{\underline{1}}$ for the image of the unit η: $J \to K$; the $\underline{\underline{1}}$ of $\underline{\underline{S}}$ is just 1. Similarly we have the notation $\phi(\psi_1 \ldots \psi_n): n(m_1 \ldots m_n) \to n'(m'_1 \ldots m'_{n'})$ for the image under θ of morphisms, where ϕ: $n \to n'$ and ψ_i: $m_i \to m'_{\phi i}$.

In accordance with the conventions of §9.3 we abbreviate

$$(10.3) \qquad 1_A[x_1 \ldots x_n]: A[X_1 \ldots X_n] \to A[X'_1 \ldots X'_n]$$

to

$$(10.4) \qquad A[x_1 \ldots x_n]: A[X_1 \ldots X_n] \to A[X'_1 \ldots X'_n]$$

where x_i: $X_i \to X'_i$; and we abbreviate

$$(10.5) \qquad a[1_{X'_{\phi 1}} \ldots 1_{X'_{\phi n}}]: A[X'_{\phi 1} \ldots X'_{\phi n}] \to A'[X'_1 \ldots X'_{n'}]$$

to

$$(10.6) \qquad a[X'_1 \ldots X'_{n'}]: A[X'_{\phi 1} \ldots X'_{\phi n}] \to A'[X'_1 \ldots X'_{n'}].$$

We make also the corresponding abbreviations with round brackets, as we did in §9.4. As in (9.22), the typical morphism (10.1) then appears as the composite

$$(10.7) \qquad A[X_1 \ldots X_n] \xrightarrow[A[x_1 \ldots x_n]]{} A[X'_{\phi 1} \ldots X'_{\phi n}] \xrightarrow[a[X'_1 \ldots X'_{n'}]]{} A'[X'_1 \ldots X'_{n'}].$$

The reader will have no trouble expressing (9.24) and (9.26) in this expanded notation.

Since the special case $CAT/\underline{P}$ has been treated in detail in [5] and [6], and since there is not much to change, I shall give only an outline in this section. There are some mild notational differences from [5] and [6]; there ϕ was always an isomorphism and (10.2) was replaced by x_i: $X_{\phi^{-1} i} \to X'_i$, which is impossible here. That, however, had the effect of making (10.6) conformable with the convention that X' stands for $1_{X'}$; here we have to sacrifice something and we choose it to be this.

<u>10.2</u> What makes the present case specially simple is the discreteness of the images ΓA under the augmentation; in fact much of what we say below would apply equally well to CAT/set, but we stick to the most practical case of finitary operations.

Let us agree that for this section "club" means "club in $CAT/\underline{\underline{S}}$ ". From the above discreteness it follows that if K is a club, so is its set of objects $|K|$, with of course the same "multiplication" $T[S_1, \dots, S_n] \mapsto T(S_1, \dots, S_n)$ for objects and the same identity $\underline{\underline{1}}$; and it follows also that the inclusion $|K| \to K$ is a club map in $CAT/\underline{\underline{S}}$.

A club which is discrete as a category is called a <u>discrete club</u>; it may be regarded as a $\circ$-monoid in $SET/\underline{\underline{N}}$. (Of course there are clubs in $CAT/\underline{\underline{N}}$ that are <u>not</u> discrete). The further consequence of the above is that, in constructing a club K from generators and relations, we can construct $|K|$ first and settle the objects before starting on the morphisms. It is this that is false for CAT/cat_0, and it is this that forms the basis of all that we do in this section.

<u>10.3</u> Let any object (P, Γ) of $SET/\underline{\underline{N}}$ be given. It is easy to construct the free discrete club (Q, Γ) it generates - which is also automatically the free club it generates. We construct the set Q inductively by

(10.8) $\underline{\underline{1}} \in Q$;

(10.9) if $P \in P$ with $\Gamma P = n$ and if $Q_1, \dots, Q_n \in Q$ then $P(Q_1 \dots Q_n) \in Q$.

We embed P in Q by identifying P with $P(\underline{\underline{1}} \dots \underline{\underline{1}})$, and extend Γ to Q by $\Gamma\underline{\underline{1}} = 1$, $\Gamma P(Q_1 \dots Q_n) = n(\Gamma Q_1, \dots, \Gamma Q_n)$. To save repetition, let us agree that until further notice "club map" means one in $CAT/\underline{\underline{S}}$ (i.e. <u>over</u> $\underline{\underline{S}}$), and not a general one in $CAT/\!\!\int^*\underline{\underline{S}}$. Let us also agree that all augmentations are denoted by Γ if no other name is given. It is clear that for any club L and any morphism $P \to L$ over $\underline{\underline{S}}$, there is a unique extension to a club map $Q \to L$.

Now let ρ be a relation on $\mathcal{Q}$ such that $T\rho S$ implies $\Gamma T = \Gamma S$, and consider club-maps $M : \mathcal{Q} \to \mathcal{L}$ such that $T\rho S$ implies $MT = MS$. It is clear that these factorize uniquely into club maps $\mathcal{Q} \to \mathcal{S} \to \mathcal{L}$, where $\mathcal{S}$ is the quotient discrete club of $\mathcal{Q}$ obtained by first enlarging ρ to a congruence $\bar{\rho}$ with respect to the operation $T(S_1,\ldots,S_n)$ of substitution, and then setting $\mathcal{S} = \mathcal{Q}/\bar{\rho}$. We can call $\mathcal{S}$ the discrete club with generators $\mathcal{P}$ and relations ρ.

A model of $\mathcal{P}$ is a category A together with, for each $P \in \mathcal{P}$, a functor $|P| : A^n \to A$, where $n = \Gamma P$. This is clearly the same thing as a morphism $\mathcal{P} \to \{A, A\}$ over $\underline{\underline{S}}$; the endo-internal-hom $\{A, A\}$ is of course always canonically a club; and so a model of $\mathcal{P}$ is the same thing as a club-map $\mathcal{Q} \to \{A, A\}$, or again as an action $\mathcal{Q} \circ A \to A$; so it is just a $\mathcal{Q}$-algebra or $\mathcal{Q}$-category.

A model of $\mathcal{P},\rho$ is such a model A of $\mathcal{P}$ satisfying $|T| = |S| : A^n \to A$ whenever $T\rho S$. It is therefore an $\mathcal{S}$-category for the discrete club $\mathcal{S}$.

The most obvious example is that of a strict monoidal category A; here $\mathcal{P}$ has objects $\otimes$ and I with $\Gamma\otimes = 2$ and $\Gamma I = 0$, and $\mathcal{Q}$ is subjected to relations ρ identifying $\otimes(\otimes,\underline{\underline{1}})$ with $\otimes(\underline{\underline{1}}, \otimes)$, $\otimes(I,\underline{\underline{1}})$ with $\underline{\underline{1}}$, and $\otimes(\underline{\underline{1}},I)$ with $\underline{\underline{1}}$; the corresponding discrete club $\mathcal{S}$ is $\underline{\underline{N}}$ itself, with identity augmentation.

10.4 Given a discrete club $\mathcal{S}$, whether by generators and relations or directly, let us be given a graph $\mathcal{D}$ (in the usual sense of graph - objects and arrows - not that of [5] and [6]) with object-set $|\mathcal{D}| = \mathcal{S}$, and let us be given a morphism of graphs $\Gamma : \mathcal{D} \to \underline{\underline{S}}$ extending $\Gamma : \mathcal{S} \to \underline{\underline{S}}$. We might as well now replace $\mathcal{S}$ by $|\mathcal{D}|$. Consider for a club $\mathcal{L}$ those graph-morphisms $\mathcal{D} \to \mathcal{L}$ over $\underline{\underline{S}}$ such that the composite $|\mathcal{D}| \to \mathcal{D} \to \mathcal{L}$ is a club-map. I assert that there is a universal such map $\mathcal{D} \to \mathcal{M}$ into a club $\mathcal{M}$, so that any such $\mathcal{D} \to \mathcal{L}$ factors through a unique club-map $\mathcal{M} \to \mathcal{L}$; we can call $\mathcal{M}$ the club generated by $\mathcal{S}$ and $\mathcal{D}$. In particular a <u>model</u> of $(S,\mathcal{D})$ means an S-category A together with, for each $f : T \to S$ in $\mathcal{D}$, a <u>natural transformation</u> $|f| : |T| \to |S|$ <u>of type</u> Γf, in the sense of (9.20). So such a model A is just an M-category for the club M.

To construct M we begin by enlarging $\mathcal{D}$ twice, without changing its objects; first we define an <u>instance</u> of an $f : T \to S$ in $\mathcal{D}$ to be a formal expression

(10.10) $$f(R_1 \ldots R_{n'}) : \; T(R_{\phi 1} \ldots R_{\phi n}) \to S(R_1 \ldots R_{n'})$$

where $\phi = \Gamma f$, and where the objects $T(R_{\phi 1} \ldots R_{\phi n})$ etc. are of course not formal but are the objects of $|\mathcal{D}|$ so denoted. We identify f with $f(\underline{\underline{1}} \ldots \underline{\underline{1}})$, and so the instances form a bigger graph $\mathcal{D}'$ extending $\mathcal{D}$, to which we also extend Γ.

We next, for an instance $g : T \to S$ in $\mathcal{D}'$, define an <u>expansion</u> of g, i.e. an <u>expanded instance</u> of some f, to be an expression

(10.11) $$P(1 \ldots g \ldots 1) : P(\underline{\underline{1}} \ldots T \ldots \underline{\underline{1}}) \to P(\underline{\underline{1}} \ldots S \ldots \underline{\underline{1}})$$

for some $P \in |\mathcal{D}|$. We identify g itself with $\underline{\underline{1}}(g)$, and now have a still bigger graph $\mathcal{D}''$, to which we again extend Γ.

We then pass to $\mathcal{D}'''$, the free category generated by the graph $\mathcal{D}''$, to which we again extend Γ. The category $\mathcal{D}'''$ is not yet a club, but a <u>partial</u> structure of $\circ$-monoid can be defined on it; and it is clear that any morphism $\mathcal{D} \to L$ over $\underline{\underline{S}}$ into a club L, whose restriction $|\mathcal{D}| \to L$ is a club-map, extends uniquely to a functor $\mathcal{D}''' \to L$ which respects the partial club-structure on $\mathcal{D}'''$. There are two sorts of diagrams in $\mathcal{D}'''$ which are necessarily sent into commutative diagrams by $\mathcal{D}''' \longrightarrow L$; they are typified by

$$(10.12)\qquad \begin{array}{ccc} T(S_1\ldots S_i\ldots S_j\ldots S_n) & \xrightarrow{T(1\ldots h\ldots 1\ldots 1)} & T(S_1\ldots S_i'\ldots S_j\ldots S_n) \\ \Big\downarrow{\scriptstyle T(1\ldots 1\ldots g\ldots 1)} & & \Big\downarrow{\scriptstyle T(1\ldots 1\ldots g\ldots 1)} \\ T(S_1\ldots S_i\ldots S_j'\ldots S_n) & \xrightarrow[T(1\ldots h\ldots 1\ldots 1)]{} & T(S_1\ldots S_i'\ldots S_j'\ldots S_n) \end{array}$$

and by

$$(10.13)\qquad \begin{array}{ccc} T(S_{\phi 1}\ldots\ \ldots\ \ldots\ S_{\phi n}) & \xrightarrow{f(S_1\ldots S_i\ldots S_{n'})} & T'(S_1\ldots S_i\ldots S_{n'}) \\ \Big\downarrow{\scriptstyle T(1\ldots k\ldots k\ldots 1)} & & \Big\downarrow{\scriptstyle T'(1\ldots k\ldots 1)} \\ T(S_{\phi 1}\ldots\ \ldots\ \ldots\ S_{\phi n}) & \xrightarrow[f(S_1\ldots S_i'\ldots S_{n'})]{} & T'(S_1\ldots S_i'\ldots S_{n'}). \end{array}$$

Here (10.13) needs some explanation. The left vertical arrow stands for either leg of a diagram like (10.12); or the corresponding thing when there are more than two maps k in it; and these k's occur at all those indices j for which $\phi j = i$. The diagram (10.12) says in effect that T is "functorial", and (10.13) that f is "natural". It suffices in fact to impose these diagrams not for all $f,g,h,k \in \mathcal{D}'''$ but for $f \in \mathcal{D}''$, $g,h,k \in \mathcal{D}'$; the more general cases then follow automatically. All this is rather more complicated than in the $\mathcal{CAT}/\underline{\underline{P}}$ case of [6].

If we write *Funct* for the class of diagrams of the form (10.12) and *Nat* for those of the form (10.13), and $\mathcal{D}'''/(Funct + Nat)$ for the quotient category with the same objects got by imposing these relations, it turns out that this last is the desired club $\mathcal{M}$; for it indeed admits a club-structure in which, in analogy with (10.7), $f(g_1\ldots g_n)$ is defined to be the composite

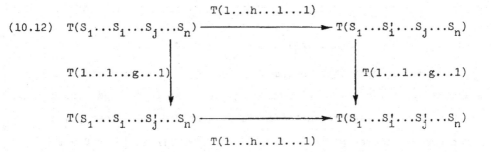

$$(10.14)\quad T(S_1 \dots S_n) \xrightarrow{T(g_1 \dots g_n)} T(S'_{\phi 1} \dots S'_{\phi n}) \xrightarrow{f(S'_1 \dots S'_{n'})} T'(S'_1 \dots S'_{n'}).$$

10.5 Now suppose that, besides the above, we are given a set σ of diagrams $f,g : T \to S$ in M with $\Gamma f = \Gamma g$; or, what is perhaps more usual, a set σ of such diagrams in $\mathcal{D}'''$; and we consider only those graph-morphisms $\mathcal{D} \to L$ over $\underline{\underline{S}}$ into a club L, still requiring $|\mathcal{D}| \to \mathcal{D} \to L$ to be a club-map, for which, in the induced $\mathcal{D}''' \to L$ or $M \to L$, the images of f and g coincide for each $f,g: T \to S$ in σ. There is again a universal such L, say K, a quotient-category of M, and itself a club, such that any such map factorizes as $\mathcal{D} \to M \to K \to L$ for a unique club-map $K \to L$. In fact K is obtained from M by imposing as extra relations all the expanded instances of the relations σ; or better, is obtained as

$$(10.15)\qquad K = \mathcal{D}'''/(Funct + Nat + Imp),$$

where Imp, standing for the imposed relations, consists of all the expanded instances of the relations σ.

We may in fine call K the club generated by the function-generators P, with relations ρ, and the natural-transformation-generators $\mathcal{D}$, with the relations σ.

A model A of $(P,\rho,\mathcal{D},\sigma)$ is of course a model of $(P,\rho,\mathcal{D})$ satisfying the relations σ; it is the same thing as a K-category A for the club K.

A typical example is that of symmetric monoidal categories, where P consists of $\otimes$ and I with $\Gamma\otimes = 2$ and $\Gamma I = 0$; where ρ is vacuous; where $\mathcal{D}$ consists of

$a : \otimes(\otimes,\underline{\underline{1}}) \to \otimes(\underline{\underline{1}},\otimes)$, $\quad \bar{a} : \otimes(\underline{\underline{1}},\otimes) \to \otimes(\otimes,\underline{\underline{1}})$,

$r : \otimes(\underline{\underline{1}},I) \to \underline{\underline{1}}$, $\quad \bar{r} : \underline{\underline{1}} \to \otimes(\underline{\underline{1}},I)$, and

$c : \otimes \to \otimes$, with Γa, $\Gamma\bar{a}$, Γr, $\Gamma\bar{r}$ identities and with Γc the non-identity permutation of 2; and where σ consists of the usual coherence axioms, together with $a\bar{a} = 1$, $\bar{a}a = 1$, $r\bar{r} = 1$, $\bar{r}r = 1$, establishing $\bar{a},\bar{r}$ as inverses of a and r. Here the club K is in fact in $CAT/\underline{\underline{P}}$;

other examples leading to clubs in $CAT/\underline{\underline{S}}$ are suggested in §1.4 above.

<u>10.6</u> It should now be clear that a complete solution to the "coherence problem" for such a structure (P,ρ,D,σ) consists in the explicit determination of the club K. To find the objects of K is already a word-problem, but usually easy, and trivial if ρ is vacuous. To find the morphisms is a word-problem that is usually much harder. We know the generators, and it is a question of knowing when $f,g : T \to S$ in D''' coincide in K because of the relations in (10.15); that is, of knowing "which diagrams commute". A necessary condition is of course $\Gamma f = \Gamma g$; this is the condition for the writability of a diagram, in terms of generic components of the "natural transformations" f and g. For example, in the symmetric-monoidal-category case, we have $c,1 : \otimes \to \otimes$ with $\Gamma c \neq \Gamma 1$; and the generic components

(10.16)

$$A \otimes B \xrightarrow{\;c\;} B \otimes A \qquad A \otimes B \xrightarrow{\;1\;} A \otimes B$$

do not form a closed diagram. The "all diagrams commute" case is that where $\Gamma f = \Gamma g$ is also <u>sufficient</u> for $f = g$, i.e. where Γ is faithful. It is of course a rare case.

In (10.16) the <u>special</u> components $c,1 : A \otimes A \to A \otimes A$ form a closed diagram, but this does not commute in a typical model, and makes no sense in the club K. If one wants to know which such <u>specialized</u> diagrams commute, one has only to look at the free model $K \circ A$ on A, which of course we also know completely when we know K.

<u>10.7</u> The considerations of §§10.3 - 10.5 assert that a "map" $(P,\rho,D,\sigma) \to L$ of the kind described there leads to a club map $K \to L$ in our current sense of the word, that is to say, in $CAT/\underline{\underline{S}}$. This situation may be recognized even when L itself is given only by generators and relations (P',ρ',D',σ'). Thus we are able without calculation to conclude for example the existence of a club map $K \to L$

over $\underline{S}$ when K is the club for monoidal categories and L that for symmetric monoidal categories; or when K is that for symmetric monoidal categories and L is that for two symmetric monoidal structures together with a distributive law, as in (1.1) above. This is often very handy, in view of the observation in §9.5 that not <u>all</u> club-maps $K \to L$ for clubs K,L in $CAT/\underline{S}$ themselves lie in $CAT/\underline{S}$. We may also observe in [8] below that we have to leave $CAT/\underline{S}$ for $CAT\!\int\!*\underline{S}$ when considering a <u>distributive law</u> $K \circ L \to L \circ K$ between clubs that themselves lie in $CAT/\underline{S}$.

<u>10.8</u>   A further application of the considerations of §§10.3 - 10.5 is in connection with K-morphisms $(\phi,\bar{\phi}) : A \to A'$ between K-categories A and A'; that is, for (lax) D-morphisms, in the sense of [12]§3.5 in this volume, where D is the doctrine $K \circ -$ for a club K in $CAT/\underline{S}$.

Denote by FUN the 2-category which is like $[\underline{2}, CAT]$ except that the morphisms are not lax n.t.'s but op-lax n.t.'s; if we took $[\underline{2}, CAT]$ itself we should get op-K-morphisms instead of K-morphisms. So an object of FUN is a functor $\phi : A \to A'$, and a morphism is θ,θ',α of the form

(10.17)

$$
\begin{array}{ccc}
A & \xrightarrow{\theta} & B \\
\phi \downarrow & \overset{\alpha}{\Longrightarrow} & \downarrow \psi \\
A' & \xrightarrow[\theta']{} & B'
\end{array}
\quad ,
$$

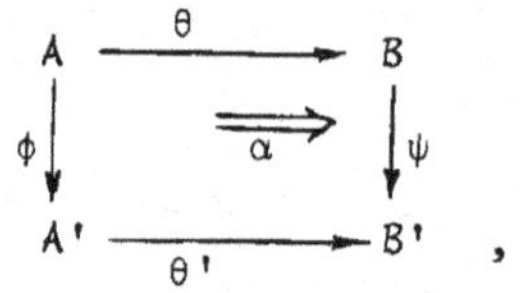

while a 2-cell is γ,γ' where $\gamma : \theta \Rightarrow \bar{\theta}$ and $\gamma' : \theta' \Rightarrow \bar{\theta}'$ satisfy

(10.18)

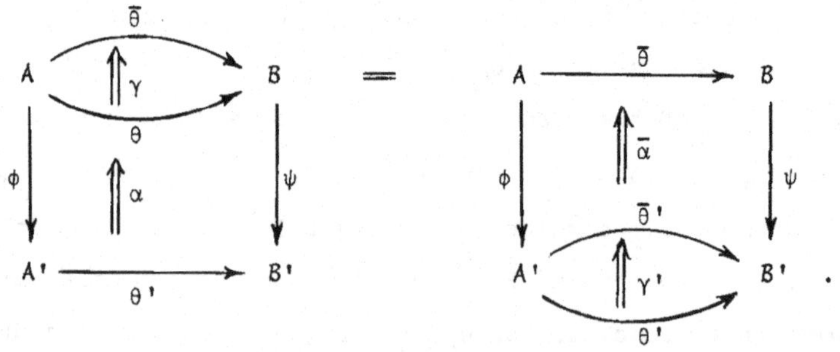

There is an evident 2-functor

(10.19) $\quad \hat{\circ} : \mathit{CAT}/\underline{\underline{S}} \times \mathit{FUN} \to \mathit{FUN}$

sending K, ϕ to $K\circ\phi$, sending $T : K \to K'$ and $(\theta, \theta', \alpha) : \phi \to \phi'$ to $(T\circ\theta,\ T\circ\theta',\ T\circ\alpha)$, and sending $\eta : T \Rightarrow \bar{T}$ and $(\gamma,\gamma') : (\theta,\ \theta',\alpha) \Rightarrow (\bar{\theta},\bar{\theta}',\bar{\alpha})$ to $(\eta\circ\gamma,\ \eta\circ\gamma')$. So there is really no need to distinguish notationally $\hat{\circ}$ from $\circ$. It is moreover clear that (10.19) provides an <u>action</u> of $\mathit{CAT}/\underline{\underline{S}}$ on FUN, in the sense that $(K\circ L)\hat{\circ}\ \phi \cong K\hat{\circ}(L\hat{\circ}\phi)$, etc. It is finally clear that we have a 2-adjunction

(10.20) $\quad \mathit{FUN}(K\hat{\circ}\phi,\psi) \cong \mathit{CAT}/\underline{\underline{S}}\ (K,<\phi,\psi>)$,

where $<\phi,\psi>$ denotes the comma object

(10.21) $\quad <\phi,\psi> \ = \{\phi,1\}/\{1,\psi\}$

in $\mathit{CAT}/\underline{\underline{S}}$, as in

(10.22)
$$\begin{array}{ccc} <\phi,\psi> & \xrightarrow{\ d_1\ } & \{A,B\} \\ {\scriptstyle d_0}\downarrow & \overset{\lambda}{\nearrow\!\!\!\nearrow} & \downarrow{\scriptstyle \{1,\psi\}} \\ \{A',B'\} & \xrightarrow[\ \{\phi,1\}\ ]{} & \{A,B'\} \end{array} \quad .$$

Thus FUN is exhibited as a tensored $\mathit{CAT}/\underline{\underline{S}}$ - category.

An <u>action</u> (θ,θ',α): $K\hat{\circ}\phi \to \phi$ of the club K in $\mathit{CAT}/\underline{\underline{S}}$ on $\phi \in \mathit{FUN}$ is clearly the same thing, by §3.5 of [12], as actions $\theta : K\circ A \to A$ and $\theta' : K\circ A' \to A'$, making A and A' into K-categories, together with an α,

(10.23)
$$\begin{array}{ccc} K\circ A & \xrightarrow{\ \theta\ } & A \\ {\scriptstyle K\circ\phi}\downarrow & \overset{\Longrightarrow}{\alpha} & \downarrow{\scriptstyle \phi} \\ K\circ A' & \xrightarrow[\ \theta'\ ]{} & A' \end{array} ,$$

making of (ϕ,α) a K-functor $A \to A'$. Such an action then is a club-map $K \to <\phi,\phi>$ over $\underline{\underline{S}}$ where $<\phi,\phi>$ has its evident club-structure.

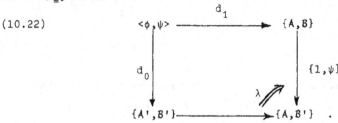

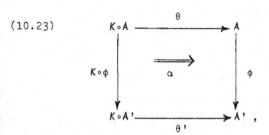

To give α when θ,θ' are already given is to give such a $K \to \langle\phi,\phi\rangle$ whose composites with d_0 and d_1 are the given club-maps $K \to \{A',A'\}$ and $K \to \{A,A\}$.

Such an α has components

(10.24)
$$\alpha_{T[A_1\ldots A_n]} : T(\phi A_1\ldots\phi A_n) \to \phi\, T(A_1\ldots A_n)$$

which are to be natural in T and in the A_i, and the axioms (3.17), (3.18) of [12] become

(10.25)
$$\alpha_{T(S_1\ldots S_n)[A_1\ldots A_m]} = \alpha_{T[S_1(A_1\ldots A_{m_1})\ldots S_n(\ldots A_m)]} \cdot T(\alpha_{S_1[A_1\ldots A_{m_1}]}\ldots\alpha_{S_n[\ldots A_m]});$$

(10.26)
$$\alpha_{\underline{1}[A]} = 1.$$

If now K is given by generators and relations (P,ρ,D,α), the description of the structure as a club-map $K \to \langle\phi,\phi\rangle$ shows how to give α in terms of these generators and relations. First, one has only to give $\alpha_T = \alpha_{T[A_1\ldots A_n]}$ for $T \in P$, and it is of course to be natural in the A_i. One then gets the general α_T by (10.25) and (10.26). To a relation $T\rho S$ one must impose the axiom $\alpha_T = \alpha_S$. It remains to impose as axioms the naturality of α_T in T only for morphisms $f : T \to S$ in D, and one is finished.

In this way one sees for instance that, when K is the club either for monoidal categories or for strict monoidal categories, a K-functor is a monoidal functor in the usual sense (not a strict one in the strict monoidal category case; the axioms from the identifications $\otimes(\otimes,\underline{1}) = \otimes(\underline{1},\otimes)$ in this case are the same as the axioms expressing naturality in a: $\otimes(\otimes,\underline{1}) \to \otimes(\underline{1},\otimes)$ in the non-strict case, etc.) Similarly for symmetric monoidal functors, monad functors, and other well-known cases.

In the same way we see that for a K-natural-transformation $\eta\colon (\phi,\alpha) \Rightarrow (\psi,\beta)\colon \;A \to A'$, we need impose on the natural transformation $\eta\colon \;\phi \Rightarrow \psi$ the axiom

(10.27)
$$\begin{array}{ccc} T(\phi A_1 \ldots \phi A_n) & \xrightarrow{\alpha_{T[A_1\ldots A_n]}} & \phi T(A_1\ldots A_n) \\ {\scriptstyle T(\eta_{A_1}\ldots\eta_{A_n})}\downarrow & & \downarrow{\scriptstyle \eta_{T(A_1\ldots A_n)}} \\ T(\psi A_1 \ldots \psi A_n) & \xrightarrow[\beta_{T[A_1\ldots A_n]}]{} & \psi T(A_1\ldots A_n) \end{array} ,$$

representing the axiom (3.19) of [12], only for those objects T in P; as we commonly do for monoidal natural transformations.

<u>10.9</u> We conclude by noticing some other properties that seem to be special to the case of clubs in $CAT/\underline{\underline{S}}$, or perhaps CAT/set_0.

Let K then be a club in $CAT/\underline{\underline{S}}$. By a <u>closed</u> K-category we mean a K-category A such that, for each $T \in K$, with $\Gamma T = n$ say, and for each i, the functor $T(A_1\ldots A_{i-1} - A_{i+1}\ldots A_n)\colon \; A \to A$ has a right adjoint for each choice of the $A_j \in A$. Thus symmetric monoidal closed category, monoidal biclosed category, cartesian closed category, etc. The reader may like to check what such a thing is when a K-category is a category A bearing a monad, or a comonad.

I showed in [7] that, when K actually lies in $CAT/\underline{\underline{P}}$, the closed K-categories were the algebras for a club L of the mixed-variance kind of §1.2 above; in particular they are monadic over CAT. I left open there the question whether the canonical $K \to L$ in this case was faithful.

This question can be answered by using a Yoneda embedding, as announced in [11]. In fact, it is to clubs K in $CAT/\underline{\underline{S}}$ that the work of Day in [1] seems to extend.

For if A is a K-category for a K in $CAT/\underline{\underline{S}}$, we can extend the K-structure on A to one on $B = [A^{op}, Set]$, and indeed uniquely (to within isomorphism) if we ask B to be a closed K-category; for then the realization $|T|: B^n \to B$ of $T \in K$ must be the left Kan extension of its realization $|T|: A^n \to A$. This solves the above question when K lies in $CAT/\underline{\underline{P}}$; for if we take B to be $[K^{op}, Set]$ there is a unique strict L-map $L \to B$ taking $\underline{\underline{1}} \in L$ to the image under Yoneda of the $\underline{\underline{1}}$ of K; the strict K-map $K \to L \to B$ then coincides with the Yoneda embedding, since they have the same effect on $\underline{\underline{1}}$; whence $K \to L$ is faithful since the Yoneda embedding is so. (We are using of course that $K \cong K \circ I$ is the free K-category on I, i.e. on $\underline{\underline{1}} \in K$).

Note that the above Yoneda extension does not work for an arbitrary doctrine D, or even for a club K in say CAT/cat_0. For I is always a D-category, and yet $[I^{op}, Set] \cong Set$ need not be. But Set, or at any rate set, is certainly a K-category for a club K in CAT/set, since there is a club-map $\Gamma: K \to set$.

Of course the above is a little loose, since the lack of strict associativity for colimits in Set causes $B = [A^{op}, Set]$ to be only a pseudo-K-algebra in general; either we take account of this by comparing pseudo-algebras with algebras, as we shall to some extent do in [8] below; or we replace Set by an equivalent category admitting strictly associative colimits - I have not thought out the question of whether set has these.

REFERENCES

[1] B.J. Day, On closed categories of functors, Lecture Notes in Math. 137 (1970), 1-38.

[2] S. Eilenberg and G.M. Kelly, A generalization of the functorial calculus, Jour. of Algebra 3 (1966), 366-375.

[3] J.W. Gray, The categorical comprehension scheme, Lecture Notes in Math. 99 (1969), 242-312.

[4] J.R. Isbell, Math. Reviews 44 #278 (1972).

[5] G.M. Kelly, Many-variable functorial calculus. I., Lecture Notes in Math. 281 (1972), 66-105.

[6] G.M. Kelly, An abstract approach to coherence, Lecture Notes in Math. 281 (1972), 106-147.

[7] G.M. Kelly, A cut-elimination theorem, Lecture Notes in Math. 281 (1972), 196-213.

[8] G.M. Kelly, Coherence theorems for lax algebras and for distributive laws, in this volume.

[9] G.M. Kelly and S. Mac Lane, Coherence in closed categories, Jour. Pure and Applied Algebra 1 (1971), 97-140.

[10] G.M. Kelly and S. Mac Lane, Closed coherence for a natural transformation, Lecture Notes in Math. 281 (1972), 1-28.

[11] G.M. Kelly and R. Street (Editors), Abstracts of the Sydney Category Theory Seminar 1972, mimeographed (originally by School of Mathematics, Univ. of New South Wales, 1972; second printing (1973) available from either editor at his present address).

[12] G.M. Kelly and R. Street, Review of the elements of 2-categories, in this volume.

[13] A. Kock, Limit monads in categories, Aarhus Univ. Mat. Inst. Preprint Series 1967/68 No. 6 (1967).

[14] M.L. Laplaza, Coherence for distributivity, Lecture Notes in Math. 281 (1972), 29-65.

[15] M.L. Laplaza, A new result of coherence for distributivity, Lecture Notes in Math. 281 (1972), 214-235.

[16] F.W. Lawvere, Ordinal sums and equational doctrines, Lecture Notes in Math. 80 (1969), 141-155.

[17] G. Lewis, Coherence for a closed functor, Lecture Notes in Math. 281 (1972), 148-195.

[18] G. Lewis, Coherence for a closed functor, Ph.D. Thesis, Univ. of New South Wales, 1974.

[19] S. Mac Lane, Natural associativity and commutativity, Rice Univ. Studies 49 (1963), 28-46.

[20] V. Zöberlein, Doktrinen auf 2-Kategorien, manuscript (Math. Inst. der Univ. Zürich, 1973).

DOCTRINAL ADJUNCTION

by

G. M. Kelly

This paper deals with adjunctions $\eta,\varepsilon\colon f \dashv u\colon A \to A'$ where A, A' are categories with structure; its purpose is to unify and simplify various isolated observations in the literature, at the same time extending them widely.

(a) First, and trivially, there is the classical result that a left adjoint f preserves colimits.

(b) If A and A' are monoidal categories, perhaps closed, and if $f \dashv u$ is merely an adjunction in *CAT*, it is not in general the case that f "preserves tensor products" in the sense that $f(X \otimes' Y) \cong fX \otimes fY$ and $fI' \cong I$. For instance, let $A = A' = \mathit{Set}$, and let $f(X) = X + X$, so that $u(A) = A \times A$. Yet it is striking how often in natural examples f <u>does</u> preserve tensor products: a whole class of cartesian closed examples is provided by geometric morphisms of topoi, but there are innumerable others, of which perhaps the best known is that where u is the forgetful functor from $(Ab, \otimes)$ to $(\mathit{Set}, \times)$.

One can dig out from the proof of an old theorem of mine ([3] §§5.1 and 5.2), which was not itself directed to this question, a <u>sufficient</u> condition for f to preserve tensor products; namely that A and A' be symmetric monoidal closed, that the adjunction $f \dashv u$ lie not merely in *CAT* but in *Sym Mon CAT*, and finally that u be <u>normal</u> (i.e. commute with the canonical symmetric monoidal functors $A \to \mathit{Set}$ and $A' \to \mathit{Set}$).

(c) Two theorems of Street deal with adjunctions for categories with structure: Theorem 9 of [8] asserts that the left adjoint of a monad-functor is an op-monad-functor, and Theorem 1 of [9] asserts that the left adjoint of a lax natural transformation is an op-lax natural transformation.

(d) Let V be a symmetric monoidal closed category and let T be a monoidal monad (= commutative monad) on V. Then provided V admits equalizers, the category V^T of T-algebras is closed, in the original "internal-hom" sense of [2] (cf. Kock [7]); it has been shown to have a tensor product, making it monoidal closed, only when it is cocomplete; and cocompleteness has only been demonstrated under highly restrictive hypotheses on V and T. In this context Wolff [11] has looked at certain adjunctions $V^T \to V^{T'}$ (arising in fact from a distributive law), and shown them by direct calculations to be closed adjunctions. Perhaps the right approach here is to change universes, recovering the missing tensor product, and then to use the easier "monoidal" methods; but it might also be of value to have a simple criterion, directly in terms of internal-hom, for an adjunction to be closed.

(e) Finally there is Day's result [1] giving sufficient conditions for a full reflective subcategory A of a biclosed monoidal category A' to admit itself a biclosed monoidal structure, in such a way that the adjunction $f \dashv u$, where u is now the inclusion, becomes a monoidal adjunction.

It turns out that there are some simple general results that illuminate all of the above situations. We start in §1 with a doctrine D, which could be on any 2-category at all, but which we take to be a doctrine on **CAT**, purely because the nomenclature is there more vivid. (For what we need here about doctrines we refer to [6] above in this volume, especially §3.5.) We suppose that A and A' are D-categories, and that we are given an adjunction $\eta,\varepsilon\colon\ f \dashv u\colon\ A \to A'$ in **CAT**. Our first result is that there is a bijection between enrichments of u to a D-functor $(u,\bar{u})$ and enrichments of f to an op-D-functor $(f,\bar{f}')$.

Our second result is that if the adjunction $f \dashv u$ admits enrichment to an adjunction $(f,\bar{f}) \dashv (u,\bar{u})$ in D-*CAT*, then $\bar{f}$ is the inverse of the above $\bar{f}'$, so that $(f,\bar{f})$ and $(f,\bar{f}')$ are <u>strong</u>; moreover that if the enrichment $\bar{u}$ is given, $(u,\bar{u})$ has a left adjoint in D-*CAT* precisely when $\bar{f}'$ is an isomorphism, the left adjoint then being $(f,\bar{f})$ where $\bar{f}$ and $\bar{f}'$ are inverse; and finally that if the enrichment $\bar{f}$ is given, $(f,\bar{f})$ has a right adjoint in D-*CAT* precisely when it is strong, the right adjoint being $(u,\bar{u})$ where $\bar{u}$ corresponds by our first result to the inverse $\bar{f}'$ of $\bar{f}$.

The first result encompasses the two theorems of Street in (c) above - for the second of these the relevant 2-category is not *CAT* but a suitable *CAT*/Λ. The second result gives a proper answer to the question raised in (b) above: to say that f preserves tensor products is to say that it is strong, so that this happens precisely when the adjunction is one in *Mon CAT*, which can also be expressed as a condition on $\bar{u}$, here represented by its generators $\tilde{u}$: $uA \otimes' uB \to u(A \otimes B)$ and u°: $I' \to uI$ (cf. §10.8 of [4] above in this volume). This shows how wide of the mark the sufficient condition given in (b) above really is: that A and A' should be closed, or even symmetric, is totally irrelevant; as for the normality of u, we shall see that it is a <u>consequence</u> of the adjunction's lying in *Mon CAT*, and not an independent condition at all. The observation (a) that left adjoints preserve colimits can also be seen as a trivial case of this second result, by taking D as the doctrine whose algebras are categories-with-colimits: <u>any</u> functor gives a comparison of colimits, and is hence canonically a D-functor, so that the adjunction is <u>always</u> in D-*CAT* and f is <u>always</u> strong, which here means colimit-preserving.

In §2 we look at the case where D is the doctrine whose algebras are monoidal categories, but supposing A, A' to be actually <u>monoidal closed</u>. Then the giving of $\tilde{u}$: $uA \otimes' uB \to u(A \otimes B)$ is equivalent to the giving of a certain $\hat{u}$: $u[A,B] \to [uA,uB]'$ where the square brackets are the internal-homs; and the conditions on $\tilde{u}$, u^o for the adjunction to be monoidal translate into conditions on $\hat{u}$ and u^o. That on u^o is equivalent to the normality of u (whether the categories are closed or not); that on $\hat{u}$ is independent (showing again how mixed-up is the "sufficient condition" given in (b) above). Similarly the conditions on $\tilde{f}$, f^o (namely that they be isomorphisms) translate into conditions on $\hat{f}$ and f^o. Now, however, that the conditions are expressed entirely on the internal-hom level, we can ask whether they are still the conditions for the adjunction to be a closed one, even in the absence of the tensor products, so as to provide the "simple criterion" desired in (d) above. (Recall that for monoidal closed categories, monoidal functor = monoidal closed functor = closed functor, and similarly for natural transformations.) The answer is that they are indeed still the precise conditions. I suppose an ideal proof would be in terms of pro-D-structures; but the theory of these has not been developed, and in view of the marginal interest of non-monoidal closed categories I merely give an independent <u>ad hoc</u> proof and leave it at that - the role here of the general results of §1 is merely to suggest the right conditions.

Finally in §3 we return to the case of a general doctrine D and turn to the analogue of Day's result in (e) above. So we now suppose A to be a full reflective subcategory of A', or equivalently we suppose the counit ε: $fu \Rightarrow 1$ to be the identity. Now only A' is <u>given</u> as a D-category; we seek conditions under which it is possible to give a D-structure to A too, at the same time enriching the adjunction to one in D-CAT. Our results in §1 show that f <u>must</u> be strong; this at once shows that the enrichment is essentially unique if it exists, and gives a necessary condition; we show that the condition is roughly sufficient. By "roughly" I mean that we may have to make do with a <u>pseudo</u>-D-<u>structure</u> on A (axioms satisfied only to within isomorphism). We can get an honest D-structure in two cases: for any D, if we actually ask f to be strict (accepting the correspondingly stronger necessary condition); and for certain "flexible" D, where a pseudo-D-algebra is "the same thing" as an honest one. The D for monoidal categories is "flexible"; that for strict monoidal categories is not. In the monoidal case, then, our condition is necessary and sufficient, and it reduces to Day's. This shows two things about Day's result. First, biclosedness is irrelevant; the result is essentially a "monoidal" one; we just get as a bonus that A is closed or biclosed if A' is. Secondly, since u must be normal by our §2, the restriction to normal u, which Day imposes upon himself in the first paragraph of the introduction of [1], is in fact no restriction at all.

1. The main results

1.1 We refer heavily to [6] above in this volume, both for its results on adjunction and for all terminology not otherwise explained.

We suppose given a doctrine D = (D,m,j) on some 2-category. Everything we say is quite independent of the 2-category in question, and so, merely for notational simplicity, we write as if the 2-category were *CAT*. We suppose that A and A' are D-categories, with actions n: DA → A and n': DA' → A'. We further suppose given an adjunction η,ε: f ⊣ u: A → A' in *CAT*.

We consider pairs of natural transformations $\bar{u}, \bar{f}'$ as in

(1.1)
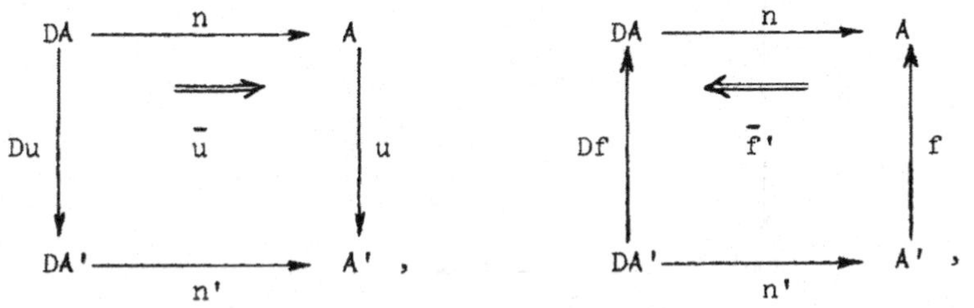

which are mates ([6] §2.2) under the adjunctions Df ⊣ Du and f ⊣ u. We record from [6] Proposition 2.1 the value of $\bar{f}'$ in terms of $\bar{u}$, to wit

(1.2) 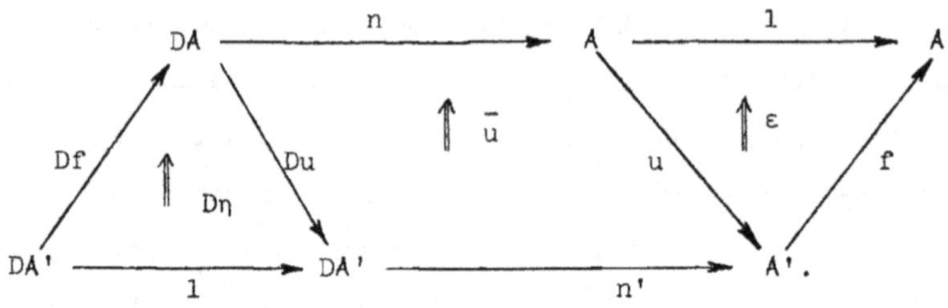

Denote by I, II the axioms for $(u,\bar{u})$ to be a D-functor, namely (3.17) and (3.18) of [6] with f replaced by u. Denote by I', II' the corresponding axioms for $(f,\bar{f}')$ to be an op-D-functor — the same (3.17) and (3.18) of [6] with $\bar{f}$ reversed in sense and re-named $\bar{f}'$, and with A,A' interchanged.

<u>Lemma 1.1</u> <u>Axiom</u> I <u>is equivalent to</u> I' <u>and</u> II <u>to</u> II'.

<u>Proof</u>. Proposition 2.5 of [6] shows that the mates of the left squares of the left sides of I and II are identities; Proposition 2.4 of [6] shows that the mate of the left square of the right side of I is $D\bar{f}'$; Proposition 2.2 of [6] completes the proof. □

<u>Theorem 1.2</u> <u>There is a bijection between enrichments of</u> u <u>to a</u> D-<u>functor</u> $U = (u,\bar{u}): A \to A'$ <u>and enrichments of</u> f <u>to an op</u>-D-<u>functor</u> $F' = (f,\bar{f}'): A' \to A$. <u>This bijection is given by taking</u> $\bar{u}$ <u>and</u> $\bar{f}'$ <u>to be mates under the adjunctions</u> $Df \dashv Du$ <u>and</u> $f \dashv u$. □

<u>1.2</u> Now consider a further natural transformation $\bar{f}$ as in

(1.3) 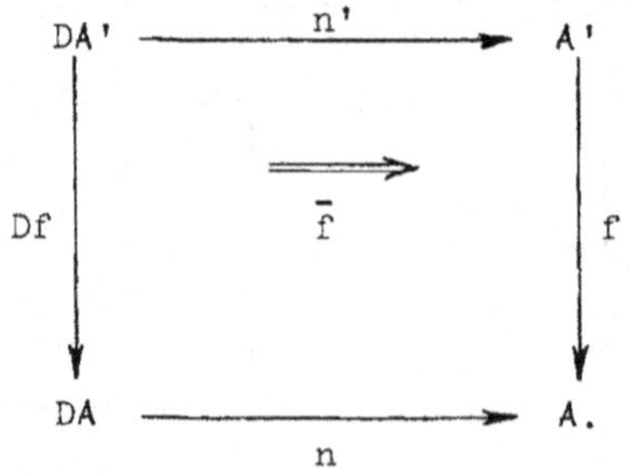

Let us call I", II" the axioms for $(f,\bar{f})$ to be a D-functor F. If these are satisfied, the condition for $\varepsilon\colon\ fu \Rightarrow 1$ to be a D-natural transformation $\varepsilon\colon\ FU \Rightarrow 1$ is, by (3.19) of [6],

(1.4)

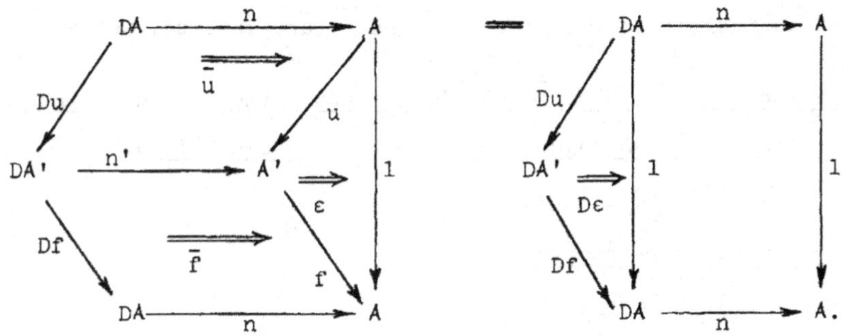

We get an equivalent condition by pasting Dη to each side along the edge Du, which is an invertible process by (2.1) and (2.2) of [6]: it comes to the same thing to say that we pass to mates under the adjunctions $Df \dashv Du$ and $1 \dashv 1$. The mate of the right side of (1.4) is then the identity natural transformation of n.Df; the mate of the left side is

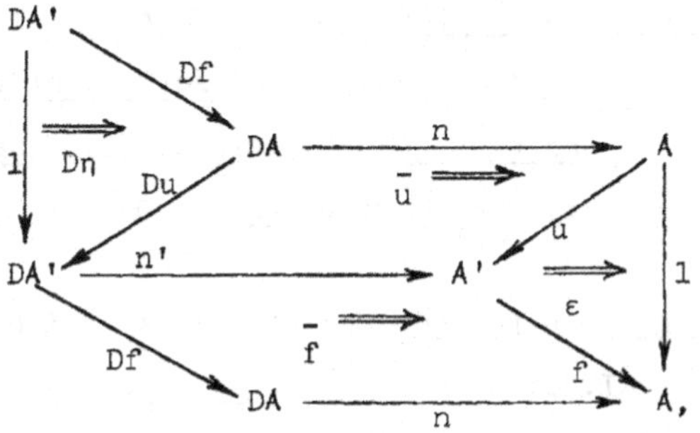

which in view of (1.2) above can be rewritten as

(1.5)

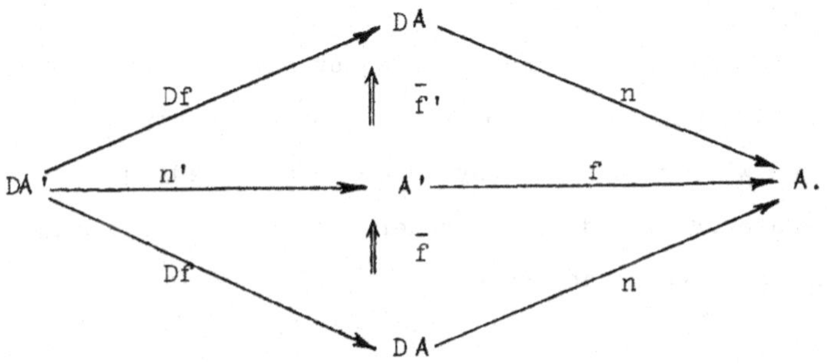

We conclude that, if $F = (f,\bar{f})$ is a D-functor, then ε is D-natural if and only if $\bar{f}'.\bar{f} = 1$, the identity of n.Df. An exactly similar argument shows that η: $1 \Rightarrow UF$ is D-natural if and only if $\bar{f}.\bar{f}' = 1$. Since I", II" are equivalent to I',II' when $\bar{f}$ and $\bar{f}'$ are inverse, we have:

Proposition 1.3 *Given $\bar{f}$ and $\bar{f}'$, and hence the mate $\bar{u}$ of the latter, η and ε constitute an adjunction in D-CAT between D-functors $U = (u,\bar{u})$ and $F = (f,\bar{f})$ if and only if*

(i) $(u,\bar{u})$ *is a D-functor;*

(ii) $(f,\bar{f})$ *is a D-functor; and*

(iii) $\bar{f}$ *and* $\bar{f}'$ *are mutually inverse.*

Moreover (i) *is equivalent to*

(i') $(f,\bar{f}')$ *is an op-D-functor;*

and (i') *and* (ii) *are equivalent in the presence of* (iii). □

Immediate consequences are:

Theorem 1.4 *In order that a D-functor $U = (u,\bar{u})$: $A \to A'$ have a left adjoint $F = (f,\bar{f})$ in D-CAT, it is necessary and sufficient that u have the left adjoint f in CAT, and that the $\bar{f}'$ given by* (1.2) *be an isomorphism. Then $\bar{f} = \bar{f}'^{-1}$, and F is necessarily strong.* □

Theorem 1.5 *In order that a D-functor $F = (f,\bar{f})$: $A' \to A$ have a right adjoint $U = (u,\bar{u})$ in D-CAT, it is necessary and sufficient that f have the right adjoint u in CAT, and that F be strong. Then $\bar{u}$ is the mate of $\bar{f}' = \bar{f}^{-1}$, in the sense of* (1.1). □

1.3 We leave the reader to formulate the obvious dual theorems obtained by replacing the doctrine D by the opposite doctrine D*, where $D^*A = (DA^{op})^{op}$; he will get in this way theorems about adjunction in the 2-category of D-categories and op-D-functors; here it is the right adjoint that must be strong; and so on.

2. The monoidal and closed cases

2.1 Now let D be either the doctrine whose algebras are monoidal categories, or else that whose algebras are symmetric monoidal categories. These doctrines arise from clubs in $CAT/\underline{P}$, and it follows from §10.8 of [4] above in this volume that a D-functor is the same thing as a monoidal functor or a symmetric monoidal functor, as the case may be. In both cases $\bar{u}$ is determined by its components $\tilde{u}$: $uA \otimes' uB \to u(A\otimes B)$ and u°: $I' \to uI$, subject to the usual axioms ([2] page 473 in the monoidal case, with the extra axiom of [2] page 513 in the symmetric monoidal case). It further follows from §10.8 of [4] that $\bar{u}$ is an isomorphism precisely when $\tilde{u}, u^{\circ}$ are isomorphisms. Similarly a D-natural transformation is, in both cases, a monoidal natural transformation.

Theorems 1.4 and 1.5 therefore apply to an adjunction $(f,\tilde{f},f^{\circ}) \dashv (u,\tilde{u},u^{\circ})$ either in *Mon CAT* or in *Sym Mon CAT*, via the intermediary of an op-monoidal functor $(f,\tilde{f}',f^{\circ\prime})$. The condition in Theorem 1.4 is now that $\tilde{f}'$ and $f^{\circ\prime}$ be isomorphisms; that in Theorem 1.5 is now that $\tilde{f}$ and f° be isomorphisms. There is no difference between the non-symmetric and symmetric cases; it is just automatic that $\tilde{u}$ satisfies the extra axiom in the symmetric case if and only if $\tilde{f}$ or $\tilde{f}'$ does. We get the explicit expressions for $\tilde{f}'$: $f(X\otimes' Y) \to fX \otimes fY$ and for $f^{\circ\prime}$: $fI' \to I$ in terms of $\tilde{u}$ and u° from (1.2), giving D its value in terms of the appropriate club. We find for $f^{\circ\prime}$ and $\tilde{f}'$ respectively the composites

$$fI' \xrightarrow[fu^{\circ}]{} fuI \xrightarrow[\varepsilon]{} I; \tag{2.1}$$

$$f(X\otimes' Y) \xrightarrow[f(\eta\otimes\eta)]{} f(ufX\otimes' ufY) \xrightarrow[f\tilde{u}]{} fu(fX\otimes fY) \xrightarrow[\varepsilon]{} fX\otimes fY. \tag{2.2}$$

Given a monoidal functor $U = (u,\tilde{u},u^{o})$: $A \to A'$, write V for $A(I,-)$: $A \to \mathit{Set}$ and V' for $A'(I',-)$: $A' \to \mathit{Set}$ with their canonical monoidal-functor structures ([2] page 504). There is a canonical monoidal natural transformation $V \Rightarrow V'U$ with A-component

$$A(I,A) \xrightarrow{u_{IA}} A'(uI,uA) \xrightarrow{A'(u^{o},1)} A'(I',uA),$$

as shown in [2] page 510; we call U normal if this is a natural isomorphism $V \cong V'U$ (we do not require it, as we did in [2], to be actually an equality). If we use the adjunction $f \dashv u$ we find at once:

Proposition 2.1 *U is normal precisely when (2.1) is an isomorphism. In particular normality of U is a necessary condition for the existence of an adjunction $F \dashv U$ in $\mathit{Mon}\ \mathit{CAT}$.* □

One can similarly transform the condition that (2.2) be an isomorphism, but I see no special meaning in the thus-transformed condition, and omit it.

2.2 Now suppose that the monoidal categories A and A' are closed, so that $-\otimes B$ and $-\otimes' Y$ have right adjoints $[B,-]$ and $[Y,-]'$. Then instead of giving $\tilde{u}$: $UA \otimes' uB \to u(A\otimes B)$ we may equally well give its mate under the adjunctions $-\otimes B \dashv [B,-]$ and $-\otimes' uB \dashv [uB,-]'$; which is a natural transformation

$$(2.3) \qquad \hat{u}\colon\ u[B,C] \longrightarrow [uB,uC]'.$$

It is shown on page 487 of [2] that the monoidal-functor axioms for $\tilde{u},u^{o}$ on page 473 of [2] translate into the closed-functor axioms for $\hat{u},u^{o}$ on page 434 of [2]; and similarly for monoidal natural transformations and closed natural transformations.

We leave the reader to verify that the condition that (2.2) be an isomorphism translates into the condition that

$$(2.4) \qquad u[fY,C] \xrightarrow{\hat{u}} [ufY,uC]' \xrightarrow{[\eta,1]} [Y,uC]'$$

be an isomorphism. Moreover, if $\tilde{f}$ is given, and corresponds to $\hat{f}$, then the condition that $\tilde{f}$ be an isomorphism translates into the condition that

$$(2.5) \qquad [Y,uC]' \xrightarrow[\eta]{} uf[Y,uC]' \xrightarrow[u\hat{f}]{} u[fY,fuC] \xrightarrow[u[1,\varepsilon]]{} u[fY,C]$$

be an isomorphism. So $(u,\hat{u},u^{o})$ has a left adjoint when u does if and only if (2.1) and (2.4) are isomorphisms; and $(f,\hat{f},f^{o})$ has a right adjoint when f does if and only if f^{o} and (2.5) are isomorphisms.

We omit the proofs of these reductions precisely because we are now going to give direct proofs of the last two statements that apply even to non-monoidal closed categories. Our motive for doing this was given in the introduction. As the reader will see, the proofs that follow largely parallel those of §1, and would doubtless be best seen in the context of pro-D-structures, if the theory of these had been worked out.

2.3 Our definition of (non-monoidal) closed category is an inessential modification of that of [2]; a category A, a functor $[\ ,\ ]: A^{op} \times A \to A$, an object I of A, and natural transformations $L: [B,C] \to [[A,B],[A,C]]$, $j: I \to [A,A]$, $i: A \cong [I,A]$, this last being an isomorphism; subjected to the axioms CC1 - CC4 of [2] page 429, together with the axiom that the $A(A,B) \to A(I,[A,B])$ induced by j be an isomorphism.

We suppose given as before an adjunction η,ε: $f \dashv u: A \to A'$ in CAT, but now we suppose A and A' to be closed, and not necessarily monoidal. It is still the case that there is a bijection between morphisms $u^{o}: I' \to uI$ and morphisms $f^{o\prime}: fI' \to I$, where $f^{o\prime}$ is given in terms of u^{o} by (2.1): this is of course immediate by the adjunction $f \dashv u$. But in place of the bijection between natural transformations $\tilde{u}$ and natural transformations $\tilde{f}'$ that we had in the monoidal case, we get something more complicated. Precisely, natural transformations $\hat{u}$ as in (2.3) are in bijection with natural transformations

$$u_1 \colon \; u[fY,C] \longrightarrow [Y,uC]',$$

u_1 being given in terms of $\hat{u}$ as the composite (2.4); and natural transformations $\hat{f}\colon \; f[Y,Z]' \to [fY,fZ]$ are in bijection with natural transformations

$$f_1 \colon \; [Y,uC]' \longrightarrow u[fY,C],$$

f_1 being given in terms of $\hat{f}$ as the composite (2.5). To see that $\hat{u} \mapsto u_1$ and $\hat{f} \mapsto f_1$ are indeed bijections, we have only to write (2.4) and (2.5) in 2-dimensional form as, respectively,

(2.6)

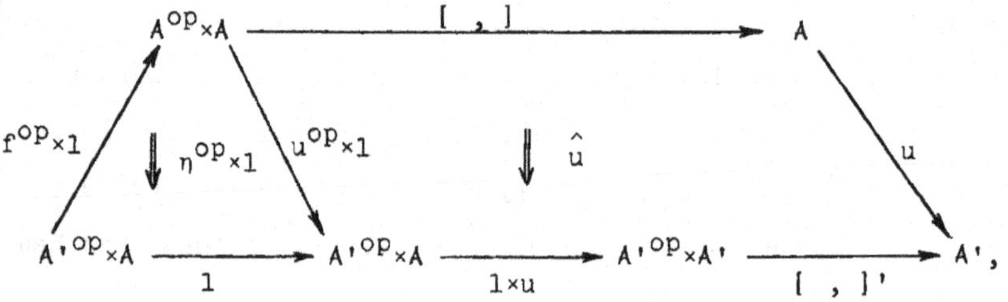

(2.7)

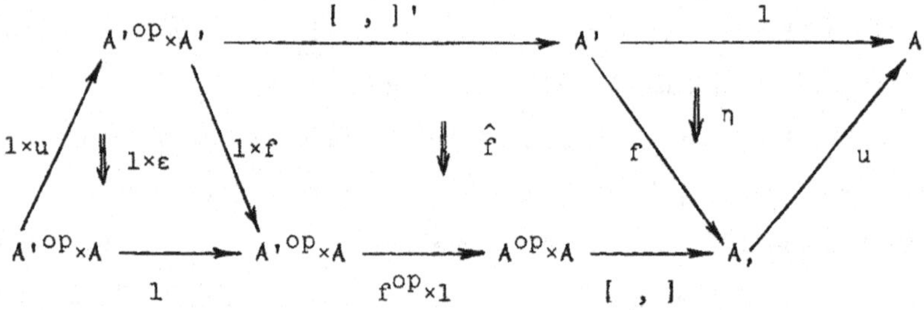

exhibiting u_1 as the mate of $\hat{u}$ under certain adjunctions, and f_1 as the mate of $\hat{f}$ under others, in the sense of [6] §2.2.

A closed functor is of course said to have an adjoint if it has one in the 2-category of closed categories, closed functors, and closed natural transformations. The main result of this section, suggested by and extending the assertions of §2.2 above, is:

Theorem 2.2 *A closed functor* $U = (u,\hat{u},u^o)\colon A \to A'$ *has a left adjoint* $F = (f,\hat{f},f^o)$ *if and only if* u *has the left adjoint* f *and moreover* $f^{o\prime}$ *and* u_1*, given by* (2.1) *and* (2.4) *respectively, are* *isomorphisms. Then* f^{o} *is the inverse of* $f^{o\prime}$ *and* $\hat{f}$ *corresponds by* (2.5) *to* $f_1 = u_1^{-1}$.

A closed functor $F = (f,\hat{f},f^{o})$: $A' \to A$ *has a right adjoint* $U = (u,\hat{u},u^{o})$ *if and only if* f *has the right adjoint* u *and moreover* f^{o} *and* f_1, *the latter given by* (2.5), *are isomorphisms. Then* u^{o} *corresponds by* (2.1) *to* $f^{o\prime} = f^{o-1}$, *and* $\hat{u}$ *corresponds by* (2.4) *to* $u_1 = f_1^{-1}$.

Proof Take all the data $\hat{u},u^{o},\hat{f},f^{o}$ as given, with the mates $u_1,f^{o\prime},f_1$ of the first three, without supposing yet that the closed-functor axioms CF1-CF3 of [2] page 434 are necessarily satisfied by U or by F. This last in no way prevents us from mechanically forming the composites $FU = \Phi = (\phi,\hat{\phi},\phi^{o})$ and $UF = \Psi = (\psi,\hat{\psi},\psi^{o})$ as on [2] page 434.

Our first observation is that η: $1 \Rightarrow \Psi$ and ε: $\Phi \Rightarrow 1$ satisfy the axioms CN1 and CN2 for closed natural transformations, on page 441 of [2], if and only if f^{o} is inverse to $f^{o\prime}$ and f_1 to u_1. In fact ε satisfies CN2 if and only if $f_1u_1 = 1$; η satisfies CN2 if and only if $u_1f_1 = 1$; ε satisfies CN1 if and only if $f^{o\prime}f^{o} = 1$; and η satisfies CN1 if and only if $f^{o}f^{o\prime} = 1$.

We prove only the first of these assertions, leaving the others to the reader. Express $f_1u_1 = 1$ by pasting (2.6) on top of (2.7). In the resulting diagram the 2-cells $\hat{f}$ and $\hat{u}$ have a common edge, and the result of pasting them along this edge is just $\hat{\phi}$. The other three 2-cells are the triangles $\eta^{op} \times 1$, η, $1 \times \varepsilon$. Get rid of the first two of these by pasting on an $\varepsilon^{op} \times 1$ and an ε to cancel them out, as in (2.1) and (2.2) of [6]; the expression of $f_1u_1 = 1$ is now that $\hat{\phi}$ with a $1 \times \varepsilon$ pasted on is equal to the identity of $u[\ ,\ ](f^{op} \times 1)$ with an $\varepsilon^{op} \times 1$ and an ε pasted on in suitable places. But this is just CN2 for ε in 2-dimensional form.

This establishes the "only if" part of both the assertions of the theorem. For the "if" part of the first assertion, we use the prescription of the theorem to <u>define</u> $\hat{f}$ and f^{o}; and it remains to show that $F = (f, \hat{f}, f^{o})$ satisfies the closed-functor axioms; similarly for the "if" part of the second assertion.

We shall show here that F (resp. U) satisfies the axiom CF3 of [2] page 434 when U (resp. F) does, and when the hypothesis is verified that u_1 and f_1 are inverse; we leave the easier axioms CF1, CF2 to the reader. Write the two legs of CF3, say for Ψ, as

$$p_1, p_2 : \quad \psi[Y,Z] \longrightarrow [\psi[X,Y], [\psi X, \psi Z]] ;$$

we are now for simplicity writing $[Y,Z]$ for $[Y,Z]'$. We similarly use p_1, p_2 for the two legs of CF3 for U or for F.

We do not, <u>a priori</u>, know CF3 for Ψ; but we do know that $\eta: 1 \Rightarrow \Psi$ satisfies CN2, and by an easy diagram-filling-in argument we get that

$$(2.8) \qquad [Y,Z] \xrightarrow[\eta]{} \psi[Y,Z] \xrightarrow[p_i]{} [\psi[X,Y], [\psi X, \psi Z]]$$

is independent of i. Now give ψ its value uf, and express the p_i for Ψ in terms of those for U and for F, as is done in the big diagram on page 435 of [2], which we shall call Δ.

First suppose we know CF3 for U. Then (2.8) and Δ give that

$$(2.9) \qquad \begin{array}{ccc} [Y,Z] \xrightarrow{\eta} uf[Y,Z] \xrightarrow{up_i} u[f[X,Y],[fX,fZ]] & & \\ \downarrow{\scriptstyle \hat{u}} & & \\ [uf[X,Y], u[fX,fZ]] \xrightarrow[{[1,\hat{u}]}]{} [uf[X,Y],[ufX,ufZ]] & & \end{array}$$

is independent of i. But since (2.4) is the isomorphism u_1, the morphisms $\hat{u}$ and $[1,\hat{u}]$ in (2.9) are both coretractions and can be cancelled: the composite of the first two morphisms in (2.9) is already independent of i. This implies that the top leg of the diagram

$$\begin{array}{ccccc}
f[Y,Z] & \xrightarrow{f\eta} & fuf[Y,Z] & \xrightarrow{fup_i} & fu[f[X,Y],[fX,fZ]] \\
 & \searrow^{1} & \downarrow{\scriptstyle\varepsilon} & & \downarrow{\scriptstyle\varepsilon} \\
 & & f[Y,Z] & \xrightarrow[p_i]{} & [f[X,Y],[fX,fZ]]
\end{array}$$

is independent of i. But the square commutes by naturality, and the triangle by one of the adjunction-equations; hence p_i is independent of i, which is CF3 for F.

Now suppose instead that we know CF3 for F and seek it for U. We conclude from (2.8) and Δ that

(2.10)
$$\begin{array}{ccccccc}
[Y,Z] & \xrightarrow{\eta} & uf[Y,Z] & \xrightarrow{u\hat{f}} & u[fY,fZ] & \xrightarrow{p_i} & [u[fX,fY],[ufX,ufZ]] \\
 & & & & & & \downarrow{\scriptstyle [u\hat{f},1]} \\
 & & & & & & [uf[X,Y],[ufX,ufZ]]
\end{array}$$

is independent of i. Write this for Z = uC, compose on the end with $[1,[1,u\varepsilon]] : [uf[X,Y], [ufX,ufuC]] \to [uf[X,Y],[ufX,uC]]$, and use naturality to move this last morphism back through $[u\hat{f},1]$ and p_i; in the process it becomes $u[1,\varepsilon] \; : \; u[fY,fuC] \to u[fY,C]$; so that the first three morphisms in the thus-transformed composite (2.10) form the isomorphism f_1 of (2.5) and can be removed. What we now have is that

(2.11) $$u[fY,C] \xrightarrow[p_i]{} [u[fX,fY],[ufX,uC]] \xrightarrow[[u\hat{f},1]]{} [uf[X,Y],[ufX,uC]]$$

is independent of i. Now set Y = uB, compose with $u[\varepsilon,1] : u[B,C] \to u[fuB,C]$ on the front, and with $[\eta,1] : [uf[X,uB],[ufX,uC]] \to [[X,uB],[ufX,uC]]$ on the end. Using naturality to move the first morphism through the second, and observing that the last three morphisms then give an instance of the isomorphism $[f_1,1]$, we have that

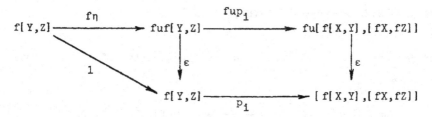

$$(2.12) \qquad u[B,C] \xrightarrow[p_i]{} [u[fX,B],[ufX,uC]]$$

is independent of i. The "extraordinary" naturality of p_i in A gives the commutativity of

$$\begin{array}{ccc}
u[B,C] & \xrightarrow{p_i} & [u[fuA,B],[ufuA,uC]] \\
{\scriptstyle p_i}\downarrow & & \downarrow{\scriptstyle [u[\varepsilon,1],1]} \\
[u[A,B],[uA,uC]] & \xrightarrow[{[1,[u\varepsilon,1]]}]{} & [u[A,B],[ufuA,uC]] \; ;
\end{array}$$

since by (2.12) the top leg is independent of i, so is the bottom leg; but $[1,[u\varepsilon,1]]$ is a coretraction and hence cancellable, $u\varepsilon$ being a retraction by one of the adjunction equations. Thus the p_i on the left is independent of i, which is CF 3 for U. □

3. Reflective subcategories

3.1 We return now to the case of an arbitrary doctrine D, and use the results of §1 to generalize Day's result in [1].

So once again we are given an adjunction η,ε: $f \dashv u$: $A \to A'$ in the 2-category on which D acts, but now we suppose that we are in the special case where

$$(3.1) \qquad fu = 1 \quad \text{and} \quad \varepsilon = 1: \; fu \Rightarrow 1.$$

The results being independent of the 2-category, we again write as if this were CAT; then (3.1) may be expressed by calling A a full reflective subcategory of A', with inclusion u and reflexion f. The adjunction equations ((2.1) and (2.2) of [6] above) now become

$$(3.2) \qquad f\eta = 1: \; f \Rightarrow fuf; \qquad \eta u = 1: \; u \Rightarrow ufu.$$

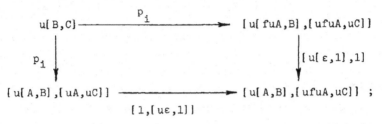

We suppose A', but not A, to be given with a D-structure, with action n': DA' → A'. By an <u>enrichment</u> of this situation we mean the giving of a D-structure to A, with action n: DA → A say, and the enriching of u and f to D-functors $U = (u,\bar{u})$ and $F = (f,\bar{f})$ in such a way that η,ε provide an adjunction $F \dashv U$ in D-CAT.

By Theorem 1.5 to give an enrichment is to give a D-structure to A and to give an isomorphism $\bar{f}$ as in

(3.3)

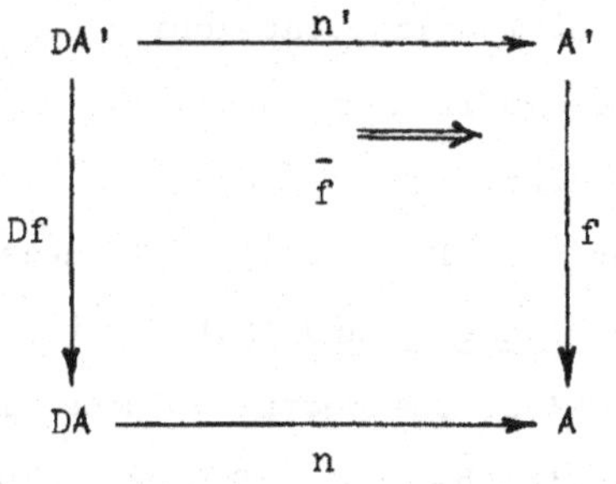

that makes of $(f,\bar{f})$ a strong D-functor. If such an enrichment exists at all, it is unique to within a suitable isomorphism; for (3.3) gives $n.Df \cong fn'$, while (3.1) gives $Df.\ Du = 1$, so that n is effectively determined by

(3.4) $n \cong f.n'.Du.$

Now consider the composite

(3.5)

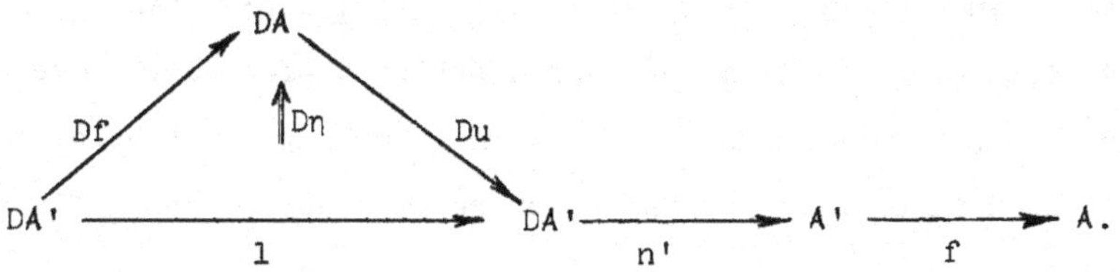

If an enrichment exists, (3.5) is isomorphic by (3.3) to n.Df.Dη, which is the identity by (3.2). Thus a <u>necessary</u> condition for an enrichment to exist is that (3.5) be an isomorphism. In the case where D is the doctrine for monoidal categories, this clearly reduces to the single condition

(3.6) $f(\eta\otimes'\eta)$: $f(X\otimes'Y) \longrightarrow f(ufX\otimes'ufY)$ is an isomorphism;

and Day [1] has shown that in this case (3.6) is also sufficient. However the invertibility of (3.5) can hardly be sufficient for a general D; if D were the doctrine for strict monoidal categories, the invertibility of (3.5) would still reduce to (3.6), but the monoidal structure constructed by Day on $\mathbb{A}$, given by $A\otimes B = f(uA\otimes'uB)$, would not in general be strict even though that on $\mathbb{A}'$ was strict. As we said in the Introduction, the doctrine for monoidal categories has a certain "flexibility" - an ability to absorb isomorphisms - which the doctrine for strict monoidal categories lacks.

It turns out that the invertibility of (3.5) is necessary even for the existence of a pseudo-enrichment; and that for this it is also sufficient. To be precise, let D' be the doctrine whose algebras are the normalized pseudo D-algebras in the sense of Street [10] §2 in this volume. That is to say, a D'-category differs from a D-category in that the associativity axiom for the action is satisfied only to within a (prescribed) isomorphism, subjected to suitable axioms; the unitary axiom however (this is what "normalized" means) is satisfied on the nose. Then a D'-functor [resp. strict D'-functor] is what Street calls a lax homomorphism [resp. strict homomorphism] of pseudo-D-algebras. Since every D-category is a fortiori a D'-category, and every strict D-functor is a fortiori a strict D'-functor, there is a doctrine-map p: $D' \Rightarrow D$; and it is shown in my paper [5] below in this volume, which also justifies the other remarks above, that p is a retraction.

3.2 We can now give the main result of this section.

Theorem 3.1 *Let the reflexion* $\eta, \varepsilon: f \dashv u$: $\mathbb{A} \to \mathbb{A}'$ *be given as above, and let* $\mathbb{A}'$ *be a D-category, hence also a D'-category. Then the reflexion admits a D'-enrichment if and only if* (3.5) *is an isomorphism.*

Proof First, the necessity. Consider the composite of (3.5) with pA': $D'A' \to DA'$. By the 2-naturality of p, the morphism pA' can be moved past the triangle $D\eta$ in (3.5), turning the latter into $D'\eta$. But then pA' composed with n': $DA' \to A'$ is just the action of D' on A'. Thus the composite of (3.5) with pA' is just the analogue of (3.5) with D' replacing D. We already know that <u>this</u> must be invertible if a D'-enrichment is to exist. But then (3.5) itself must be invertible, since pA' is a retraction.

We turn to the sufficiency. Guided by (3.4), we <u>define</u> n: $DA \to A$ by

(3.7) $n = f.n'.Du.$

We can then write (3.5) as an isomorphism

(3.8) $\phi = f.n'.D\eta : fn' \Rightarrow n.Df,$

and we define the $\bar{f}$ of (3.3) to be the inverse of ϕ.

We check that n satisfies the unitary law for an action. We have

$$\begin{aligned} n.jA &= f.n'.Du.jA && \text{by (3.7)} \\ &= f.n'.jA'.u && \text{by the naturality of } j \\ &= fu && \text{by the unitary law for } n' \\ &= 1 && \text{by (3.1)}. \end{aligned}$$

As regards the associative law, we have

$$\begin{aligned} n.mA &= f.n'.Du.mA && \text{by (3.7)} \\ &= f.n'.mA'.D^2u && \text{by the naturality of } m \\ &= f.n'.Dn'.D^2u && \text{by the associative law for } n'; \end{aligned}$$

while

$$n.Dn = n.Df.Dn'.D^2u \qquad \text{by (3.7)}.$$

Therefore the isomorphism

(3.9) $\mu = \phi.Dn'.D^2u$: $f.n'.Dn'.D^2u \Rightarrow n.Df.Dn'.D^2u$

is an isomorphism

(3.10) μ: $n.mA \Rightarrow n.Dn$,

and we define ν to be μ^{-1}.

It remains to verify that n and ν constitute a normalized pseudo-action of D on A, and that f and $\bar{f}$ constitute a "lax homomorphism of pseudo-D-algebras" in Street's language. Then in terms of D', we have a D'-action on A and a strong D'-functor $(f,\bar{f})$: $A' \to A$, whence the desired D'-enrichment follows by Theorem 1.5.

The axioms to be verified are the axioms (1), (2), (3) of [10] §2 for ν, and the axioms (5), (6) of [10] §2 for $\bar{f}$. The notations are rather different: Street's i is our j, his c is both our m and our n, as well as our n'; his ζ is 1 in our case (normality); his θ is our ν, and also our ν' which is 1 (A' being a honest D-category); and his θ_f is our $\bar{f}$. In verifying these axioms we replace ν and $\bar{f}$ by their inverses, μ and ϕ respectively, inverting the arrows accordingly.

Axiom (1) reads $\mu.jDA = 1$. But

$$\begin{aligned} \mu.jDA &= \phi.\ Dn'.D^2u.\ jDA && \text{by (3.9)} \\ &= \phi.\ jA'.\ n'.Du && \text{by the naturality of } j. \end{aligned}$$

This will therefore follow from axiom (5), which reads $\phi.jA' = 1$. But

$$\begin{aligned} \phi.jA' &= f.n'.\ D\eta.jA' && \text{by (3.8)} \\ &= f.n'.jA'.\eta && \text{by the 2-naturality of } j \\ &= f\eta && \text{by the unitary axiom for } n' \\ &= 1 && \text{by (3.2).} \end{aligned}$$

Axiom (2) reads $\mu.DjA = 1$. But

$$\begin{aligned} \mu.DjA &= \phi.Dn'.D^2u.DjA && \text{by (3.9)} \\ &= \phi.Dn'.DjA'.Du && \text{by the naturality of } j \\ &= \phi.Du && \text{by the unitary axiom for } n' \\ &= f.n'.D\eta.Du && \text{by (3.8)} \end{aligned}$$

$= 1$ since $\eta u = 1$ by (3.2).

This leaves us with axioms (3) and (6). The first of these reduces to the second, via (3.9). For the left side of axiom (3) is the vertical composite of $n.D\mu$ with $\mu.DmA$, and the right side is the vertical composite of $\mu.D^2n$ with $\mu.mDA$. Now using (3.9)

$$n.D\mu = n.D\phi.D^2n'.D^3u;$$

$$\mu.DmA = \phi.Dn'.D^2u.DmA$$
$$= \phi.Dn'.DmA'.D^3u \quad \text{by the naturality of } m$$
$$= \phi.Dn'.D^2n'.D^3u \quad \text{by the associativity axiom for } n';$$

$$\mu.D^2n = \mu.D^2f.D^2n'.D^3u \quad \text{by (3.7);}$$

$$\mu.mDA = \phi.Dn'.D^2u.mDA$$
$$= \phi.mA'.D^2n'.D^3u \quad \text{by the naturality of } m.$$

But the left side of axiom (6) is the vertical composite of $n.D\phi$ with $\phi.Dn'$, and the right side is the vertical composite of $\mu.D^2f$ with $\phi.mA'$. So axiom (3) follows from axiom (6) on composing with $D^2n'.D^3u$.

It remains then to verify axiom (6). Now

$$n.D\phi = f.n'.Du.Df.Dn'.D^2\eta \quad \text{by (3.7) and (3.8);}$$

$$\phi.Dn' = f.n'.D\eta.Dn' \quad \text{by (3.8);}$$

$$\mu.D^2f = f.n'.D\eta.Dn'.D^2u.D^2f \quad \text{by (3.9) and (3.8);}$$

$$\phi.mA' = f.n'.D\eta.mA' \quad \text{by (3.8)}$$
$$= f.n'.mA'.D^2\eta \quad \text{by the 2-naturality of } m$$
$$= f.n'.Dn'.D^2\eta \quad \text{by the associativity axiom for } n'.$$

The vertical composite of the first two of these is indeed equal to that of the second two, both being

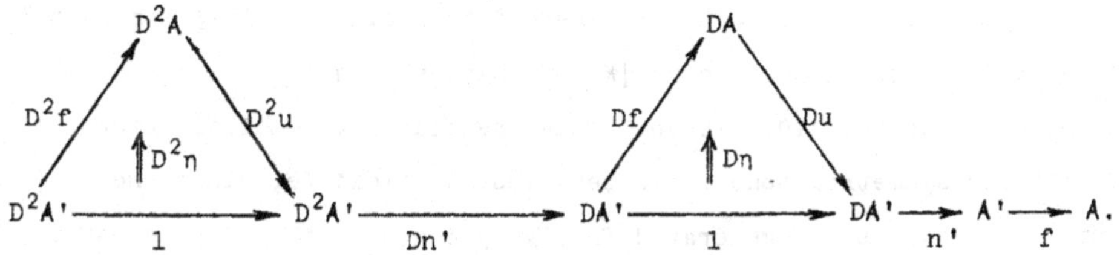

This completes the proof. □

3.3 As we have said in the Introduction, there are two cases in which we can actually get a D-enrichment. The first is that where (3.5) is not merely an isomorphism but an identity; which corresponds to asking $F = (f, \bar{f})$ to be not merely strong (as it must be by Theorem 1.5) but strict. This may seem rather an artificial case; but it often arises in nature, as the paper [5] below will show, and I have found it extremely useful in various (unpublished) considerations related to coherence. To state it formally:

Theorem 3.2 Let the reflexion $\eta, \varepsilon\colon f \dashv u : A \to A'$ be given as above, and let A' be a D-category. Then the reflexion admits a D-enrichment with F strict if and only if (3.5) is the identity (which includes the assertion that $f.n'.Du.Df = f.n'$).

Proof For the necessity, we are to have (3.3) with $\bar{f} = 1$, so that we must have $f.n' = n.Df$. So (3.5) is $n.Df.D\eta$, which is the identity since $f\eta = 1$ by (3.2).

The sufficiency is contained in the proof of Theorem 3.1, since in the present case we have $\phi = 1$ and $\mu = 1$, or equally $\bar{f} = 1$ and $\nu = 1$. □

We said towards the end of §3.1 that the doctrine-map $p\colon D' \Rightarrow D$ is a retraction; that is, there is a 2-natural transformation $q\colon D \Rightarrow D'$ with $pq = 1$. In general we cannot suppose q to be a doctrine-map. If we can find a doctrine-map q with $pq = 1$, we call the doctrine D flexible. In the paper [5] below, we show that for any doctrine D the doctrine D' is flexible; and we also show that D is flexible if D is of the form $K\circ -$, where K is a club in $\mathbf{CAT}/\underline{\underline{S}}$ (or even in $\mathbf{CAT}/\mathcal{set}$) whose discrete club $|K|$ of objects is a free discrete club in the sense of [4] §10.3 above. Thus the club for monoidal categories, or that for symmetric monoidal categories, is flexible, since the functor-operations are generated freely by ⊗ and I; the club for strict monoidal categories escapes the result since $\otimes$ and I are subjected to axioms like $\otimes(\otimes,\underline{1}) = \otimes(\underline{1},\otimes)$.

By applying the theorem below in the case where D is replaced by D', we see that Theorem 3.1 remains true even when A' is originally given as a D'-category rather than a D-category.

Theorem 3.3 *Let the reflexion* η,ε: $f \dashv u$: $A \to A'$ *be given as above, and let* A' *be a* D-*category, where* D *is flexible. Then the reflexion admits a* D-*enrichment if and only if* (3.5) *is an isomorphism*.

Proof The necessity was shown in §3.1. For the sufficiency, first use Theorem 3.1 to get a D'-enrichment. Then apply the 2-functor q-CAT: D'-CAT $\to$ D-CAT (cf. [6] §3.6 above) to get a D-enrichment. The only point at issue is whether the D-structure A' now has is that it started with; this is ensured by pq = 1. □

REFERENCES

[1] B.Day (=B.J.Day), A reflection theorem for closed categories, Jour. Pure and Applied Algebra 2 (1972), 1-11.

[2] S. Eilenberg and G.M. Kelly, Closed categories, Proc. Conf. on Categorical Algebra (La Jolla 1965), Springer-Verlag 1966.

[3] G.M. Kelly, Adjunction for enriched categories, Lecture Notes in Math. 106 (1969), 166-177.

[4] G.M. Kelly, On clubs and doctrines, in this volume.

[5] G.M. Kelly, Coherence theorems for lax algebras and for distributive laws, in this volume.

[6] G.M. Kelly and R. Street, Review of the elements of 2-categories, in this volume.

[7] A. Kock, Monads on symmetric monoidal closed categories, Arch. Math. 21 (1970), 1-10.

[8] R. Street, The formal theory of monads, Jour. Pure and Applied Algebra 2 (1972), 149-168.

[9] R. Street, Two constructions on lax functors, Cahiers de Top. et Géom. Diff. XIII 3 (1972), 217-264.

[10] R. Street, Fibrations and Yoneda's lemma in a 2-category, in this volume.

[11] H. Wolff, Commutative distributive laws, to appear in Jour. Australian Math. Soc.

COHERENCE THEOREMS FOR LAX ALGEBRAS AND FOR DISTRIBUTIVE LAWS

by

G.M. Kelly

1.1 We prove two main results. First, let D be a doctrine, say on *Cat*. (For the most basic notions about doctrines, see [13] above, familiarity with which we assume.) Let D^* be that doctrine whose (honest) algebras are the lax algebras for D, in the sense of [24] above. Since a D-algebra is a fortiori a D^*-algebra, there is a doctrine-map $s\colon D^* \to D$. We prove the existence, in the 2-category [*Cat*,*Cat*] of endo-2-functors, of a morphism $h\colon D \to D^*$ (not a doctrine-map) such that $sh = 1$, and of a 2-cell $\eta\colon 1 \Rightarrow hs$ such that $s\eta = 1$ and $\eta h = 1$. Borrowing the terminology usual in the 2-category *Cat*, we may say that this exhibits s as a "reflexion" of D^* onto the "full reflective subobject" D. Moreover, if we replace "lax" above by "pseudo", the same is still true, but now η is an isomorphism; so that in this case s is an equivalence of endo-2-functors. Part of this result, along with some refinements we shall also prove, has been applied in my paper [10] above.

For our second result, let $p\colon D'D \to DD'$ be a distributive law between two doctrines ——— or rather a pseudo-distributive-law; for strict ones in the original sense of Beck [2] are rare in natural examples at the doctrine level (just as $A \otimes B = B \otimes A$ is hopelessly rare for categories, while $A \otimes B \cong B \otimes A$ is common). A category provided with actions both of D and of D', and for which the D-action is given the structure of a (non-strict!) D'-morphism, is an algebra for some doctrine D^*. Were the distributive law a strict one,

DD' would itself be a doctrine, and _its_ algebras would be categories with a D-action and a D'-action, the D-action being a _strict_ D'-morphism; this is the classical situation of Beck [2]. In that case there would clearly be a doctrine-map s: $D^* \to DD'$. In our pseudo-distributive case, DD' is no longer a doctrine, but is at any rate a D^*-algebra; so we can define s: $D^* \to DD'$ as the unique strict D^*-morphism whose composition with $j^*: 1 \to D^*$ is jj': $1 = 11 \to DD'$. Our result is once again that s is a reflexion of D^* onto a full reflective subobject DD'; and that it is actually an equivalence of endo-2-functors if we require the D-action to be a _strong_ (although still not a _strict_) D'-morphism.

Kock [14] distinguished, among the enriched monads on a symmetric monoidal closed category, those that he called the _commutative_ monads — now more often called _monoidal_ monads. Once again, strictly commutative doctrines are rare, but _pseudo-commutative_ ones are easily found. One such is the doctrine for symmetric monoidal categories; let us now call _this_ D'. Then if D is any pseudo-commutative doctrine, there is a pseudo-distributive-law p: $D'D \to DD'$, and the above result applies. In particular it applies when D too is taken to be this particular D'. In that case our result contains that of Laplaza [19], who considers a category with two symmetric monoidal structures $\otimes$ and $\oplus$ together with a "distributivity" $A\otimes(B\oplus C) \to (A\otimes B)\oplus(A\otimes C)$.

The similarity of our two main results suggests that each is a special case of something very general, relating the doctrines for a more lax and a less lax situation. I have not attempted to formulate this: the present proofs have in common only some general principles collected in §2 and §3 below, and are otherwise distinct.

The rest of this introduction is devoted to general comments on coherence problems, including the relation between such results as we have stated and results about the commutativity of diagrams.

1.2 To be told, as in the above situations, what are the algebras for some doctrine D^* (along with the strict morphisms of these, and the 2-cells between the strict morphisms) is to be given D^* implicitly. By the "coherence problem" for D^* I understand primarily that of determining D^* explicitly from this information. A complete solution of this is often beyond our powers. By a "coherence theorem" I understand a complete or partial solution of this problem: a result which tells us something at least about D^* qua endo-2-functor of *Cat*. For instance, that it is equivalent to some known endo-2-functor D; or even that it contains a known D, in some definite way, as a full reflective subobject. Results about certain diagrams commuting may be part of such a theorem, or may be among its important consequences; but such results can, I believe, no longer be seen as constituting the essence of a coherence theorem.

I shall return later to the matter of commuting diagrams. First, I want to observe that the last paragraph has been oversimplified in various ways for the sake of its polemical point.

(a) The word "doctrine" (= 2-monad) may have to be replaced by "monad". Categories endowed with some equational structure are the algebras for a doctrine on *Cat* when the structural functors are all covariant; but they are the algebras for a mere monad when, as in the case of symmetric monoidal closed categories, functors of mixed variance, such as the internal-hom, are involved.

(b) One takes much too narrow a view if one considers only doctrines on *Cat*. An equational structure may be borne by a Λ-indexed family of categories, and these are algebras for a doctrine on the 2-category Cat/Λ. It is easy to conceive of doctrines on V-*Cat* for a symmetric monoidal closed V. But then too there are things like "lax doctrines"; an $E \in [Cat, Cat]$ with multiplication $E^2 \to E$ and unit $1 \to E$ satisfying the "doctrine axioms" only to within certain 2-cells, themselves subjected to so-called "coherence axioms";

such an E is an algebra for some doctrine D^* on the 2-category $[\mathcal{Cat},\mathcal{Cat}]$. Clearly we should take D^* to be a doctrine (or possibly a mere monad) on an arbitrary 2-category K; although it may be reasonable to suppose that K has some of the good properties of $\mathcal{Cat}$, such as being locally presentable.

(c) It is reasonable - and probably very common - to use the term "coherence theorem" for a result which, having found out something about D^* from a knowledge of its algebras, goes back to these algebras and deduces something useful about them. For instance, Mac Lane's original coherence theorem [23] may be interpreted as saying that the doctrine-map s: $D^* \to D$ is an equivalence of endo-2-functors, where D^* is the doctrine for symmetric monoidal categories and D that for <u>strict</u> symmetric monoidal categories. This in fact gives a complete determination of D^* (since D^* is $P\circ -$ for a club P which is wholly determined by Mac Lane's result). But it has the further consequence that every symmetric monoidal category is symmetric-monoidally equivalent to a strict one. [This is well known although I do not recall any precise statement in print. Beck [3] takes it for granted and proceeds to refine the equivalence in this case, and in one other, via categories of fractions. It is a special case of the assertion of Isbell in [7], but this assertion has been withdrawn by its author; it asserts more than is true, and he discovered a radical error in the proof (at the Oberwolfach conference "Kategorien 1973").] How generally the existence of such an equivalence s: $D^* \to D$ implies that every D^*-algebra is D^*-equivalent to a D-algebra has not to my knowledge been discussed; nor do we pursue it in this paper - but it cries out for investigation, and would be many people's ideal of what a coherence theorem should be.

1.3 We return now to the matter of commuting diagrams. In some cases we know at once from the description of the D^*-algebras that D^* is $C\circ-$ for some club C, and that C is given by specified generators and relations; see §1 and §10 of [9] above, and the further references given there. The word-problem involved in finding the objects of C is typically easy - often indeed they are freely generated - while that involved in finding the morphisms is typically much harder. It is a matter of finding the category generated by a certain graph subject to certain relations - a generalization of the word-problem for monoids, inasmuch as a category is a "monoid with many objects". Deciding when two words in the generators represent the same morphism of C is deciding which diagrams commute in C. This is the most common sense in which solving a coherence problem, completely or partially, may involve proving that certain diagrams commute. The typical example is that of symmetric monoidal closed categories, the club for which was determined in part by Kelly-Mac Lane [11], and then more fully by Voreadu ([27],[28],[29]), but is still not known completely.

(Other examples where the club is partially but not completely determined are: a symmetric monoidal closed category V, two V-categories, two V-functors, and a V-natural transformation (Kelly-Mac Lane [12]); two symmetric monoidal closed categories and a symmetric monoidal functor (Lewis [21],[22]); a category with two symmetric monoidal structures $\otimes$ and $\oplus$ along with a "distributivity" $A\otimes(B\oplus C) \to (A\otimes B)\oplus(A\otimes C)$ that is not required to be an isomorphism (Laplaza [18] and [19]). Some examples where the club is completely determined are: a monoidal or symmetric monoidal category (Mac Lane [23]); a category with a tensor-product and a non-isomorphic $a\colon (A\otimes B)\otimes C \to A\otimes(B\otimes C)$ (Laplaza [17]); a category with a monad (Lawvere [20]; Lambek [16]) - the doctrine is just $\underline{\underline{\Delta}}\times-$; two symmetric monoidal categories and a symmetric monoidal functor (Lewis [21] and

[22]). I do not suggest that the above results are expressed in terms of clubs by their authors; but this is what they amount to. The last-mentioned result of Lewis is so expressed, and indeed must be: for it sets out to determine C completely in a case where the augmentation Γ: $C \to T$ is not faithful. Finally, this list is not meant to be exhaustive.)

One could also regard the determination of D^* as the problem of determining D^*A for every object A; that is, of determining the free D^*-algebras. This is the view taken by Lambek ([15] and [16]), who sets up generators and relations for D^*A, and who attacks the problem of which diagrams in D^*A commute when A is a discrete category. To this end he has introduced a brilliant adaptation of Gentzen's work on cut-elimination. Yet when D^* is $C\circ-$ for a club C, as it is in the examples considered by Lambek, the consideration of $D^*A = C\circ A$ in place of C is an unnecessary complication - for any knowledge of C immediately gives corresponding knowledge of $C\circ A$, and this for any A, discrete or not; while the cut-elimination techniques work perfectly well at the level of C. In fact, those results of Lambek of the form "equi-generality implies commutativity" turn out to be re-phrasings of " Γ: $C \to T$ is faithful"; and some of them are then seen to be false (cf. [8] §6). When, however, D^* does not come from a club, Lambek's idea of setting up generators and relations for D^*A, and studying commutativity of diagrams at this level, may well be the best way of getting at D^*. An example is that where a D^*-algebra is a cartesian closed category, which has been investigated by Szabo ([25] and [26]).

1.4 Thus proving diagrams commutative may be a tool in establishing a coherence theorem, or even a way of stating it; but it need not be the only way of getting some grip on D^*, as is shown by our results stated in §1.1. On the other hand I now point out that such results imply theorems about diagrams commuting, at least when

D,D' come from clubs: and imply them wholesale, in so far as these clubs are arbitrary. In this way I believe one clinches the argument for calling them "coherence theorems".

In the first of those results it is easy to see that if D is $\mathcal{D}\circ-$ for a club $\mathcal{D}$ in $CAT/\underline{\underline{S}}$, then D^* is $\mathcal{D}^*\circ-$ for another such club $\mathcal{D}^*$, and that s,h,η all arise from things in $CAT/\underline{\underline{S}}$; so that we may as well change notation and write s: $\mathcal{D}^* \to \mathcal{D}$, h: $\mathcal{D} \to \mathcal{D}^*$, η: $1 \Rightarrow hs$, $sh = 1$, $s\eta = 1$, $\eta h = 1$. If f,g: $T \to S$ is a diagram in $\mathcal{D}^*$, a necessary condition for its commutativity is $sf = sg$. This is not in general sufficient, for it implies only that $\eta_S f = \eta_S g$, where η_S: $S \to hsS$. It is however sufficient if η_S is an isomorphism, as it always is in the "pseudo" case, and as it is in the "lax" case when S is in the image of h, i.e. when S belongs to the full subcategory $\mathcal{D}$ of $\mathcal{D}^*$. Since s is a club-map, sf can be calculated explicitly and directly; so we have an effective test for the commutativity of any diagram in the "pseudo" case, and of a large class in the "lax" case.

The same is true in the second result of §1.1, in the form in which it is first stated (the D-action an <u>arbitrary</u> D'-morphism). If f,g: $T \to S$ in $\mathcal{D}^*$ with S in the full reflective subcategory $\mathcal{D}\circ\mathcal{D}'$, then $f = g$ if and only if $sf = sg$. In the case where the D-action is to be a <u>strong</u> D'-morphism, it is no longer the case that D^* comes from a club in $CAT/\underline{\underline{S}}$ when D and D' do. This is for the reasons indicated in §1.5 of [9]; in the special case corresponding to Laplaza's problem of two symmetric monoidal structures and a distributivity d: $A\otimes(B\oplus C) \to (A\otimes B)\oplus(A\otimes C)$, the <u>type</u> of d lies in $\underline{\underline{S}}^{op}$, and we can take it to be in $\underline{\underline{S}}$ by passing to the opposite doctrine. But if we ask d to be an isomorphism, its inverse d^{-1} <u>already</u> has its type in $\underline{\underline{S}}$, and dd^{-1}: $(A\otimes B)\oplus(A\otimes C) \to (A\otimes B)\oplus(A\otimes C)$ has its type neither in $\underline{\underline{S}}$ nor in $\underline{\underline{S}}^{op}$; the doctrine does not come from a club in any sense we can at present give to that notion.

What we <u>can</u> do is to return to the club $\mathcal{D}^*$ in the non-strong case, and consider a model A (that is, a $\mathcal{D}^*$-algebra) in which the image of d <u>happens</u> to be an isomorphism (or more generally one in which the image of η <u>happens</u> to be a monomorphism). Then although sf = sg does not imply f = g in $\mathcal{D}^*$, it <u>does</u> imply the equality of their images in $\{A,A\}$, the "rich endo-functor category" of [9] §9, whose objects are functors $A^n \to A$ and whose morphisms are suitably general natural transformations. So for such a model A we have a criterion at the model-level for commutativity of any diagram f,g: $T \to S$ describable in $\mathcal{D}^*$ - which means (in Laplaza's case) a diagram involving d but not d^{-1}.

With this said, and with the observation that in Laplaza's case the doctrines $\mathcal{D}$ and $\mathcal{D}'$ are both the doctrine $\mathcal{P}$ for symmetric monoidal categories, which is equivalent to $\underline{\underline{P}}$ by Mac Lane's original result [23], so that s can be regarded as a map $\mathcal{D}^* \to \underline{\underline{P}}\circ\underline{\underline{P}}$, it is easy to see that s is Laplaza's "distortion" ([19] §2), and that our result includes his theorem in [19] §4. We give more details below.

<u>1.5</u> We end this introduction by observing that the <u>kind</u> of coherence theorem exemplified by the results of §1.1 is pretty common.

First, if we take Mac Lane's original result in [23], and look just at the associativity part, we see that the invertibility of a: $(A\otimes B)\otimes C \to A\otimes(B\otimes C)$ is not central to the main argument. What is really proved is that s: $\mathcal{D}^* \to \mathcal{D}$ is a reflexion, where $\mathcal{D}$ is the subcategory of the appropriate club $\mathcal{D}^*$, consisting of the objects bracketed wholly from the right. It is only because a is invertible that the reflexion is in fact an equivalence. It follows that, even when a is not an isomorphism, sf = sg is necessary and sufficient for f = g: $T \to S$ whenever S lies in $\mathcal{D}$. This is a <u>partial</u> coherence result in that case. A total one requires more work, and is given by Laplaza in [17].

To get back to the partial one, however: I say that it is virtually predictable, and in a sense automatic. The method of proof of the present paper requires a $\mathcal{D}^*$-functor h: $\mathcal{D} \to \mathcal{D}^*$ with sh = 1. Identifying $\mathcal{D}$ with the set of natural numbers >0, we define h as a functor by h(1) = $\underline{\underline{1}}$, h(n) = $\underline{\underline{1}} \otimes$ h(n-1). Enriching it to a $\mathcal{D}^*$-functor means giving $\tilde{h}$: hn $\otimes$ hm $\to$ h(n+m); in particular we need $\tilde{h}$: h2 $\otimes$ h1 $\to$ h3, and this is of course a; the other components of $\tilde{h}$ are then given by an easy induction. But (h,$\tilde{h}$) has to satisfy an axiom if it is to be a $\mathcal{D}^*$-functor: and this reduces at once to the pentagonal axiom for a. This is in itself a lightweight observation; but note that similar considerations <u>predict</u> what the "coherence axioms" <u>should be</u> when we also have a constant object I and non-isomorphic maps ℓ: I$\otimes$A $\to$ A, r: A$\otimes$I $\to$ A; this is not an artificial situation, but arises precisely in the old, unsolved, coherence problem for a non-monoidal, non-symmetric closed category.

A final comment on Mac Lane's original coherence result, say for monoidal categories. If D = $\underline{\underline{N}}\circ$- is the doctrine for strict monoidal categories, and if D' = $\mathcal{N}\circ$- is that for monoidal categories, Mac Lane's result gives a doctrine-map D' $\to$ D that is an equivalence of endo-2-functors. <u>Our</u> first result gives a similar equivalence $D^* \to D$. But Mac Lane's D' is not <u>isomorphic</u> to our D^*; although they are <u>equivalent as doctrines</u>, and not merely as endo-2-functors. This equivalence is discussed towards the end of §3 of [13] above.

As a last example, let $\mathcal{C}$ be a club in $\mathcal{CAT}/\underline{\underline{S}}$ and consider pairs h: $\mathcal{A} \to \mathcal{A}'$ of $\mathcal{C}$-categories connected by a $\mathcal{C}$-functor h. Lewis ([21] and [22]) has determined the corresponding club completely when $\mathcal{C} = \mathcal{P}$ or $\mathcal{N}$. But we can get a <u>partial</u> result for a <u>general</u> $\mathcal{C}$ with hardly any effort. Observe that the diagrams whose commutativity was needed by Eilenberg-Kelly [6] in the monoidal ($\mathcal{N}$) and symmetric monoidal ($\mathcal{P}$) cases were always rather special: they contained variables only from $\mathcal{A}$, not from $\mathcal{A}'$; they lay in $\mathcal{A}'$; and they always had codomain of the form hA. A commutativity criterion for such diagrams can be given at once, for a general C. Having variables only from A means looking not at the whole club, which is the free algebra on $A = I$ and $A' = I$ where I is the unit category, but at the free algebra on $A = I$ and A' = empty. Let this be $h\colon\ C \to C'$; it is immediate that its domain is C itself, while C' is the "unknown". Since h is free on the given generators and since $1\colon\ C \to C$ is a model, there are unique <u>strict</u> C-functors n,s rendering commutative

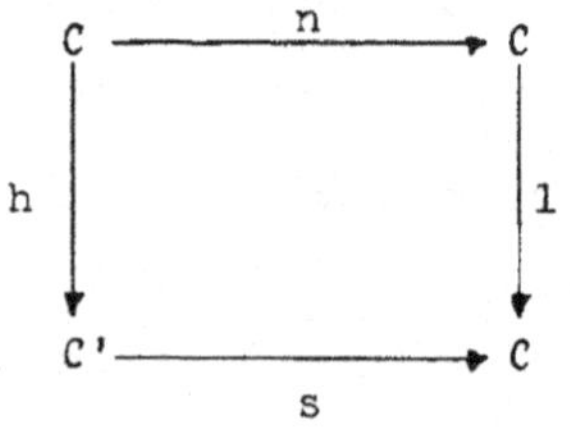

and having $n(\underline{\underline{1}}) = \underline{\underline{1}}$. But this last requirement, since C is the free C-category on $\underline{\underline{1}}$, gives $n = 1$; so that $sh = 1$. It is easy to show that s is in fact a reflexion of C' onto the full subcategory C, embedded by h; so that if S is in the image of h, the diagram $f,g\colon\ T \to S$ in C' commutes if and only if $sf = sg$. A lot less than Lewis proves for $C = P$ or N, admittedly; but much more general, very useful, and incredibly cheap.

<u>2. A method of constructing D-reflexions</u>

<u>2.1</u> Since this section is purely formal, we take (D,m,j) to be a doctrine on an arbitrary 2-category K, and we understand "D-algebra" in the wide sense: an action $a\colon\ DA \to A$ of D on a 2-functor $A\colon\ L \to K$, not necessarily on an object A of K (the special case $L = I$). As far as possible we use a,b, etc., for actions of D on A,B, etc.

We assume acquaintance with the general facts about monads and doctrines given in [13] §3 above; we recall in particular that the free D-algebra DA has action mA: $D^2A \to DA$, and that an action a: $DA \to A$ is a strict D-morphism. We add a few enrichments appropriate to the doctrine case.

First, if γ: $r \Rightarrow r'$: $A \to B$ is any 2-cell, then in

(2.1) $D\gamma$: $Dr \Rightarrow Dr'$: $DA \to DB$

not only are Dr, Dr' strict D-morphisms but $D\gamma$ is a D-2-cell. Next, for a free D-algebra DA and a D-algebra B, there is an isomorphism not only of sets but of categories

(2.2) Φ: $[\mathcal{L},\mathcal{K}](A,B) \to$ D-$\mathcal{A}\ell g_*$(DA,B).

(Here D-$\mathcal{A}\ell g_*$ has as objects the D-algebras of domain $\mathcal{L}$; its morphisms are <u>strict</u> D-morphisms, and its 2-cells are D-2-cells. We recall that we replace D-$\mathcal{A}\ell g_*$ by D-$\mathcal{A}\ell g$ when we allow <u>all</u> D-morphisms as 1-cells.) In detail,

(2.3) Φ sends 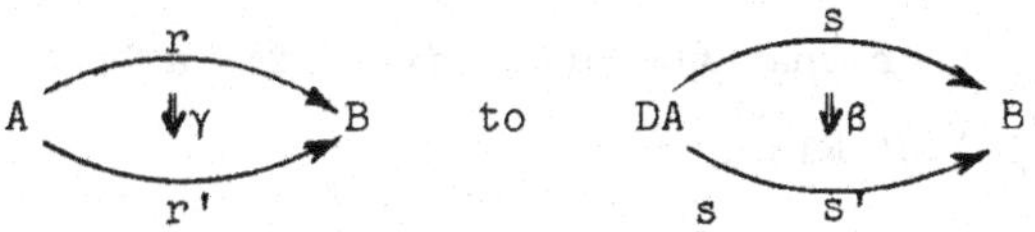

where

(2.4) 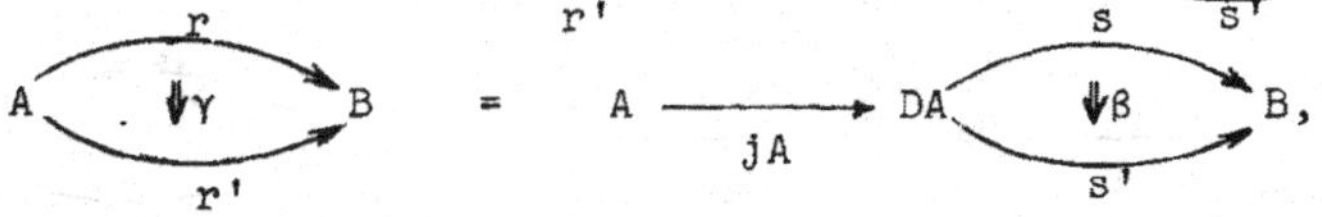

(2.5) 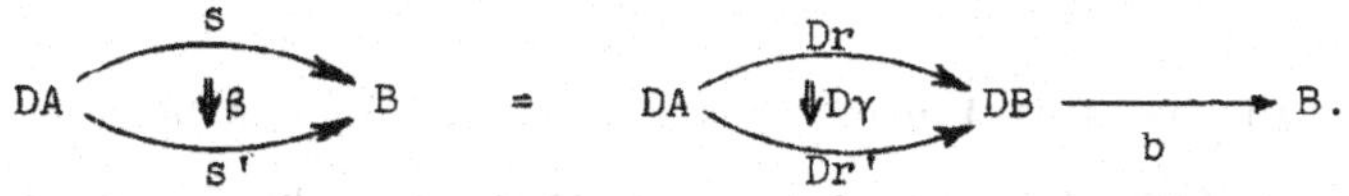

Now consider an arbitrary (i.e. not necessarily strict) D-morphism $G = (g,\bar{g})$: $A \to B$, so that $\bar{g}$ is a 2-cell

(2.6) 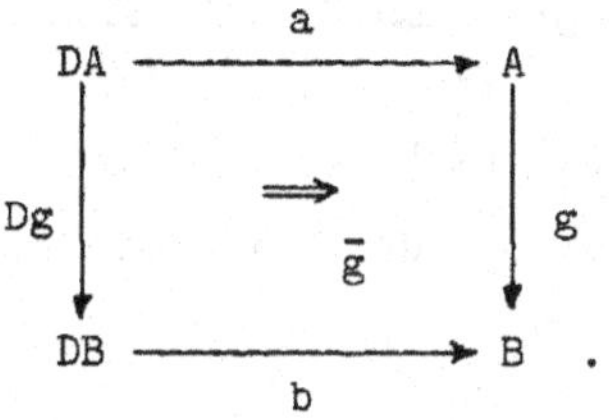

<u>Theorem 2.1</u> $\bar{g}$ <u>is a</u> D-<u>2-cell</u>

(2.7)

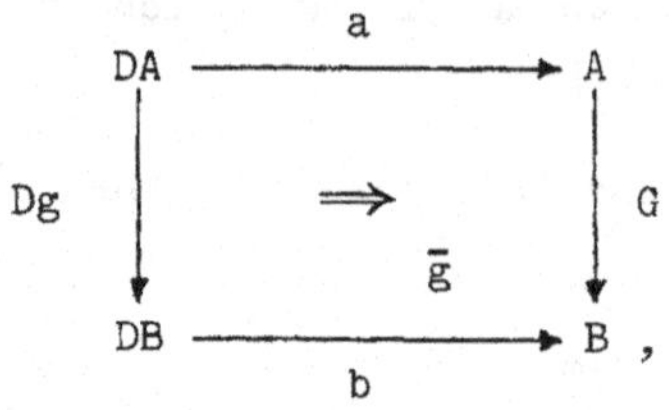

<u>all the edges now being regarded as</u> D-<u>morphisms, of which</u> Dg,a,b <u>are strict</u>.

<u>Proof</u> The D-naturality axiom (3.19) of [13] for $\bar{g}$ reduces to the associativity axiom (3.18) of [13] for a D-morphism $(g,\bar{g})$. □

Observe that, since $b.Dg = \Phi g$ by (2.5), we can also write (2.7) as a D-2-cell

(2.8) $\bar{g}: \Phi g \Rightarrow Ga.$

Observe further that the other axiom for a D-morphism, the identity axiom (3.17) of [13], gives

(2.9)

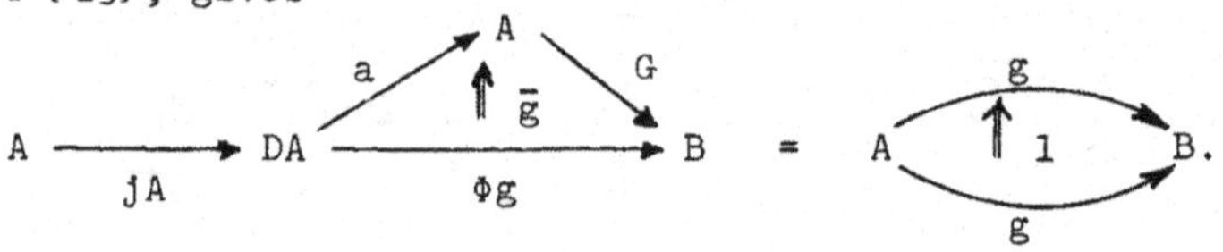

<u>2.2</u> Now, in the situation of Theorem 2.1, replace A by a <u>free</u> D-algebra DA. Define the strict D-morphism ΨG and the D-2-cell ψ_G by:

(2.10)

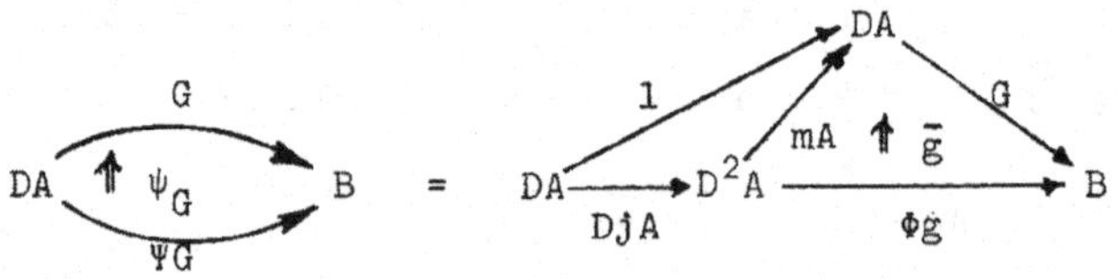

Had we replaced DjA by jDA on the right side of this definition, we should by (2.9) have got merely the identity. Since, however, we have

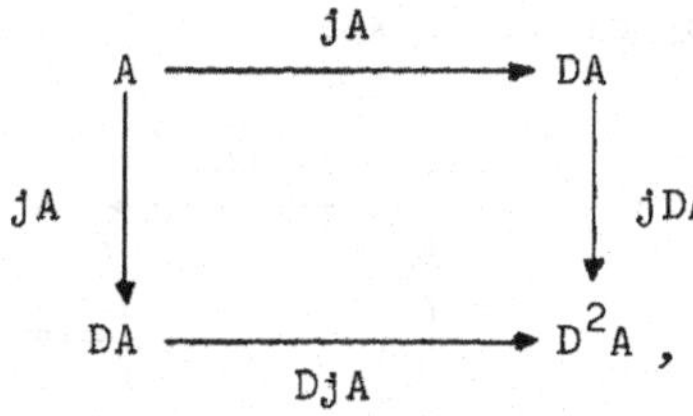

we can conclude from (2.9), not that $\psi_G = 1$, but that

(2.11)
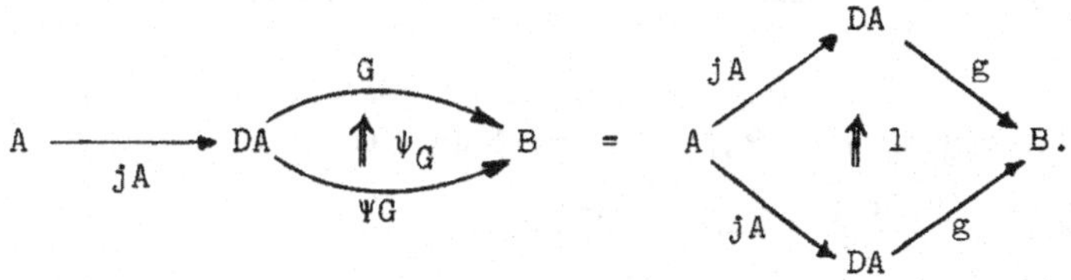

Note in particular that, by (2.4),

(2.12) $\Psi G = \Phi(g.jA)$.

Theorem 2.2 $\psi_G\colon \Psi G \Rightarrow G$ *is the coreflexion of* G *into the strict* D-*morphisms* DA $\to$ B. *That is to say, if* s: DA $\to$ B *is a strict* D-*morphism and* α: s $\Rightarrow$ G *is a* D-2-*cell, then*

(2.13) $$\alpha \;=\; s \xRightarrow[\beta]{} \Psi G \xRightarrow[\psi_G]{} G$$

for a unique D-2-*cell* β. *Explicitly,*

(2.14) $\beta = \Phi(\alpha.jA)$,

and β *is the unique* D-2-*cell satisfying*

(2.15)
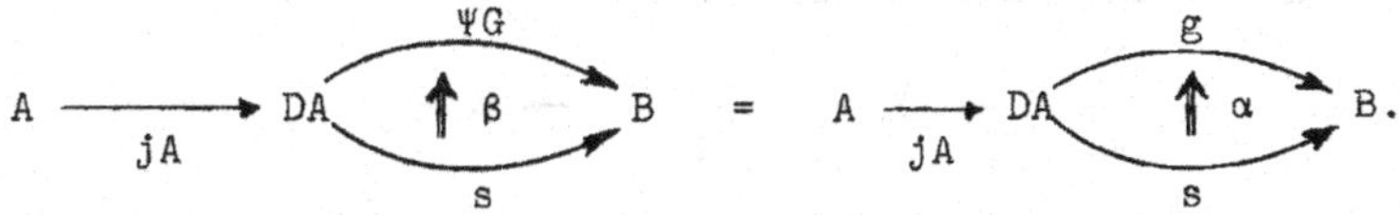

Proof Since $\psi_G.jA = 1$ by (2.11), (2.13) implies (2.15), which in turn gives (2.14) by (2.4), proving the uniqueness of β satisfying (2.13).

Define β therefore by (2.14), so that by (2.5) the explicit value of β is b. D(α.jA). Using this, the definition (2.10), and the fact that mA. DjA = 1, we get (2.13) if we compose with DjA the diagram expressing the D-naturality of α, to wit

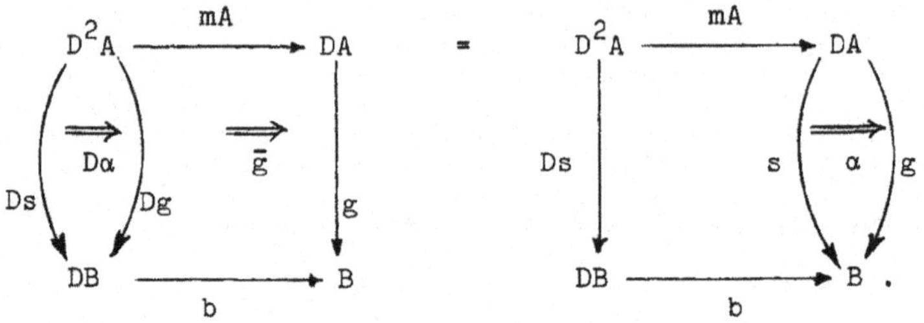

□

The naturality of Ψ, ψ is summed up in:

<u>Proposition 2.3</u> <u>Let</u> G <u>be the composite</u>

(2.16) $$DA \xrightarrow{\ s\ } C \xrightarrow{\ H\ } E \xrightarrow{\ t\ } B$$

<u>where</u> $H = (h, \bar{h})$, t <u>and</u> s <u>are strict</u>, <u>and</u> $s = \Phi r$ <u>for</u> $r\colon\ A \to C$. <u>Then</u>

(2.17)

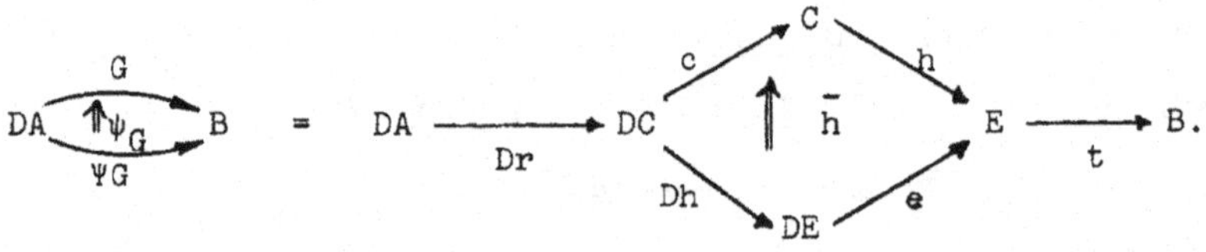

<u>Proof</u> $\psi_G = \bar{g}.\ DjA$ by (2.10)

$= t.\bar{h}.Ds.DjA$ by (2.16)

$= t.\bar{h}.\ Dr$ by (2.4). □

<u>Corollary 2.4</u> <u>Let</u> G <u>be the composite</u>

(2.18) $$DA \xrightarrow{\ Dq\ } DC \xrightarrow{\ H\ } E \xrightarrow{\ t\ } B;$$

<u>then</u>

(2.19)

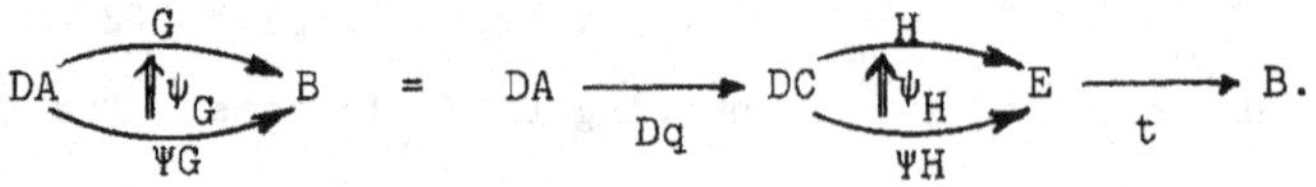

<u>Proof</u> In this case the r of Proposition 2.3 is $jC.q$. □

<u>2.3</u> In [10] §3 above we considered the problem of enriching an adjunction $s \dashv h: B \to B'$ to a D-adjunction, in the case $sh = 1$, given B' as a D-algebra. A condition must be satisfied for the enrichment to exist; when it does, s necessarily becomes a <u>strong</u> D-morphism; and we considered in particular the special case where s is to be strict.

Here we approach the same situation from a different starting point, in the case where B' is a <u>free</u> D-algebra DA.

For this section, we suppose that B is <u>given</u> as a D-algebra; that we are given a strict D-morphism $s: DA \to B$, say $s = \Phi r$ for $r: A \to B$; that we are given a D-morphism $H = (h,\bar{h}): B \to DA$; and that we have

(2.20) $sH = 1$

as D-morphisms. We ask what else we need in order to get a D-2-cell $\eta: 1 \Rightarrow Hs$ satisfying $s\eta = 1$ and $\eta h = 1$, so as to complete the data for a D-reflexion of DA onto B.

Write G for the composite Hs; then by (2.12) $\Psi G = \Phi(h.s.jA)$, which is $\Phi(hr)$ by (2.4). On the other hand $1: DA \to DA$ is a strict D-morphism, and is $\Phi(jA)$. By Theorem 2.2, there is a bijection between D-2-cells $\eta: 1 \Rightarrow Hs = G$ and D-2-cells $\zeta: 1 \Rightarrow \Psi G$; such D-2-cells $\zeta: \Phi(jA) \Rightarrow \Phi(hr)$ are in turn in bijection by (2.2) with (mere) 2-cells $\xi: jA \Rightarrow hr$, as in

(2.21)
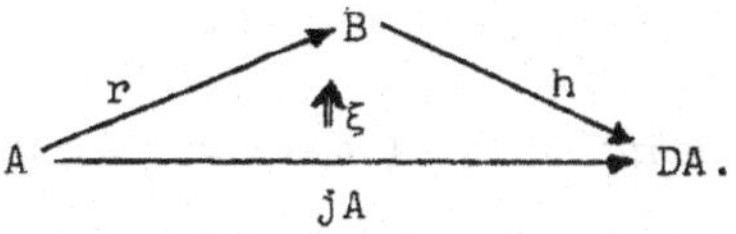

Using (2.5) to write $\zeta = \Phi\xi$, and using (2.17) (with $t = 1$) to write ΨG and ψ_G, we see by Theorem 2.2 that η is the composite

(2.22)

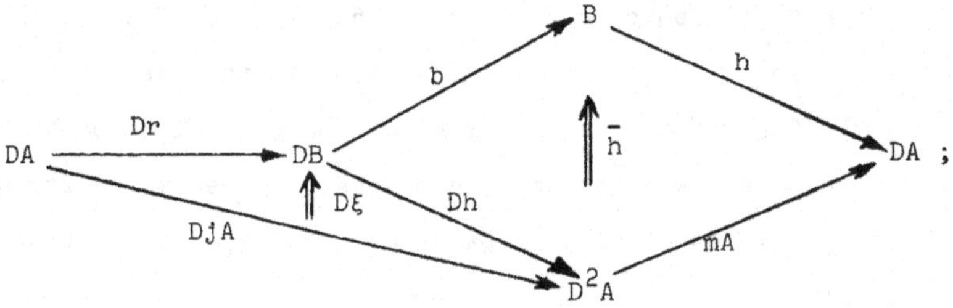

of course the top leg of this is hs since b.Dr = s by (2.5), and the bottom leg is 1 since mA.DjA = 1.

Now consider sη. By (2.20), $s\bar{h} = 1$, so

$$
\begin{aligned}
s\eta &= s.mA.D\xi \\
&= b.\ Ds.\ D\xi \qquad \text{since } s \text{ is strict} \\
&= b.D(s\xi) \\
&= \Phi(s\xi) \quad \text{by } (2.5)\ .
\end{aligned}
$$

Since Φ is an isomorphism of categories, sη = 1 if and only if sξ = 1. Summing all thus up:

Theorem 2.5 *Given* D-*morphisms* s: DA → B *and* H: B → DA *with* s = Φr *strict and with* sH = 1, *there is a bijection as in* (2.22) *between* D-2-*cells* η: 1 ⇒ Hs *and* 2-*cells* ξ: jA ⇒ hr; *and* sη = 1 *if and only if* sξ = 1. □

To get our desired reflexion, we also need ηh = 1; I see no general way of expressing this, and have it verify it by an *ad hoc* method in each case.

There is an important special case which, while it occurs in only one of the two problems studied in this paper, does occur widely in similar problems. It is that where ΨG = 1. Since ΨG = Φ(hr) and 1 = Φ(jA), this is the case where

(2.23)

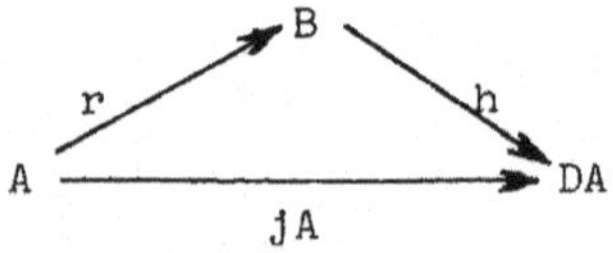

commutes. The condition $s\xi = 1$ is of course automatically satisfied if we now choose $\xi = 1$. On the other hand this choice is forced on us if we want $\eta h = 1$; for then $\eta hr = 1$ and (2.23) gives $\eta . jA = 1$; since $\psi_G . jA = 1$ by (2.11), this gives $\zeta . jA = 1$; which is $\xi = 1$ by (2.4). (Even in this case, $\xi = 1$ is not sufficient for $\eta h = 1$; this still requires ad hoc verification.) Summing up this special case gives:

Theorem 2.6 If the data are as in Theorem 2.5, and if moreover $hr = jA$, there is at most one D-2-cell $\eta: 1 \Rightarrow Hs$ with $s\eta = 1$ and $\eta h = 1$. It must be the D-2-cell $\eta = \bar{h}.Dr$ obtained by setting $\xi = 1$ in (2.22), and this automatically satisfies $s\eta = 1$. It satisfies $\eta h = 1$ if and only if $\bar{h}.Dr.h = 1$. □

3. D-algebras and D-morphisms as doctrine-maps

3.1 We are going to obtain our results announced in §1.1 by applying Theorem 2.5 and its special case Theorem 2.6. So we are seemingly going to get more than we asserted in §1.1; the adjunction $s \dashv h$ is actually going to be a D^*-adjunction. Once given, however, that s is a strict D^*-morphism, as it is in both the results of §1.1, it is automatic by Theorem 1.5 of [10] above that any such adjunction enriches to a D^*-adjunction. So our working with D^*-adjunctions is a requirement not of our results but of our method of proof.

What this method of proof does require, however, is that we can recognize and deal with non-strict D^*-morphisms, when all we are told originally about the unknown doctrine D^* is the 2-category $D^*\text{-}\mathcal{A}lg_*$ of D^*-algebras, strict D^*-morphisms, and D^*-2-cells. Moreover I here mean $D^*\text{-}\mathcal{A}lg_*$ in the narrow sense: a D^*-algebra is an action of D^* on an object of K, not on a 2-functor $L \to K$. However in saying that we are given $D^*\text{-}\mathcal{A}lg_*$, I do mean to imply that we are also given the forgetful functor from $D^*\text{-}\mathcal{A}lg_*$ to K; and it is classical that all this determines D^* to within isomorphism. The purpose of this section is to introduce the simple machinery that allows us to recognize D^*-morphisms from the implicit knowledge of D^*.

3.2 We now assume once for all that the 2-category K has "small hom-categories" - that is, that each $K(A,B)$ is a small category; and that K is 2-complete, i.e. *Cat*-complete. We recall from [4] that this means (a) that K admits all small limits; (b) that these limits are preserved by the representables $K(A,-)$: $K \to Cat$; and (c) that K admits cotensor products $[n,A]$ for $n \in Cat$, $A \in K$. As is well known, we can then construct comma objects in K, using pullbacks and the cotensor product $[\underline{2},A]$ where $\underline{2}$ is the arrow category $0 \to 1$; and these are then 2-comma-objects (the universal property extends to 2-cells). The 2-category *Cat* itself has these properties, as does Cat/Λ for a set Λ.

It is true that many of the common doctrines on *Cat* extend canonically to doctrines on CAT; certainly all those of the form $\mathcal{D}\circ -$ for a small club $\mathcal{D}$ do so - one can form $\mathcal{D}\circ A$ independently of the size of A. For that matter we can always extend D on *Cat* to CAT by right Kan extension. However we do not pursue here the properties of these extensions, which add nothing new in the club case. What is important for us is that D takes small categories to small categories.

We write E for the 2-category $[K,K]$ of endo-2-functors. It is a strict monoidal category with composition as tensor product and 1 as identity. It fails to be a closed category; only for <u>certain</u> S,R in E is it the case that $E(TS,R) \cong E(T,V)$ for some V, namely when the right Kan extension of R along S exists; then this right Kan extension is V.

However E acts on K, in the sense that there is a 2-functor $E\times K \to K$, sending (D,A) to DA, and satisfying the associativity and identity laws $(ED)A = E(DA)$, $1A = A$(along with the corresponding things for 1-cells and 2-cells). Moreover we have 2-naturally

(3.1) $\mathcal{K}(DA,B) \cong \mathcal{E}(D,\{A,B\})$,

where $\{A,B\}$ is the right Kan extension of B: $\mathcal{I} \to \mathcal{K}$ along A: $\mathcal{I} \to \mathcal{K}$. We know explicitly what this is, namely for C in $\mathcal{K}$

(3.2) $\{A,B\}C = [\mathcal{K}(C,A),B]$,

the cotensor product of $\mathcal{K}(C,A) \in \mathcal{C}at$ with $B \in \mathcal{K}$. The counit of the adjunction (3.1) is a 2-natural evaluation e: $\{A,B\}A \to B$, and there is similarly a 2-natural unit d: $D \to \{A,DA\}$. We have formally the same kinds of things as we have in a closed category, in particular a 2-natural multiplication μ: $\{B,C\}\{A,B\} \to \{A,C\}$ and a unit ι: $1 \to \{A,A\}$, which make $\mathcal{K}$ into an $\mathcal{E}$-category. (That the underlying 2-category of this $\mathcal{E}$-category really is $\mathcal{K}$ itself is immediate, since $\mathcal{K}(1,\{A,B\}) \cong \mathcal{K}(A,B)$ by (3.1).)

It is moreover easy to check that the $\mathcal{E}$-valued endomorphism object $\{A,A\}$, with its multiplication μ: $\{A,A\}\{A,A\} \to \{A,A\}$ and its unit ι: $1 \to \{A,A\}$, is a doctrine; and that for any doctrine D, an action a: $DA \to A$ corresponds under (3.1) precisely to a doctrine-map $D \to \{A,A\}$.

3.3 We now write $\mathcal{K}'$ for the functor 2-category $[\underline{2},\mathcal{K}]$ and $\mathcal{K}''$ for the 2-category which is like the $[\![\underline{2},\mathcal{K}]\!]$ of [9] §2.2 above except that its morphisms are not lax natural transformations but op-lax ones; it is what we called $\mathcal{FUN}$ in [9] §10.8 above in the case $\mathcal{K} = \mathcal{CAT}$, and it is described there (in the context of just such considerations as follow, but there specialized to the club case).

So $\mathcal{K}'$ and $\mathcal{K}''$ have the same objects, namely morphisms f: $A \to B$ in $\mathcal{K}$; a typical morphism in $\mathcal{K}''$ is a triple (u,v,α) of the form

(3.3)
$$\begin{array}{ccc} A & \xrightarrow{\;u\;} & A' \\ {\scriptstyle f}\downarrow & \overset{\Rightarrow}{\scriptstyle\alpha} & \downarrow{\scriptstyle f'} \\ B & \xrightarrow[\;v\;]{} & B' \end{array} \;;$$

a typical 2-cell $(u,v,\alpha) \Rightarrow (\bar{u},\bar{v},\bar{\alpha})$ is a pair of 2-cells $\rho\colon\ u \Rightarrow \bar{u}$, $\sigma\colon\ v \Rightarrow \bar{v}$ in $\mathcal{K}$ satisfying the obvious conditions with respect to α and $\bar{\alpha}$; and the various compositions are by pasting. The morphisms of $\mathcal{K}'$ are those of $\mathcal{K}''$ in which $\alpha = 1$, and are thus just pairs (u,v) making the outside of (3.3) commute; its 2-cells are those in $\mathcal{K}''$ between these morphisms.

There are evident actions of $\mathcal{E}$ on $\mathcal{K}'$ and on $\mathcal{K}''$, the first being the restriction of the second; namely (D,f) goes to Df, $(D,(u,v,\alpha))$ goes to $(Du,Dv,D\alpha)$, and in general D is applied to everything in sight, with appropriate arrangements also for the 1-cells and 2-cells of $\mathcal{E}$. Just as in the case of $\mathcal{K}$, these actions have right adjoints; we have 2-natural isomorphisms

(3.4) $\quad \mathcal{K}'(Df,f') \cong \mathcal{E}(D,[f,f'])$,

(3.5) $\quad \mathcal{K}''(Df,f') \cong \mathcal{E}(D,\langle f,f'\rangle)$,

where $[f,f']$ is the pullback

(3.6)
$$\begin{array}{ccc} [f,f'] & \xrightarrow{\ \delta_0\ } & \{A,A'\} \\ {\scriptstyle \delta_1}\downarrow & & \downarrow{\scriptstyle \{1,f'\}} \\ \{B,B'\} & \xrightarrow[\ \{f,1\}\ ]{} & \{A,B'\} \end{array}$$

and $\langle f,f'\rangle$ is the comma object

(3.7)
$$\begin{array}{ccc} \langle f,f'\rangle & \xrightarrow{\ \partial_0\ } & \{A,A'\} \\ {\scriptstyle \partial_1}\downarrow & \overset{\lambda}{\nearrow} & \downarrow{\scriptstyle \{1,f'\}} \\ \{B,B'\} & \xrightarrow[\ \{f,1\}\ ]{} & \{A,B'\} \end{array} \quad ;$$

of course there is a canonical map

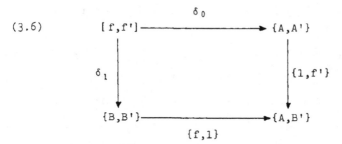

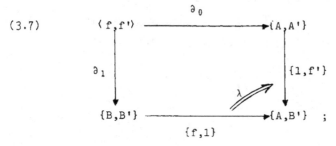

(3.8) $\quad \varepsilon\colon\ [f,f'] \to \langle f,f'\rangle$

with $\partial_0\varepsilon = \delta_0$, $\partial_1\varepsilon = \delta_1$, $\lambda\varepsilon = 1$; it is a monomorphism since δ_0 and δ_1 are jointly monomorphic.

Thus K' and K'' become E-categories, just like K, with the same kind of formal properties. In particular, taking $f' = f$, we observe that $[f,f]$ and $\langle f,f\rangle$ are doctrines; and it is easy to see in this case that δ_0, δ_1, ∂_0, ∂_1, ε are all doctrine-maps.

Moreover for any doctrine D in E, a doctrine-map $k\colon\ D \to \langle f,f\rangle$ corresponds under (3.5) to an action $(a,b,\bar{f})\colon\ Df \to f$ of D on f in K'', which when written out as

(3.9)
$$\begin{array}{ccc} DA & \xrightarrow{\quad a \quad} & A \\ {\scriptstyle Df}\downarrow & \overset{\Longrightarrow}{\bar{f}} & \downarrow{\scriptstyle f} \\ DB & \xrightarrow[\quad b \quad]{} & B \end{array}$$

is easily seen to amount to actions $a\colon\ DA \to A$ and $b\colon\ DB \to B$ of D on A and B, together with an enrichment $\bar{f}$ of f to a D-morphism $F = (f,\bar{f})\colon\ A \to B$. The action $a\colon\ DA \to A$ itself corresponds to the doctrine-map

(3.10) $\quad D \xrightarrow[k]{} \langle f,f\rangle \xrightarrow[\partial_0]{} \{A,A\},$

and similarly for b.

Similarly a doctrine-map $D \to [f,f]$ or an action of D on f in K' amounts to actions of D on A and on B such that f is a <u>strict</u> D-map; the situation of (3.9) with $\bar{f} = 1$. Since ε is a monomorphism, there is no difference between a doctrine-map $D \to [f,f]$ and a doctrine-map $D \to \langle f,f\rangle$ that happens to factorize through $[f,f]$.

<u>3.4</u> Finally, in order to describe D-2-cells as doctrine-maps, we go one step further. Consider $(K'')' = [\underline{\underline{2}},K'']$, the 2-category of morphisms of K''; so an object of $(K'')'$ is a triple (u,v,α) as in (3.3). Now write $K^{\#}$ for the <u>full</u> sub-2-category of $(K'')'$ determined by those

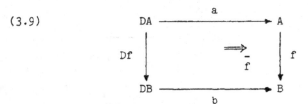

objects $(1,1,\gamma)$; that is, those of the form

(3.11)
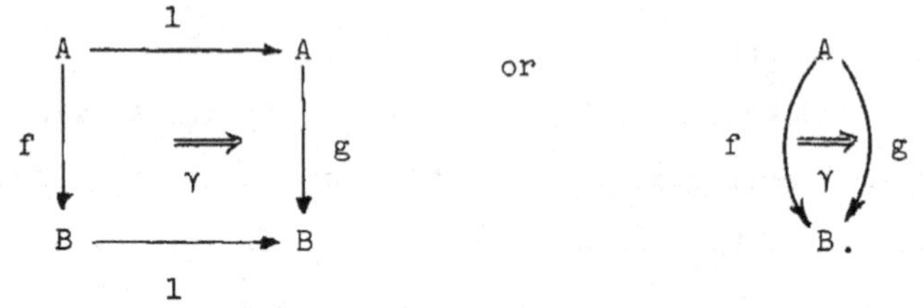

Calling this object γ for short, a morphism from γ to γ' in $K^{\#}$ consists of pairs (u,v,α): $f \to f'$ and (x,y,β): $g \to g'$ in K'' such that $(x,y,\beta)(1,1,\gamma) = (1,1,\gamma')(u,v,\alpha)$. This last implies that $x = u$, $y = v$; so finally a morphism $\gamma \to \gamma'$ in $K^{\#}$ consists of (u,v,α,β) satisfying

(3.12)
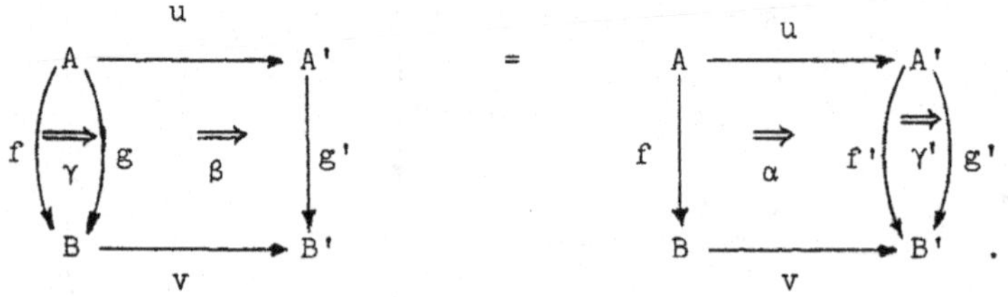

Then a 2-cell $(u,v,\alpha,\beta) \Rightarrow (\bar{u},\bar{v},\bar{\alpha},\bar{\beta})$ in $K^{\#}$ consists of a pair of 2-cells in K'', say (ρ,σ): $(u,v,\alpha) \Rightarrow (\bar{u},\bar{v},\bar{\alpha})$ and (θ,ϕ): $(u,v,\beta) \Rightarrow (\bar{u},\bar{v},\bar{\beta})$, satisfying the equation

(3.13)
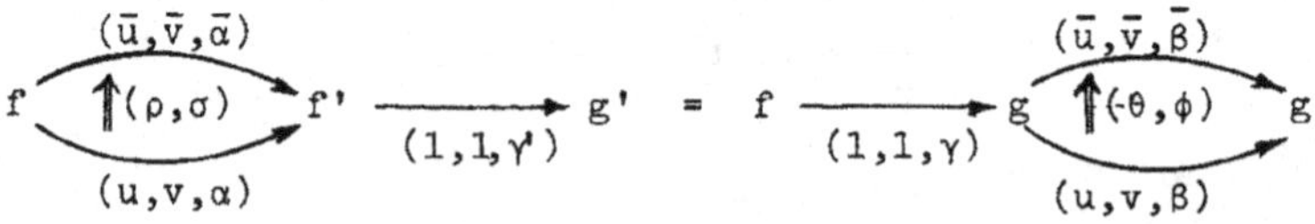

in K''. But this last equation gives $\theta = \rho$, $\phi = \sigma$; so in the end a 2-cell $(u,v,\alpha,\beta) \Rightarrow (\bar{u},\bar{v},\bar{\alpha},\bar{\beta})$ in $K^{\#}$ consists of a pair ρ: $u \Rightarrow \bar{u}$, σ: $v \Rightarrow \bar{v}$ satisfying the conditions that (ρ,σ): $(u,v,\alpha) \Rightarrow (\bar{u},\bar{v},\bar{\alpha})$ and that (ρ,σ): $(u,v,\beta) \Rightarrow (\bar{u},\bar{v},\bar{\beta})$ are 2-cells in K'', namely

(3.14)
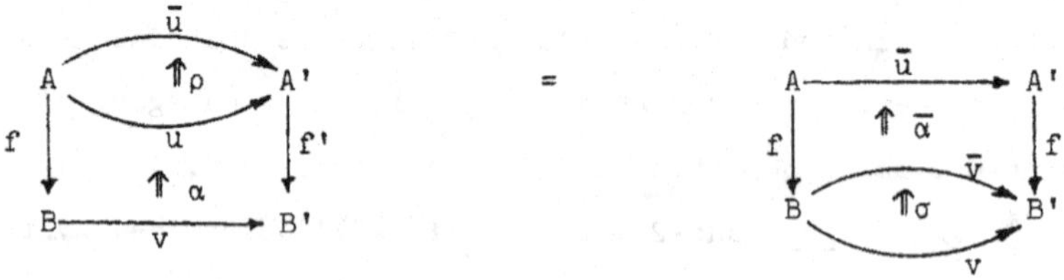

,

(3.15)

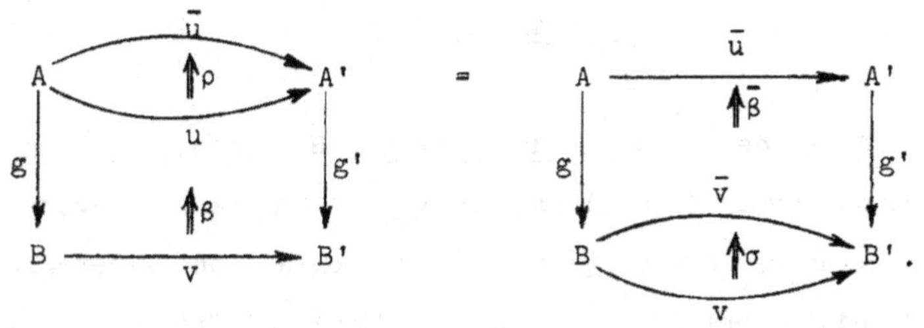

Now since E acts on K'' it also acts on $(K'')'$, and this action restricts to one on $K^\#$; and we have an evident 2-natural isomorphism

(3.16) $K^\#(D\gamma,\gamma') \cong E(D,[\![\gamma,\gamma']\!])$

where $[\![\gamma,\gamma']\!]$ is the pullback

(3.17)

$$
\begin{array}{ccc}
[\![\gamma,\gamma']\!] & \xrightarrow{\ \ \pi_0\ \ } & \langle f,f'\rangle \\[2mm]
\ \ \downarrow{\scriptstyle \pi_1} & & \ \ \downarrow{\scriptstyle \langle 1,\gamma'\rangle} \\[2mm]
\langle g,g'\rangle & \xrightarrow[\ \langle \gamma,1\rangle\ ]{} & \langle f,g'\rangle\ .
\end{array}
$$

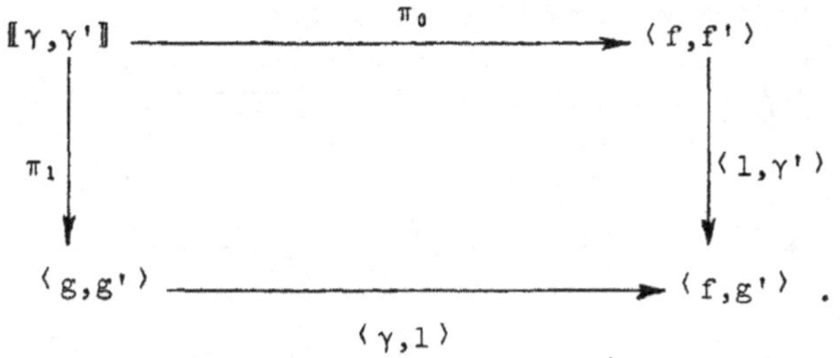

Once again when $\gamma' = \gamma$ we get a doctrine $[\![\gamma,\gamma]\!]$, and π_0,π_1 are then doctrine-maps. It is clear that a doctrine-map $D \to [\![\gamma,\gamma]\!]$, or equally an action of D on γ in $K^\#$, is just a D-2-cell $\gamma \colon F \twoheadrightarrow G$ between the D-morphisms $F = (f,\bar{f}) \colon A \to B$ and $G = (g,\bar{g}) \colon A \to B$, as defined in (3.19) of [13].

3.5   There is an analogue of (3.7) in which λ is an <u>isomorphism</u>: it can be called either the <u>strong comma object</u> or the <u>pseudo pullback</u>. Each of the main results stated in §1.1 has both a "lax" and a "pseudo" case. In order to make our treatment of these cases formally identical, we agree to use $\langle f,f\rangle$ in the strong sense, without introducing a new name for it, when handling the "pseudo" case. Then the $\bar{f}$ of (3.9) is invertible, etc.

4. Lax algebras

4.1 We first define a lax doctrine map $H = (h, \tilde{h}, h^{o})$: $D \to D^*$ between two doctrines (D,m,j) and (D^*,m^*,j^*) on the 2-category K. The most elegant way of doing it would be to take the 2-category $E = [K,K]$, and to form E'' from it as we formed K'' from K in §3.2; then to observe that E'' was a strict monoidal category, the tensor product of the objects f: $D \to E$ and f': $D' \to E'$ being ff': $DD' \to EE'$; and finally to define a lax doctrine map as an object h: $D \to D^*$ of E'' with the structure of a monoid for this tensor product. However we say the same thing in elementary terms.

A lax doctrine map, then, consists of a 2-natural transformation h: $D \to D^*$ together with modifications $h^{o}, \tilde{h}$ as in

(4.1)
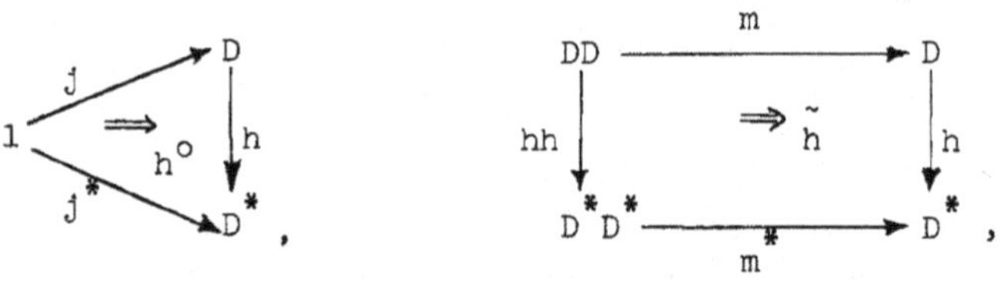

satisfying the axioms

(4.2)
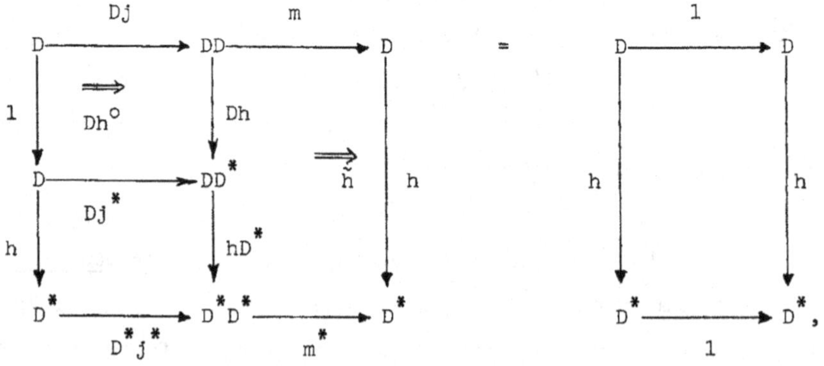

(4.3)

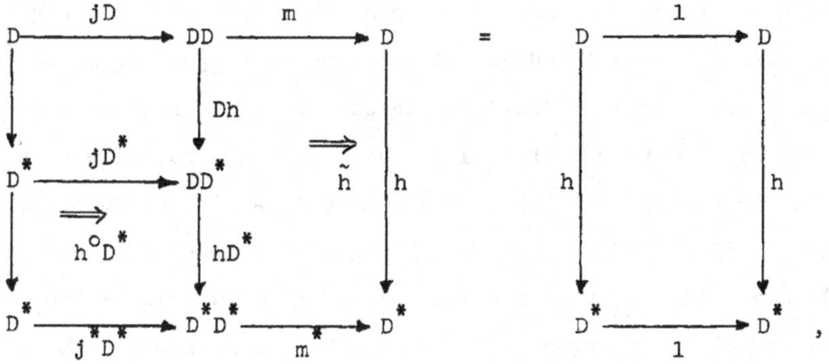

(4.4)

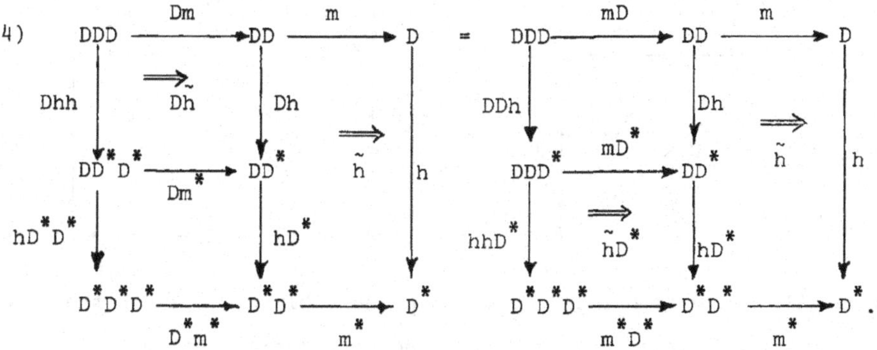

Lax doctrine maps compose by vertical pasting of diagrams (4.1); that the composite satisfies the axioms follows easily from the 2-naturality of h etc. Thus we get a category of doctrines and lax doctrine maps, which becomes a 2-category *Lax Doct* when we define a 2-cell to be a modification ρ: $h \Rightarrow k$ satisfying the obvious axioms with respect to $h^{\circ}, \tilde{h}, k^{\circ}, \tilde{k}$. It has the sub-2-category *Doct* with the same objects and the same 2-cells, but whose 1-cells are those with $\tilde{h}$ and h° identities; these are the ordinary (strict) doctrine maps, and the 2-cells are the modifications of these, as defined in [13] §3.6.

There are lots of 2-categories intermediate between *Lax Doct* and *Doct*. We call the lax doctrine map $H = (h, \tilde{h}, h^{\circ})$ a <u>pseudo</u> doctrine map if $\tilde{h}$ and h° are isomorphisms; we call it <u>normal</u> or <u>normalized</u> if

h^o is an identity. Our arguments below are expressed in terms of the relation between *Lax Doct* and *Doct*, but they go over absolutely unchanged if we replace "lax" by "pseudo" or if we require normality.

4.2 For $A \in K$, we define a lax action of the doctrine D on A to be a lax doctrine map K: $D \to \{A,A\}$ where $\{A,A\}$ is the doctrine defined in §3.2. If k: $D \to \{A,A\}$ corresponds under the isomorphism (3.1) to a: $DA \to A$, then the kind of argument familiar in the context of closed categories shows that $k^o, \tilde{k}$ correspond respectively to $\hat{a}$, $\bar{a}$ in

(4.5)

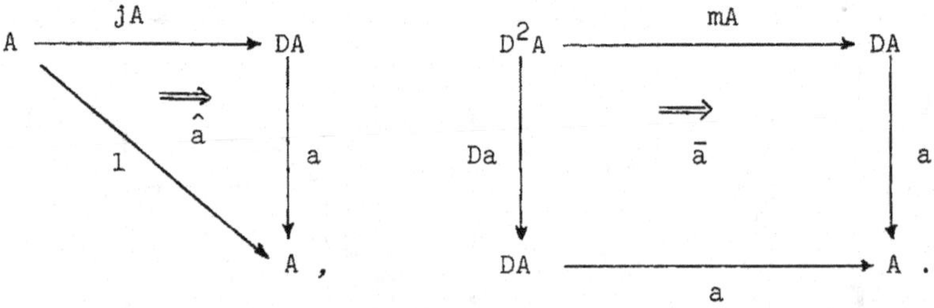

(In detail, we go from (4.1) to (4.5) by multiplying (4.1) on the right by A and composing on the tail end with the evaluation e: $\{A,A\}A \to A$. We go in the other direction by applying $\{A,-\}$ to (4.5) and composing on the front with the unit $E \to \{A,EA\}$ where $E = 1$ or D^2.) Equally simple calculations show that the axioms (4.2)-(4.4) are equivalent to the axioms (4.6)-(4.8) below:

(4.6)

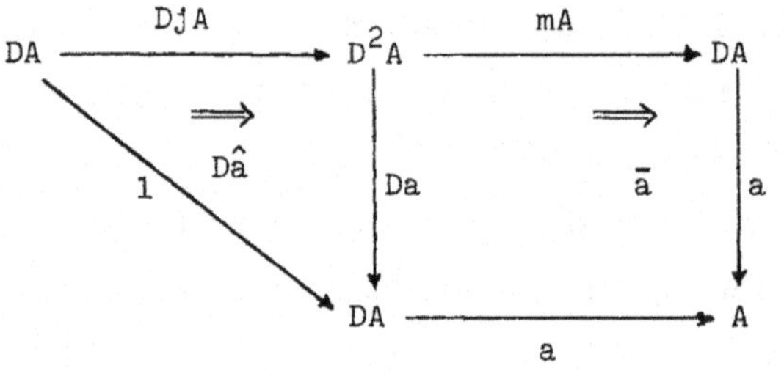

= identity,

(4.7)

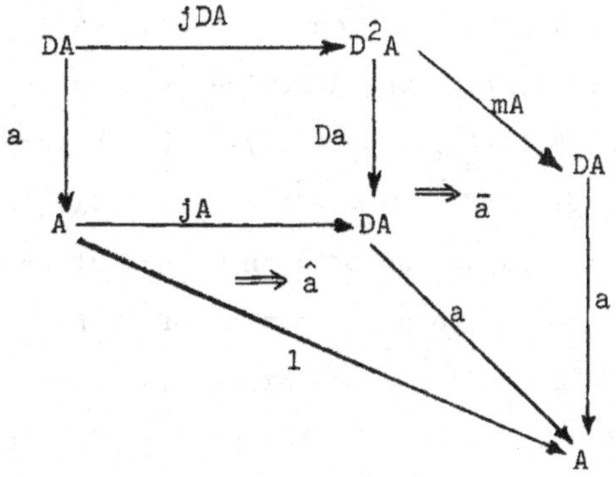

= identity,

(4.8)

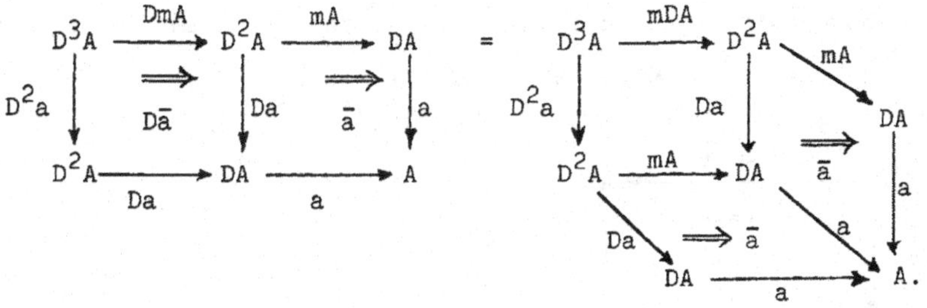

We more commonly use the term <u>lax action of</u> D <u>on</u> A not for the lax doctrine map $K = (k,\tilde{k},k^{\circ})$: $D \to \{A,A\}$ but for the above triple $(a,\bar{a},\hat{a})$ satisfying the axioms (4.6)-(4.8). This latter definition can be used even when A is not an object of K but a morphism $L \to K$, in which case the first definition fails because the right Kan extension {A,A} of A along itself may not exist; and we do use it in this extended sense. The A with such a lax action of D is called a <u>lax</u> D-<u>algebra</u>; note that this definition agrees precisely with that of Street in [24] §2 above. Moreover $\bar{a}$, for instance, is an isomorphism or an identity precisely when $\tilde{k}$ is ; so we carry over the words "pseudo", "normal" from the lax doctrine maps to the lax algebras. Finally the strict doctrine maps correspond of course to the honest D-algebras, where both $\bar{a}$ and $\hat{a}$ are identities.

<u>4.3</u> For f: A → B in K, and a doctrine D on K, consider a lax doctrine map P: D → ⟨f,f⟩ where ⟨f,f⟩ is the doctrine defined in §3.3. Because the passage from (4.1)-(4.4) to (4.5)-(4.8) is purely formal, depending only on the action of E on K satisfying (3.1), we can repeat it all at the level of the action of E on K'' satisfying (3.5). We conclude that P corresponds to a lax action of D on f in K'', given by precisely the data (4.5) and the axioms (4.6)-(4.8), but with A replaced by f and with the 1-cell a and the 2-cells $\hat{a},\bar{a}$ of K replaced by 1-cells and 2-cells of K''.

Now consider what this means. Let the morphism Df → f in K'' replacing a: DA → A in K be $(a,b,\bar{f})$ as in

(4.9) 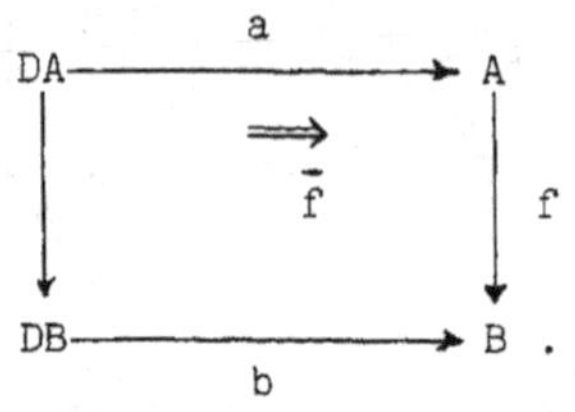

Let the 2-cells in K'' replacing the 2-cells $\hat{a},\bar{a}$ of K be the respective pairs $(\hat{a},\hat{b})$, $(\bar{a},\bar{b})$. By the definition of what a 2-cell <u>is</u> in K'', these have to satisfy

(4.10) 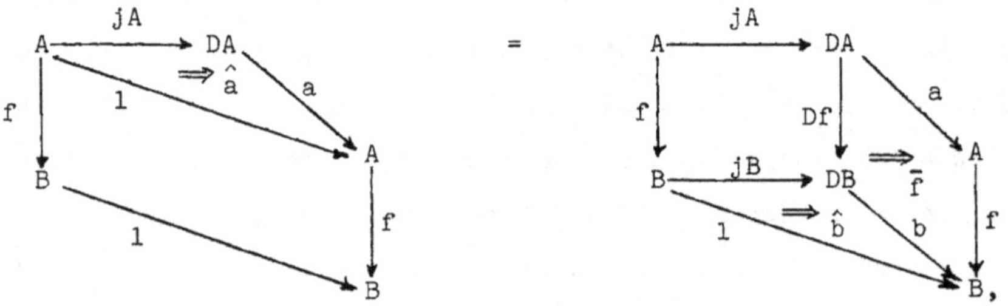

(4.11) 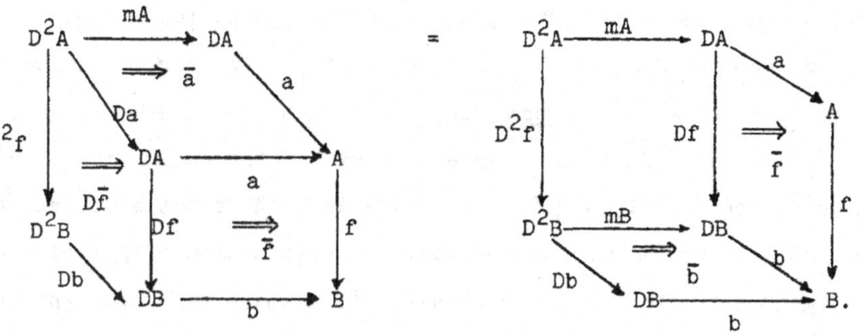

When it comes to the axioms (4.6)-(4.8), we observe that these are essentially about 2-cells in K'', which are only pairs of 2-cells in K - satisfying conditions indeed, but these latter automatic once given (4.10) and (4.11). Thus (4.6)-(4.8) for the lax action of D on f reduce to the original (4.6)-(4.8) for $\hat{a},\bar{a}$ and for $\hat{b},\bar{b}$. Putting all this together, we see that a lax action of D on f in K'' is the same thing as a lax action $(a,\bar{a},\hat{a})$ of D on A (corresponding of course to the lax doctrine map

$$D \xrightarrow[P]{} \langle f,f\rangle \xrightarrow[\partial_0]{} \{A,A\}),$$

a corresponding lax action $(b,\bar{b},\hat{b})$ of D on B, and an $\bar{f}$ as in (4.9) satisfying (4.10) and (4.11).

We call the pair $F = (f,\bar{f})$ a <u>morphism of lax D-algebras</u> $F\colon A \to B$; it is just what Street calls a "lax homomorphism" in [24] §2 above. We call it a <u>strong morphism</u> if $\bar{f}$ is an isomorphism, and a <u>strict</u> one if $\bar{f}$ is an identity. Of course an argument precisely similar to that above identifies a strict morphism $f\colon A \to B$ of lax D-algebras as a lax action of D on f in K', and hence as a lax doctrine map $Q\colon D \to [f,f]$. This shows that, for lax doctrine maps just as for strict ones, there is no difference between one into $[f,f]$, and one into $\langle f,f\rangle$ that happens to factorize through $\varepsilon\colon [f,f] \to \langle f,f\rangle$; which could alternatively have been seen by going back to the definitions of $[f,f]$ and $\langle f,f \rangle$ as a pullback and a comma object. This point needs to be kept in mind for our argument below. (Note that we get the <u>strong</u> morphisms by using the different $\langle f,f \rangle$ of §3.5.)

Finally, just as we agreed to extend the definition of lax algebra given by (4.5)-(4.8) to the case where A is not an object of K but a 2-functor $L \to K$, so we extend the definition of morphism of lax algebras to this case, by (4.9)-(4.11), where f is now 2-natural and $\bar{f}$ is a modification.

<u>4.4</u> Lastly in this hierarchy, we have to consider a lax doctrine map $R\colon\ D \to [\![\gamma,\gamma]\!]$, where $\gamma\colon\ f \Rightarrow g\colon\ A \to B$ in K as in (3.11), and $[\![\gamma,\gamma]\!]$ is the doctrine defined in §3.4. Arguing as above, we see that this is the same thing as a lax action of D on γ in $K^{\#}$. Let the analogue of the $a\colon\ DA \to A$ of §4.2 be $(a,b,\bar{f},\bar{g})\colon\ D\gamma \to \gamma$, so that as in (3.12) we have

(4.12) 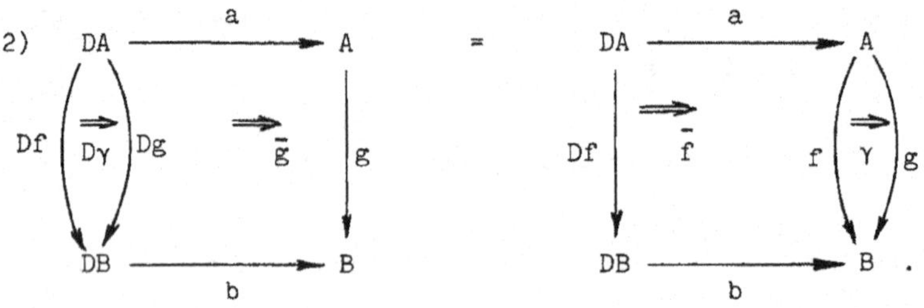

Let the analogues of the 2-cells $\hat{a},\bar{a}$ of §4.2 be the 2-cells $(\hat{a},\hat{b})$ and $(\bar{a},\bar{b})$ of $K^{\#}$. Then $\hat{a}$ and $\bar{a}$ are as in (4.5), and $(\hat{b},\bar{b})$ similarly. The axioms (3.14) and (3.15) that these must satisfy to be 2-cells in $K^{\#}$ reduce to (4.10) and (4.11), with their analogues for $g,\bar{g}$. Finally the lax-action axioms (4.6)-(4.8) for $(\hat{a},\hat{b})$ and $(\bar{a},\bar{b})$ reduce to (4.6)-(4.8) for $\hat{a},\bar{a}$ and the analogues for $\hat{b},\bar{b}$. Summing up, we have lax D-algebra structures on A and B, morphisms $F,G\colon\ A \to B$ of these where $F = (f,\bar{f})$ and $G = (g,\bar{g})$, and finally a 2-cell $\gamma\colon\ f \Rightarrow g$ satisfying (4.12) (which is identical with the (3.19) of [13], defining D-2-cells in the case of honest D-algebras).

We therefore call such a γ satisfying (4.12) a D-2-cell even in this lax case; and we extend this definition to the case where the lax D-algebras A and B are not objects of K but 2-functors $L \to K$, in which case γ is of course a modification.

Now that we have lax D-algebras, morphisms of these, and D-2-cells, we make a 2-category *Lax-D-Alg* of these elements, defining the various compositions by the evident pasting operations; if we restrict to the strict morphisms of lax D-algebras, we get the sub-2-category *Lax-D-Alg*$_*$. There is a similar restriction to the strong morphisms.

4.5 It is easy to see, although tedious to write out, that the inclusion of *Doct* in *Lax Doct* is 2-continuous: it preserves limits and cotensor products. It follows that a doctrine D, regarded as an object of *Lax Doct*, admits a 2-reflexion $H = (h, \tilde{h}, h^{\circ})$: $D \to D^*$ into *Doct*, provided that the appropriate solution-set condition is satisfied for the given D. That is to say, there is a doctrine D^* and a lax doctrine map H: $D \to D^*$ such that any lax doctrine map K: $D \to E$ is of the form K = tH for a unique strict doctrine map t: $D^* \to E$. We get the same 2-continuity if we replace "lax" by "pseudo", or if we require normality; and exactly the same considerations apply.

I do not intend to study in this paper the conditions under which, for a given D, the solution-set condition is satisfied and the reflexion D^* exists. Certainly when D is $\mathcal{D}\circ-$ for a small club $\mathcal{D}$ in *Cat*/$\underline{\underline{S}}$ it is immediate that D^* exists, as we see below in §4.10. I suppose the methods of Barr [1] or of Dubuc [5] will show D^* to exist if say K is locally presentable and D has a rank. In part I leave the question aside because I haven't thought out the details; but in part too because I suspect that it doesn't really matter: it seems to me likely that D^* always exists as a doctrine on *CAT* when D is one on *Cat*, or that D^* exists on a suitable completion of K in a bigger universe when D is given on K. The question is connected with that of extensions of D adumbrated in the second paragraph of §3.2, and really deserves a fuller treatment in its own right at another time. Our concern in this paper is with the consequences of the existence of D^*, and the club case is sufficient to show that our considerations are not vacuous.

A final point in this vein: I assumed in [10] above, for the proof of Theorem 3.1 of that paper, that D^* (there called D') does exist. If it does so only in some bigger universe, that in no way affects its utility for the proof of that theorem. And if it does not, there is no doubt that that theorem remains true, replacing " D'-category" by "normalized pseudo D-category" and so on; we should merely be denied the use of the results of [10] §1 as they stand, and should have to prove them over again in the pseudo-algebra case, which would be an awful nuisance. But no-one would doubt that such purely formal diagram-arguments, which clearly remain such when translated from D'-terms to pseudo-D-terms, would be valid independently of the "size" of D - even though the metatheorem involved is felt rather than stated.

4.6 Coming back to the main point, we identify the D^*-algebras A in $\mathcal{K}$ - something we have anticipated in the last paragraph. Such an algebra is given by a strict doctrine map $t\colon\ D^* \to \{A,A\}$; and these are in bijection under $K = tH$ with lax doctrine maps $K\colon\ D \to \{A,A\}$. Thus the D^*-algebras in $\mathcal{K}$ are just the lax D-algebras. Moreover if K corresponds as before to the lax action $(a,\bar{a},\hat{a})$, while t corresponds to the honest action $a^*\colon\ D^*A \to A$, the relation $K = tH$ translates into

$$(4.13)\qquad a = a^*.hA,\quad \bar{a} = a^*.\tilde{h}A,\quad \hat{a} = a^*.h^{\circ}A.$$

Otherwise put, every lax D-action is of the form (4.13) for a unique D^*-action a^*.

Similarly a morphism $F^* = (f,\bar{f}^*)\colon\ A \to B$ of D^*-algebras is given by a strict doctrine map $D^* \to \langle f,f\rangle$ and hence by a lax doctrine map $D \to \langle f,f\rangle$, which corresponds to a morphism $F = (f,\bar{f})\colon\ A \to B$ of lax D-algebras. The connexion between F and F^* is given by

$$(4.14)\qquad \begin{array}{ccc} DA & \xrightarrow{\;a\;} & A \\ {\scriptstyle Df}\downarrow & \overset{\Longrightarrow}{\bar{f}} & \downarrow{\scriptstyle f} \\ DB & \xrightarrow[\;b\;]{} & B \end{array} \quad = \quad \begin{array}{ccccc} DA & \xrightarrow{\;hA\;} & D^*A & \xrightarrow{\;a^*\;} & A \\ {\scriptstyle Df}\downarrow & & {\scriptstyle D^*f}\downarrow & \overset{\Longrightarrow}{\bar{f}^*} & \downarrow{\scriptstyle f} \\ DB & \xrightarrow[\;hB\;]{} & D^*B & \xrightarrow[\;b^*\;]{} & B. \end{array}$$

Again, we can say that any F is of the form (4.14) for a unique F^*. Since moreover $\varepsilon\colon\ [f,f] \to \langle f,f\rangle$ is a strict doctrine map, $D^* \to \langle f,f\rangle$ factorizes through $[f,f]$ if and only if $D \to \langle f,f\rangle$ does so; whence

$$(4.15)\qquad \bar{f} = 1 \quad \text{if and only if} \quad \bar{f}^* = 1,$$

that is, F is strict if and only if F^* is so.

Finally doctrine maps $D^* \to [\![\gamma,\gamma]\!]$ correspond to lax doctrine maps $D \to [\![\gamma,\gamma]\!]$, and we conclude that

(4.16) the D^*-2-cells $\gamma\colon\ F^* \Rightarrow G^*$ are just the D-2-cells $\gamma\colon\ F \Rightarrow G$.

Moreover the correspondences (4.14) and (4.15) respect the various kinds of composition; the only case that is not absolutely trivial is that of vertical pasting of diagrams (4.14), which clearly produces another diagram of the same kind by the naturality of h. Thus we have identified the 2-category D^*-*Alg* [resp. D^*-*Alg*$_*$], namely as essentially the 2-category *Lax-D-Alg* [resp. *Lax-D-Alg*$_*$]. Since we also know the forgetful 2-functor to *K*, we do indeed have an implicit determination of D^*. (This whole section remains valid for the stronger $\langle f,f\rangle$ of §3.5.)

<u>4.7</u> For the purposes of our proof, we need the results of §4.6 not only for algebras A that are objects of *K* but for algebras that are 2-functors A: $L \to K$ (*L* being *K* itself in our applications). Since for such an A the right Kan extension $\{A,A\}$ of A along itself need not exist, we cannot argue directly as in §4.6. Instead we have to argue object-wise, <u>using</u> the results of §4.6.

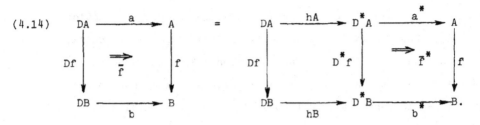

Let $A\colon \mathcal{L} \to \mathcal{K}$ then be a lax D-algebra, with lax action $(a,\bar{a},\hat{a})$. For each $X \in \mathcal{L}$ we get a lax D-action $(aX,\bar{a}X,\hat{a}X)$ on AX, and hence as in (4.13) a unique D^*-action a^*X on AX such that $aX = a^*X.hAX$, $\bar{a}X = a^*X.\tilde{h}AX$, and $\hat{a}X = a^*X.h^{\circ}AX$. What has to proved is that a^* is 2-natural, so that a^* is indeed a D^*-action on A, and of course the unique one satisfying (4.13).

To show the naturality of a^*, consider $\phi\colon\ X \to Y$ in $\mathcal{L}$. Because a is natural, we have commutativity in

(4.17)
$$\begin{array}{ccc} DAX & \xrightarrow{\quad aX \quad} & AX \\ {\scriptstyle DA\phi}\big\downarrow & & \big\downarrow{\scriptstyle A\phi} \\ DAY & \xrightarrow[\quad aY \quad]{} & AY \end{array} ;$$

and $A\phi$ is in fact a strict morphism of lax D-algebras, the axioms (4.10) and (4.11) being satisfied because $\hat{a}$ and $\bar{a}$ are modifications. Hence by (4.14) and (4.15) we have commutativity in

(4.18)
$$\begin{array}{ccc} D^*AX & \xrightarrow{\quad a^*X \quad} & AX \\ {\scriptstyle D^*A\phi}\big\downarrow & & \big\downarrow{\scriptstyle A\phi} \\ D^*AY & \xrightarrow[\quad a^*Y \quad]{} & AY \end{array} ,$$

showing the naturality of a^*. For its 2-naturality, let $\alpha\colon\ \phi \Rightarrow \psi\colon\ X \to Y$ in $\mathcal{L}$. Then $A\alpha\colon\ A\phi \Rightarrow A\psi$ is a D-2-cell by the 2-naturality of a, whence it is a D^*-2-cell by (4.16), giving the 2-naturality of a^*.

Next, let $A,B\colon\ \mathcal{L} \to \mathcal{K}$ be lax D-algebras and $F = (f,\bar{f})\colon\ A \to B$ a morphism of these. Then $FX = (fX,\bar{f}X)\colon\ AX \to BX$ is a morphism of lax D-algebras in $\mathcal{K}$, which therefore corresponds as in (4.14) to a unique morphism $F^*X = (fX,\bar{f}^*X)$ of D^*-algebras in $\mathcal{K}$, where $\bar{f}X = \bar{f}^*X.hAX$. If

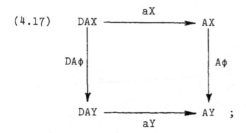

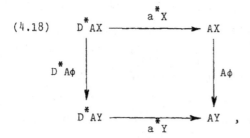

F is strict, so is F^*, and there is no more to prove; but in the general case we must show that $\bar{f}^*$ is a modification. This means the equality

$$
\begin{array}{ccc}
D^*AX & \xrightarrow{\;a^*X\;} & AX \\
{\scriptstyle D^*fX}\downarrow & \overset{\Longrightarrow}{\bar{f}^*X} & \downarrow{\scriptstyle fX} \\
D^*BX & \xrightarrow[\;b^*X\;]{} & BX \\
{\scriptstyle D^*B\phi}\downarrow & & \downarrow{\scriptstyle B\phi} \\
D^*BY & \xrightarrow[\;b^*Y\;]{} & BY
\end{array}
\quad = \quad
\begin{array}{ccc}
D^*AX & \xrightarrow{\;a^*X\;} & AX \\
{\scriptstyle D^*A\phi}\downarrow & & \downarrow{\scriptstyle A\phi} \\
D^*AY & \xrightarrow{\;a^*Y\;} & AY \\
{\scriptstyle D^*fY}\downarrow & \overset{\Longrightarrow}{\bar{f}^*Y} & \downarrow{\scriptstyle fY} \\
D^*BY & \xrightarrow[\;b^*Y\;]{} & BY \; ;
\end{array}
$$

which we have because the correspondence (4.14) respects composition, because under this correpondence $\bar{f}^*X$ corresponds to $\bar{f}X$ and (4.18) to (4.17), and because $\bar{f}$ is a modification by hypothesis.

At the next level, that of D-2-cells, there is by (4.16) nothing to prove.

We conclude that the results of §4.6, expressed in the form that every lax D-action $(a,\bar{a},\hat{a})$ is of the form (4.13) for a unique D^*-action a^*, etc., remain true when we take the algebras in the present, wider, sense.

4.8 We can now proceed rapidly to the proof of our first main result.

Theorem 4.1 *Let* D *be a doctrine on the* 2-*complete* 2-*category* $\mathcal{K}$. *Suppose that, considered as an object of* **Lax Doct**, D *admits a reflexion* $H = (h,\tilde{h},h^o)\colon\; D \to D^*$ *into* **Doct**. *Let* $s\colon\; D^* \to D$ *be the unique strict doctrine map satisfying*

(4.19) $\quad sH \;=\; 1,$

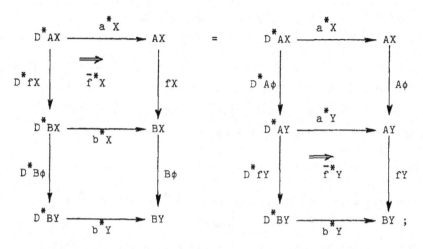

which exists because 1: $D \to D$ is a strict doctrine map and a fortiori a lax one. Then there is a modification η: $1 \Rightarrow hs$ such that $s\eta = 1$ and $\eta h = 1$. The same is true if we require normality, or if we replace "lax" by "pseudo"; in the last case η is an isomorphism.

Proof. The last sentence of the theorem is immediate from the comments at the end of §4.1, from those in the first paragraph of §4.5, and from the details of the following proof, in the light of §3.5.

Observe that, by the definition of composition of lax doctrine maps, (4.19) may be written as

(4.20) $\quad sh = 1, \quad s\tilde{h} = 1, \quad sh^{o} = 1.$

Because s is a doctrine map, $m.sD$: $D^*D \to D$ is a D^*-action on D, as in (3.8) of [13]. The lax D-action on D which corresponds to this under (4.13) is, by (4.20), the lax action (m,1,1); that is to say, the strict action m: $D^2 \to D$ of D on itself.

The D^*-action m^*: $D^* D^* \to D^*$ of D^* on itself corresponds by (4.13) to the lax action $(m^*.hD^*,\ m^*.\tilde{h}D^*,\ m^*.\ h^{o}D^*)$ of D on D^*.

The second diagram of (4.1) may be written as

(4.20)

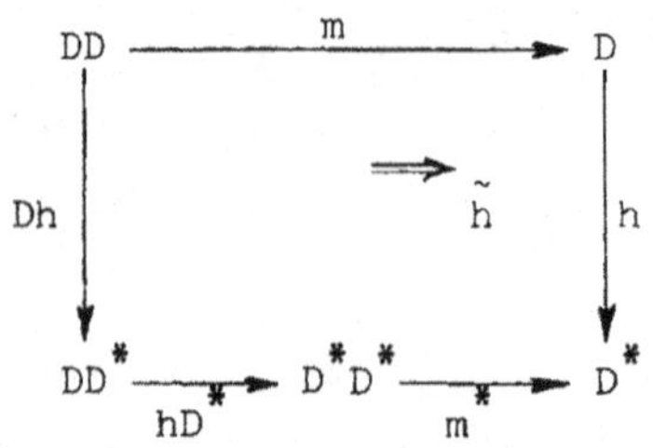

I assert that $\tilde{H} = (h,\tilde{h})$ is a morphism $D \to D^*$ of lax D-algebras. For the axiom (4.10) in this case reduces to (4.3), and the axiom (4.11) reduces to (4.4). It follows as in (4.14) that there is a unique morphism $H^* = (h,\bar{h}^*)$: $D \to D^*$ of D^*-algebras satisfying

(4.21)

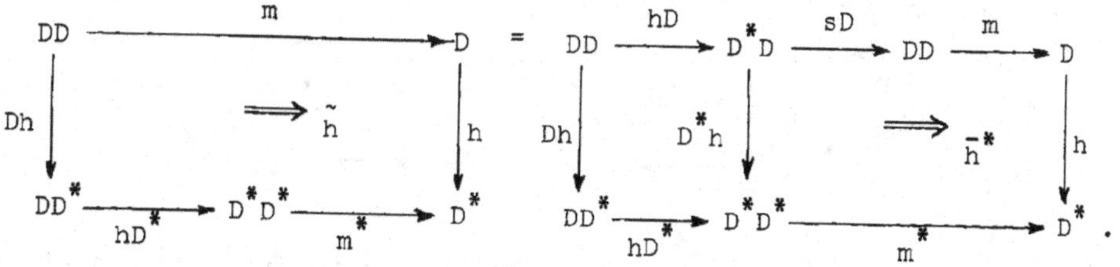

On the other hand $s\colon D^* \to D$ is a strict morphism of D^*-algebras, and hence a strict morphism of lax D-algebras. Since $s\tilde{H} = 1$ by (4.20) and since the correspondence (4.14) respects composition, we have

(4.22) $\quad sH^* = 1.$

We are now in a position to apply Theorem 2.5; with of course D replaced by D^*, DA replaced by $D^*1 = D^*$, B replaced by D, and H replaced by H^*. The r of Theorem 2.5 now becomes, by (2.4), the composite $sj^*\colon 1 \to D$, which is $j\colon 1 \to D$ since s is a doctrine map. The ξ of Theorem 2.5 is therefore to be a modification $j^* \Rightarrow hj$; and we take it to be the h^o of (4.1). Since $sh^o = 1$ by (4.20), we have $s\xi = 1$, so that by Theorem 2.5 again we have $s\eta = 1$.

According to (2.22), the $\eta\colon 1 \Rightarrow H^*s$ produced by Theorem 2.5 is the composite

(4.23)

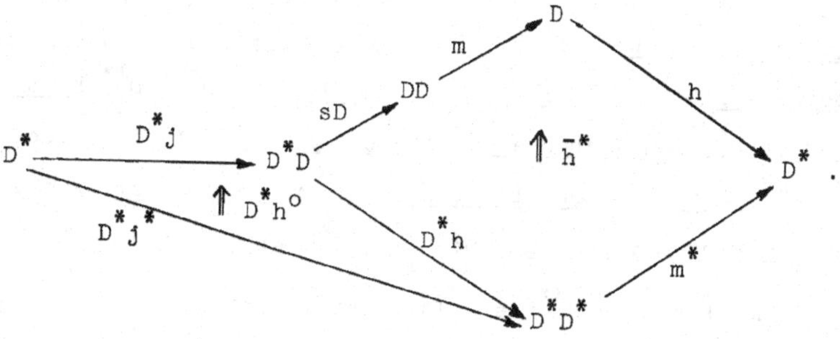

It remains only to give a proof that $\eta h = 1$. But by the 2-naturality of h we have $D^*h^o.h = hD^*.Dh^o$, so that (4.23) composed with h becomes

(4.24) 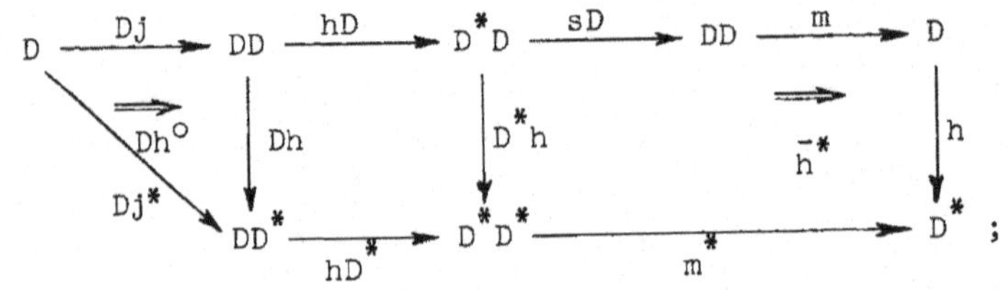

however the composite of the two rectangles on the right is $\tilde{h}$ as in (4.21), whence (4.24) is the identity by (4.2). This completes the proof. □

<u>4.9</u> In my paper [10] above in this volume, I stated at the end of §3.1, and used in the proof of Theorem 3.1, only one very small part of the above theorem: namely that $s\colon D^* \to D$ had a right inverse h, as in (4.20). The D^* in question here is that for the normalized pseudo case, which is the case that arises in [10].

In §3.3 of [10] I called the doctrine D <u>flexible</u> if $s\colon D^* \to D$ had a right inverse $t\colon D \to D^*$ that was a (strict) doctrine map; this notion was used in Theorem 3.3 of [10]. It was stated there that the doctrine D^* is always flexible: we now justify this.

The proof is elementary, not depending on Theorem 4.1, and applies equally in the lax case as in the pseudo case, with or without normality.

<u>Proposition 4.2</u> <u>Let</u> $H\colon D \to D^*$ <u>be the reflexion of</u> D, <u>considered as an object of</u> *Lax Doct*, <u>into</u> *Doct*; <u>and let</u> $K\colon D^* \to D^{**}$ <u>be the reflexion of</u> D^*, <u>considered as an object of</u> *Lax Doct*, <u>into</u> *Doct*. <u>Let</u> $q\colon D^{**} \to D^*$ <u>be the unique strict doctrine map such that</u> $qK = 1$. <u>Then there is a strict doctrine map</u> $p\colon D^* \to D^{**}$ <u>with</u> $qp = 1$.

<u>Proof</u> Since $KH\colon D \to D^{**}$ is a lax doctrine map, and since $H\colon D \to D^*$ is the reflexion, we have $KH = pH$ for a unique strict doctrine map $p\colon D^* \to D^{**}$. Then $qKH = qpH$; but again $qKH = H = 1H$ since $qK = 1$. Since qp and 1 are both strict, the uniqueness property for the reflexion H gives $qp = 1$. □

In the pseudo case, since q is an equivalence of endo-2-functors by Theorem 4.1, it follows that $pq \cong 1$. I suspect, but have not proved, that the isomorphism $pq \cong 1$ can be chosen to be a doctrine-modification, so that D^* and D^{**} are equivalent doctrines. This would be very convenient if it were true, and the question should be looked into.

4.10 We consider finally the special but important case when $D = \mathcal{D}\circ -$ for some club $\mathcal{D}$ in $\mathcal{C}at/\underline{\underline{S}}$. We again refer the reader to §1 and §10 of [9] above for general facts about clubs of this kind.

To give a lax D-algebra structure, which we shall also call a lax $\mathcal{D}$-algebra structure, on a category A, we have first to give a functor $a\colon\ \mathcal{D}\circ A \to A$, or equally a morphism $\mathcal{D} \to \{A,A\}$ in $\mathcal{C}at/\underline{\underline{S}}$; here $\{A,A\}$ is now the "rich endo-functor category" of [9] §9, and not the right Kan extension of (3.2) above. So for $T \in \mathcal{D}$ we have to give $|T|\colon\ A^n \to A$ where $\Gamma T = n$; and for $f\colon\ T \to S$ in $\mathcal{D}$ we have to give the generalized natural transformation $|f|\colon\ |T| \Rightarrow |S|$ of type Γf. Since this is to be a functor, we must require

(4.25) $|fg| = |f|\ |g|$ and $|1| = 1.$

Next we have to give $\hat{a}$ as in (4.5), a natural transformation with components

$$\hat{a}_A\colon\ A \to |\underline{\underline{1}}|(A),$$

that is, a natural transformation

$$\hat{a}\colon\ 1 \to |\underline{\underline{1}}|.$$

In the normal case we require $\hat{a} = 1$, which involves demanding that $|\underline{\underline{1}}| = 1$. In the pseudo case we have to provide $\hat{a}$ with an inverse, so we must also give a natural transformation

$$\hat{a}'\colon\ |\underline{\underline{1}}| \to 1,$$

and demand that

$$\hat{a}\,\hat{a}' = 1 \quad \text{and} \quad \hat{a}'\hat{a} = 1.$$

Similarly for the $\bar{a}$ of (4.5); it is to have components

$$\bar{a}_{T[S_1 \dots S_n][A_1 \dots A_m]} : |T|(|S_1|(A_1 \dots), \dots, |S_n|(\dots A_m))$$

$$\to |T(S_1 \dots S_n)|(A_1 \dots A_m),$$

with inverses $\bar{a}'$ provided in the pseudo case. All these data are then to satisfy the further axioms corresponding to (4.6)-(4.8).

It is clear from [9] §10 that such a lax $\mathcal{D}$-algebra is itself an algebra for a club $\mathcal{D}^*$ in $\mathcal{Cat}/\underline{\underline{S}}$, which we have in effect just described by its generators and relations. Explicitly, the objects of $\mathcal{D}^*$ are generated by objects $\bar{T}$ in bijection with the objects T of $\mathcal{D}$, and with $\Gamma\bar{T} = \Gamma T$. These are subjected to no relations except in the normal case, where we impose the relation $\bar{\underline{\underline{1}}} = \underline{\underline{1}}$. The generators for the morphisms of $\mathcal{D}^*$ consist of an $\bar{f}\colon \bar{T} \to \bar{S}$ with $\Gamma\bar{f} = \Gamma f$ for each $f\colon T \to S$ in $\mathcal{D}$; of an $\hat{a}\colon \underline{\underline{1}} \to \bar{\underline{\underline{1}}}$, with $\Gamma\hat{a} = 1$, which is to be omitted in the normal case; and of an

$$\bar{a}_{T[S_1 \dots S_n]} : \bar{T}(\bar{S}_1, \dots, \bar{S}_n) \to \overline{T(S_1, \dots, S_n)},$$

with $\Gamma\bar{a} = 1$, for each object $T[S_1, \dots, S_n]$ of $\mathcal{D}\circ\mathcal{D}$. In the pseudo case, these are to be augmented by further generators $\hat{a}'$ and $\bar{a}'_{T[S_1 \dots S_n]}$ in the reversed senses. The relations between expanded instances of these generators are $\bar{f}\bar{g} = \overline{fg}$ and $\bar{1}_T = 1_{\bar{T}}$ corresponding to (4.25); the relations $\hat{a}\hat{a}' = 1$, $\hat{a}'\hat{a} = 1$, $\bar{a}\bar{a}' = 1$, $\bar{a}'\bar{a} = 1$ in the pseudo case only; and finally the relations corresponding to (4.6)-(4.8). The first of these for instance asserts the commutativity of

$$\begin{array}{ccc}
\bar{T}(\underline{\underline{1}}\ldots\underline{\underline{1}}) & \xrightarrow{\ \bar{T}(\hat{a}\ldots\hat{a})\ } & \bar{T}(\bar{\underline{\underline{1}}}\ldots\bar{\underline{\underline{1}}}) \\
\Big\| & & \Big\downarrow \bar{a}_{T[\underline{\underline{1}}\ldots\underline{\underline{1}}]} \\
\bar{T} & = & \overline{T(\underline{\underline{1}}\ldots\underline{\underline{1}})}
\end{array} ,$$

and the other two are equally easily written down.

It is further easy to check that a $\mathcal{D}^*$-morphism is the same thing as a D^*-morphism, and a $\mathcal{D}^*$-2-cell as a D^*-2-cell; so that D^* is indeed the doctrine $\mathcal{D}^*\circ-$ for the club $\mathcal{D}^*$.

We can regard h and s as maps h: $\mathcal{D} \to \mathcal{D}^*$ and s: $\mathcal{D}^* \to \mathcal{D}$; they might a priori lie in $\mathcal{C}at\int^*\underline{\underline{S}}$, by the general principles of [9], but it is easy to see that in fact they lie in $\mathcal{C}at/\underline{\underline{S}}$. That this is so for s follows from the considerations of [9] §10.7; s is clearly defined on generators and relations by $s\bar{T} = T$, $s\bar{f} = f$, $s\,\hat{a} = 1$, $s\,\bar{a} = 1$. As for h, we have only to recall just how a $\mathcal{D}^*$-algebra gives a lax $\mathcal{D}$-algebra: clearly $hT = \bar{T}$, $hf = \bar{f}$, while the components of h° and of $\tilde{h}$ are given by $\hat{a}$ and $\bar{a}$ respectively.

It follows that h embeds $\mathcal{D}$ as a full subcategory of $\mathcal{D}^*$, onto which s reflects $\mathcal{D}^*$; with the consequences for commutativity of diagrams described in §1.4 above.

4.11 It remains only to prove the further assertion of [10] §3.3, namely that D is flexible if it is $\mathcal{D}\circ-$ for a club $\mathcal{D}$ in $\mathcal{C}at/\underline{\underline{S}}$ whose objects are freely generated. I recall that the definition of flexibility in [10] was concerned with the D^* for the normalized pseudo case; the result we shall prove is independent of normality, but does require that we be pseudo rather than lax.

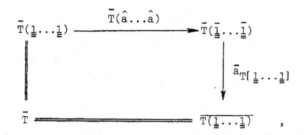

Proposition 4.3. Let $\mathcal{D}$ be a club in $\mathcal{C}at/\underline{\underline{S}}$ whose discrete club $|\mathcal{D}|$ of objects is freely generated, and let $\mathcal{D}^*$ be the club for pseudo $\mathcal{D}$-algebras, normalized or not, with $s\colon\ \mathcal{D}^* \to \mathcal{D}$ the canonical map of clubs. Then $st = 1$ for some club-map $t\colon\ \mathcal{D} \to \mathcal{D}^*$ in $\mathcal{C}at/\underline{\underline{S}}$.

Proof. As in §4.10 we already have $h\colon\ \mathcal{D} \to \mathcal{D}^*$ with $sh = 1$. Let $\mathcal{G}$ be the set of free generators of $|\mathcal{D}|$. For $G \in \mathcal{G}$, set $tG = hG$. Take $t\colon\ |\mathcal{D}| \to \mathcal{D}^*$ to be the unique club-map extending $t\colon\ \mathcal{G} \to \mathcal{D}^*$ (by club-map in this proof we always mean club-map in $\mathcal{C}at/\underline{\underline{S}}$). Then for any $T \in \mathcal{D}$ we have an isomorphism $\alpha_T\colon\ tT \to hT$ which is a composite of expanded instances of $\bar{a}$ and $\hat{a}$. Define $t\colon\ \mathcal{D} \to \mathcal{D}^*$ to be the conjugate of h under α. Since s is a club-map and since $s\bar{a} = 1$ and $s\hat{a} = 1$, we have $st = 1$. Note that t is in $\mathcal{C}at/\underline{\underline{S}}$ since h is and since $\Gamma\bar{a} = 1$, $\Gamma\hat{a} = 1$. It remains to prove that t is a club-map, that is, that we have commutativity in

$$
\begin{array}{ccc}
\mathcal{D}\circ\mathcal{D} & \xrightarrow{\quad m \quad} & \mathcal{D} \\
{\scriptstyle t\circ t}\downarrow & & \downarrow{\scriptstyle t} \\
\mathcal{D}^*\circ\mathcal{D}^* & \xrightarrow[\quad m^* \quad]{} & \mathcal{D}^* .
\end{array}
\tag{4.26}
$$

But we have forcibly made (4.26) commute on objects, by taking $t\colon\ |\mathcal{D}| \to \mathcal{D}^*$ to be a club-map. So we need only check that it commutes on morphisms. But the composite of (4.26) with $s\colon\ \mathcal{D}^* \to \mathcal{D}$ commutes, since s is a club-map with $st = 1$. Hence (4.26) itself commutes, since s is an equivalence of categories and so in particular a faithful functor. □

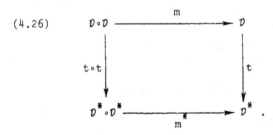

5. Pseudo distributive laws

5.1 Let (D,m,j) and (D',m',j') be doctrines on a 2-category K. To give a 2-functor $\tilde{D}_*$ rendering commutative

$$(5.1)\qquad \begin{array}{ccc} D'\text{-}\mathcal{A}\ell g_* & \xrightarrow{\tilde{D}_*} & D'\text{-}\mathcal{A}\ell g_* \\ {\scriptstyle U'_*}\downarrow & & \downarrow{\scriptstyle U'_*} \\ K & \xrightarrow[D]{} & K \end{array},$$

where U'_* is the forgetful 2-functor, and to give 2-natural $\tilde{j}_*\colon\ 1 \to \tilde{D}_*$ and $\tilde{m}_*\colon\ \tilde{D}_*{}^2 \to \tilde{D}_*$ such that $U'_*\tilde{j}_* = j\ U'_*$ and $U'_*\tilde{m}_* = m\ U'_*$, is to give a distributive law of D' over D in the classical sense of Beck [2]. Because U'_* is faithful, $\tilde{j}_*$ and $\tilde{m}_*$ are unique if they exist, and automatically make of $\tilde{D}_*$ a doctrine on $D'\text{-}\mathcal{A}\ell g_*$.

We get a much more lax notion of distributive law if we ask the doctrine D to lift not through $U'_*\colon\ D'\text{-}\mathcal{A}\ell g_* \to K$ but through its extension $U'\colon\ D'\text{-}\mathcal{A}\ell g \to K$. The pseudo distributive laws of this paper occupy an intermediate place, being only a mild extension of the classical ones.

For a pseudo distributive law we suppose that the 2-functor D is given a lifting $\tilde{D}_*$ as in (5.1). Then, as we shall show shortly, $\tilde{D}_*$ has a canonical extension to a 2-functor $\tilde{D}$ satisfying

$$(5.2)\qquad \begin{array}{ccc} D'\text{-}\mathcal{A}\ell g & \xrightarrow{\tilde{D}} & D'\text{-}\mathcal{A}\ell g \\ {\scriptstyle U'}\downarrow & & \downarrow{\scriptstyle U'} \\ K & \xrightarrow[D]{} & K \end{array}.$$

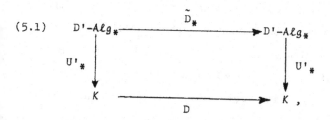

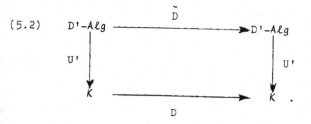

We suppose that j lifts as above to a $\tilde{j}_*$: $1 \to \tilde{D}_*$, and we observe that the components of $\tilde{j}_*$ are also those of a 2-natural $\tilde{j}$: $1 \to \tilde{D}$, satisfying $U'\tilde{j} = jU'$. What we do <u>not</u> suppose is that m lifts to an $\tilde{m}_*$. Instead, we suppose <u>given</u> an $\tilde{m}$: $\tilde{D}^2 \to \tilde{D}$ satisfying $U'\tilde{m} = mU'$; since U' is not faithful, $\tilde{m}$ is not uniquely determined by this - the morphism part of each component is so, but not the 2-cell part - so more data is involved. We further suppose that each component of $\tilde{m}$ is a <u>strong</u> D'-morphism. (If we were to replace <u>strong</u> here by <u>strict</u>, we should be back in the classical case: this then is what our relaxation consists in.) Finally we suppose that $(\tilde{D},\tilde{m},\tilde{j})$ satisfy the axioms for a doctrine; these are no longer automatic as in the classical case.

Examples of such pseudo distributive laws will be given in §6.

<u>5.2</u> Beck showed in [2] that to give a classical distributive law amounted to giving a p: $D'D \to DD'$ satisfying four axioms. We now carry out a corresponding analysis of a pseudo distributive law.

It turns out that to give a pseudo distributive law we must first give a 2-natural p: $D'D \to DD'$ satisfying three of Beck's axioms, to wit

(5.3)
$$\begin{array}{ccc} D & \xrightarrow{\;j'D\;} & D'D \\ & \searrow^{Dj'} & \downarrow{\scriptstyle p} \\ & & DD', \end{array}$$

(5.4)
$$\begin{array}{ccccc} D'D'D & \xrightarrow{\;D'p\;} & D'DD' & \xrightarrow{\;pD'\;} & DD'D' \\ {\scriptstyle m'D}\downarrow & & & & \downarrow{\scriptstyle Dm'} \\ D'D & & \xrightarrow{\qquad\quad p \qquad\quad} & & DD' \end{array} ,$$

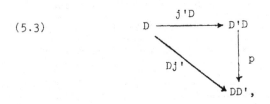

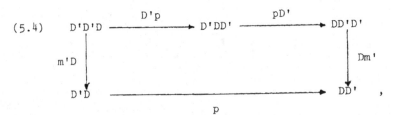

(5.5)

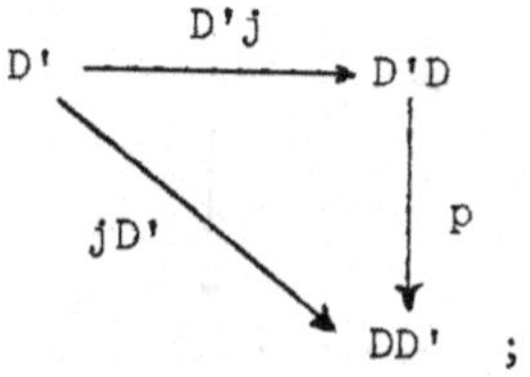

;

then we must give an invertible modification π sitting in what, if π were 1, would be Beck's fourth axiom:

(5.6)

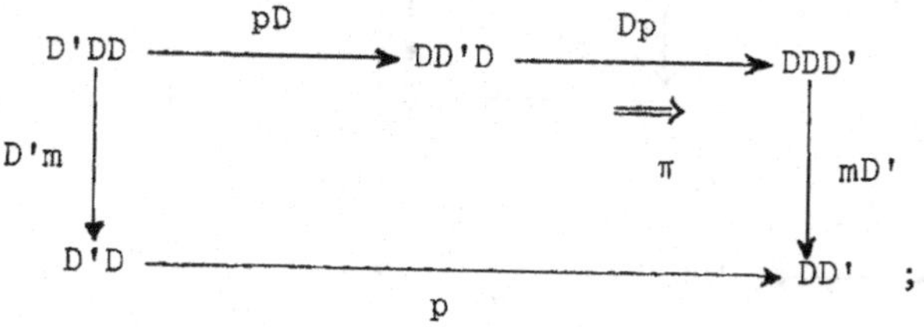

;

and finally we must subject π to five axioms; the reader will note that in each of these equality at the level of 1-cells is automatic by (5.3)-(5.5) and naturality:

(5.7)

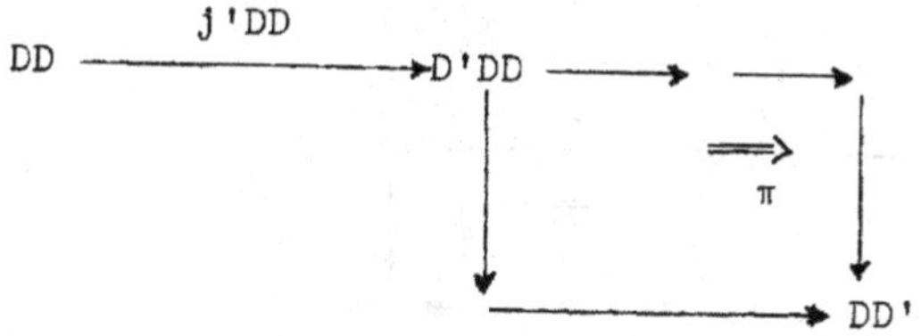

= identity,

(5.8)

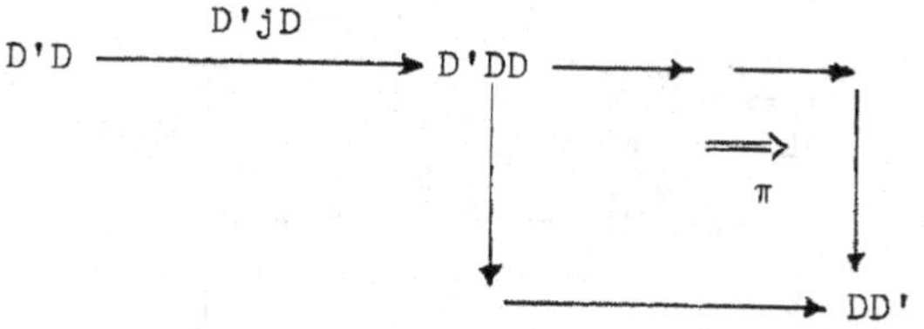

= identity,

(5.9)

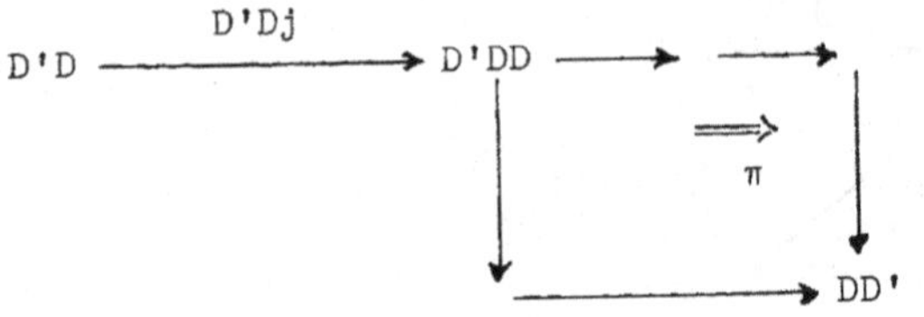

= identity,

(5.10)

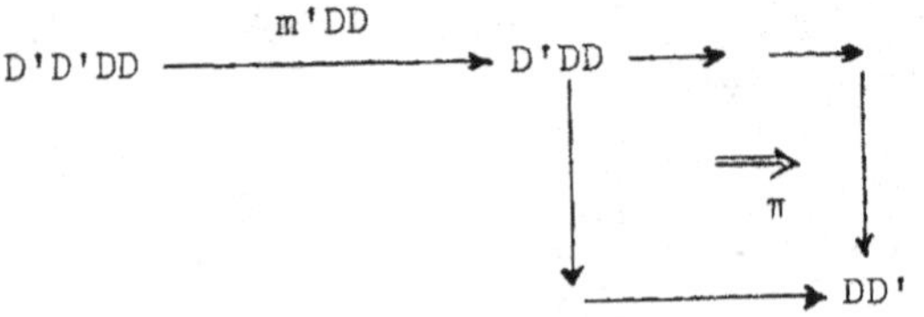

=

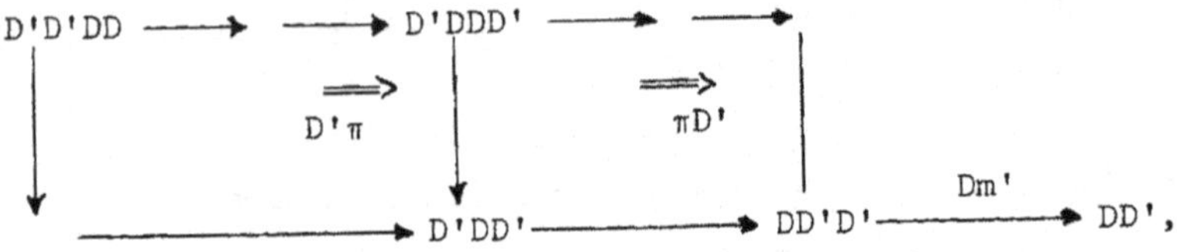

(5.11)

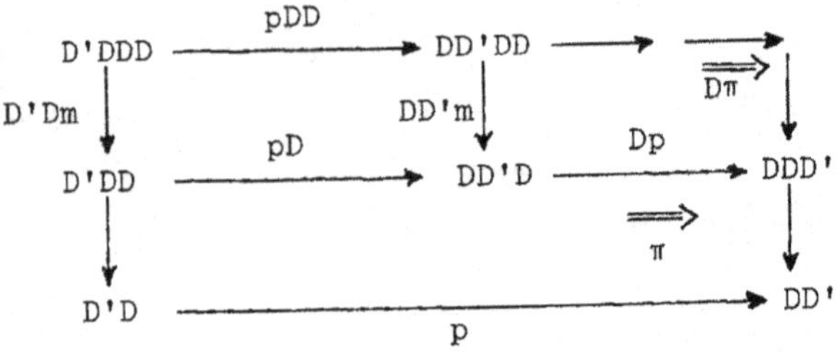

=

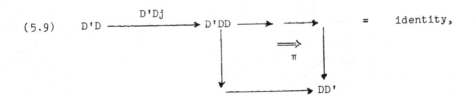

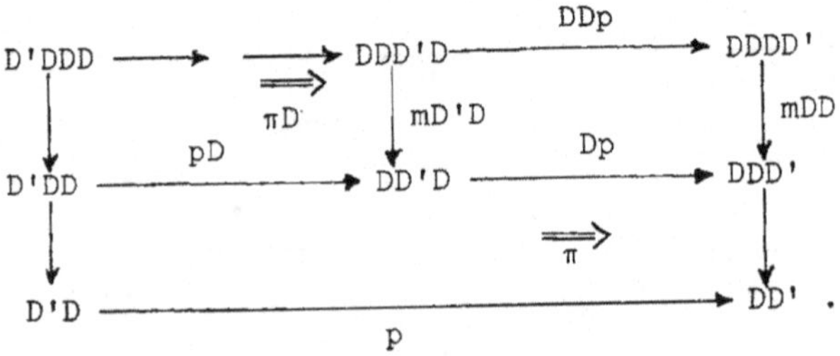

We now justify the above assertion. First, to give $\tilde{D}_*$ as in (5.1) is by [13] §3.3 to give an <u>action</u> q: $D'DU'_* \to DU'_*$ of D' on DU'_*. By [13] Proposition 2.1 to give a 2-natural q as above is to give a 2-natural p: $D'D \to D'U'_*F'_*$ where F'_* is the left adjoint of U'_*; that is, to give a 2-natural p: $D'D \to DD'$. The conditions for q to be an <u>action</u> translate at once into (5.3) and (5.4).

We give the explicit form of $\tilde{D}_*$ in terms of p; but we may as well give at the same time its canonical extension $\tilde{D}$, and write $\tilde{D}A$ rather than $\tilde{D}_*A$, etc. In doing so we use A both for an object of K and for a D'-algebra (A,a') consisting of the object A together with the D'-action a': $D'A \to A$. We then have

(5.12) $\tilde{D}A = (DA,\ Da'.pA)$,

the D'-algebra with object DA and with D'-action

(5.13) $$D'DA \xrightarrow[pA]{} DD'A \xrightarrow[Da']{} DA.$$

For a morphism $F' = (f,\bar{f}')$: $A \to B$ of D'-algebras, we have a D'-morphism

(5.14) $\tilde{D}F' = (Df,\ D\bar{f}'.pA)$

as in

(5.15) $$\begin{array}{ccccc} D'DA & \xrightarrow{pA} & DD'A & \xrightarrow{Da'} & DA \\ {\scriptstyle D'Df}\downarrow & & \downarrow{\scriptstyle DD'f} & \overset{\Longrightarrow}{D\bar{f}'} & \downarrow{\scriptstyle Df} \\ D'DB & \xrightarrow[pB]{} & DD'B & \xrightarrow[Db']{} & DB \end{array} .$$

Finally a D'-2-cell γ: $F' \Rightarrow G'$ gives a D'-2-cell $\tilde{D}\gamma$: $\tilde{D}F' \Rightarrow \tilde{D}G'$ directly by

$$(5.16) \qquad \tilde{D}\gamma = D\gamma .$$

The 2-functor $\tilde{D}_*$ is what we get by restricting $\tilde{D}$ to strict D'-morphisms; observe that if f: $A \to B$ is a strict D'-morphism then $\tilde{D}_*f = \tilde{D}f$ is just the strict D'-morphism Df: $DA \to DB$. Since what we have said about $\tilde{D}_*$ is all justified by our general remarks above, all that needs separate verification is that (5.15) really is a D'-morphism when F' is not strict, that (5.16) really is a D'-2-cell when F',G' are not strict, and that $\tilde{D}$ really is a 2-functor: all this is easy.

The next thing we wanted was $\tilde{j}_*$: $1 \to \tilde{D}_*$ with $U'_*\tilde{j}_* = jU'_*$. The last requirement forces the component $\tilde{j}_*A$: $A \to \tilde{D}_*A$ to be jA: $A \to DA$; the desired $\tilde{j}_*$ exists precisely when jA: $A \to DA$ is indeed a strict D'-morphism, and this condition reduces to (5.5). Once we have this, it is immediate that jA: $A \to DA$ is 2-natural not only for <u>strict</u> D'-morphisms $A \to B$ but for all; so we in fact get a 2-natural $\tilde{j}$: $1 \to \tilde{D}$ with the same components:

$$(5.17) \qquad \tilde{j}A = \tilde{j}_*A = jA;$$

of course we have $U'\tilde{j} = jU'$.

Then we wanted $\tilde{m}$: $\tilde{D}^2 \to \tilde{D}$ with $U'\tilde{m} = mU'$. The last requirement forces the morphism-part of the component $\tilde{m}A$: $D^2A \to DA$ to be mA. Thus

$$(5.18) \qquad \tilde{m}A = (mA, \bar{m}'A)$$

for some invertible 2-cell $\bar{m}'A$ (since $\tilde{m}A$ is to be <u>strong</u>). The form of $\bar{m}'A$ is determined by the requirement that $\tilde{m}$ be 2-natural. Indeed, mere naturality in A of $\tilde{m}A$, and that merely for <u>strict</u> D'-morphisms $f\colon\ A \to B$, suffices to fix the form of $\bar{m}'A$. First, taking $f\colon\ A \to B$ to be $a'\colon\ D'A \to A$, we easily see that $\bar{m}'A$ is of the form

$$(5.19)\qquad \begin{array}{ccccccc} D'DDA & \xrightarrow{pDA} & DD'DA & \xrightarrow{DpA} & DDD'A & \xrightarrow{DDa'} & DDA \\ {\scriptstyle D'mA}\downarrow & & \overset{\Longrightarrow}{\scriptstyle \pi A} & & \downarrow{\scriptstyle mD'A} & & \downarrow{\scriptstyle mA} \\ D'DA & & \xrightarrow{\qquad\quad pA \qquad\quad} & & DD'A & \xrightarrow{Da'} & DA \end{array}$$

for an invertible πA; next, taking $f\colon\ A \to B$ to be $D'f\colon\ D'A \to D'B$, we easily see that π must be a modification, as in (5.6). Then no more is needed to make $\tilde{m}$ 2-natural: the verification is simple.

The further things we <u>do</u> need are that $\tilde{m}A$ satisfy the axioms for a D'-morphism - these reduce to (5.7) and (5.10) - and that $(\tilde{D},\tilde{m},\tilde{j})$ satisfy the axioms for a doctrine; these reduce to (5.8), (5.9), and (5.11) (automatic in the classical case $\pi = 1$, but not here).

This completes our justification of the above analysis of a pseudo distributive law in terms of p and π.

<u>5.3</u> We now look at the 2-category $\tilde{D}$-$\mathcal{A}\ell g$ and describe its elements in elementary terms involving D, D', p, π and elements of K.

A $\tilde{D}$-algebra A is a D'-algebra A (with D'-action $a'\colon\ D'A \to A$) together with a $\tilde{D}$-action $\tilde{a}\colon\ \tilde{D}A \to A$. The latter is to be a D'-morphism $\tilde{a} = (a,\bar{a}')\colon\ DA \to A$, which in view of (5.13) has the form

$$(5.20)\qquad \begin{array}{ccccc} D'DA & \xrightarrow{pA} & DD'A & \xrightarrow{Da'} & DA \\ {\scriptstyle D'a}\downarrow & & & \overset{\Longrightarrow}{\scriptstyle \bar{a}'} & \downarrow{\scriptstyle a} \\ D'A & & \xrightarrow{\qquad\quad a' \qquad\quad} & & A \end{array} .$$

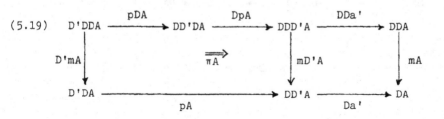

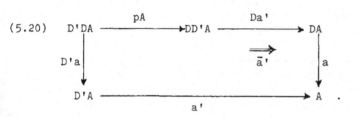

There are four axioms to be satisfied: the first two say that $\tilde{a}$ really is a D'-morphism, and the second two say that it is a D-action:

(5.21) 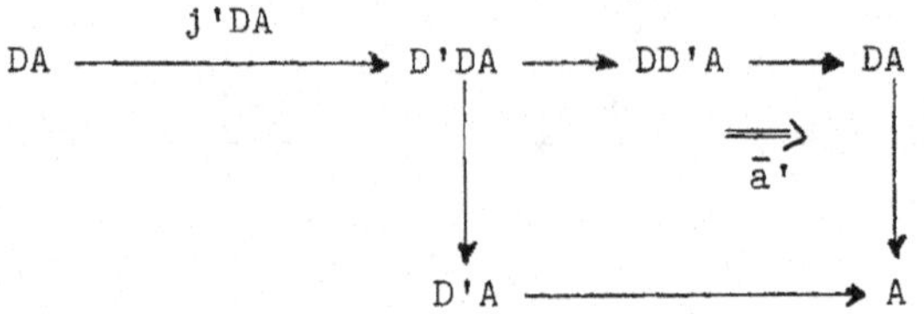

= identity,

(5.22) 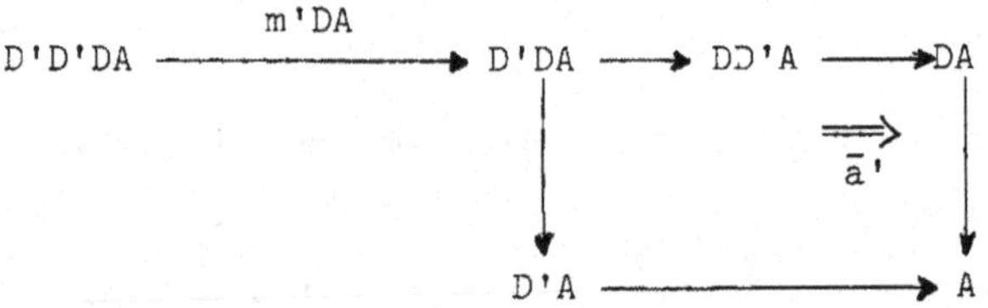

=

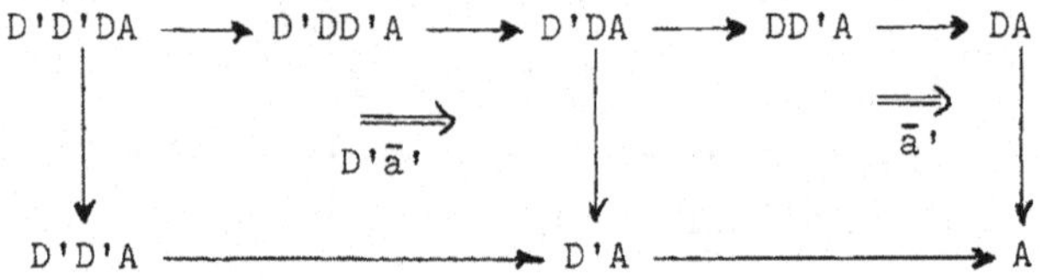

,

(5.23) 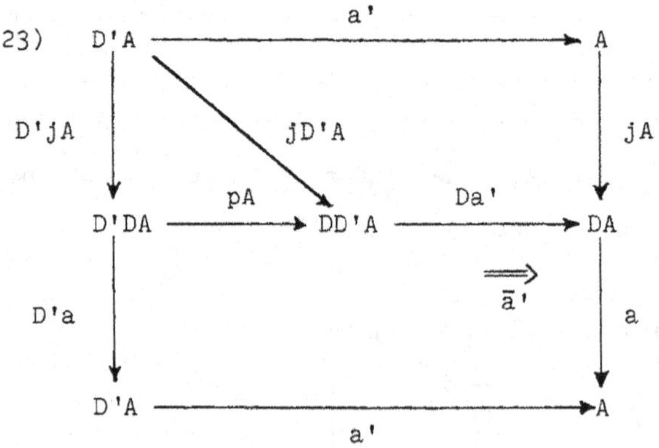

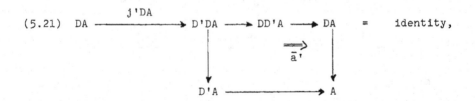

= identity,

(5.24) 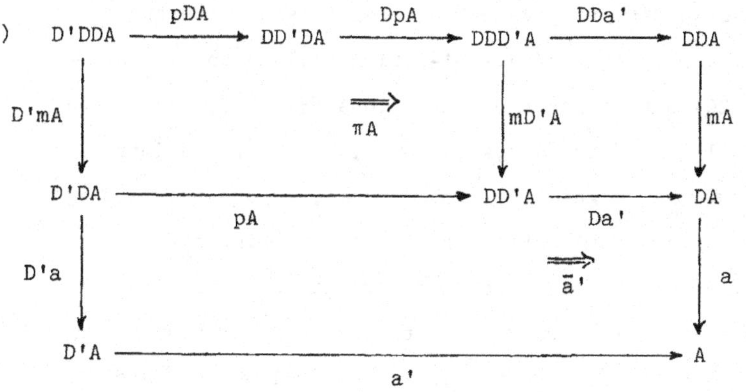

=

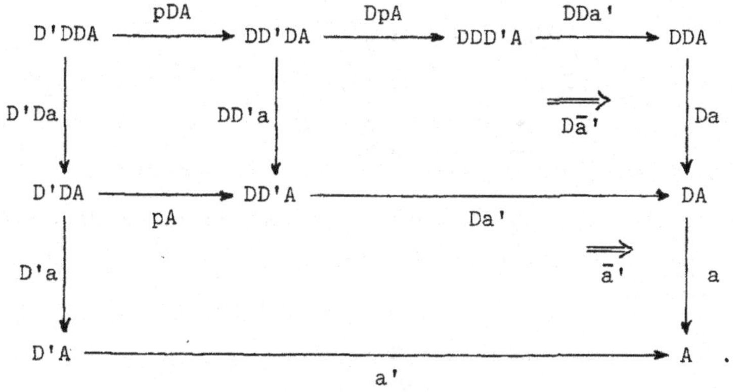

Of these, (5.21) and (5.22) are straightforward; (5.23) expresses $\tilde{a}.\tilde{j}A = 1$ using (5.20) and (5.17); (5.24) expresses $\tilde{a}.\tilde{m}A = \tilde{a}.D\tilde{a}$, using (5.20), (5.19), and (5.15). Note that at the level of 1-cells the equalities (5.21), (5.22) are automatic, since a' was given as a D'-action. The same is not true of (5.23) and (5.24), where equality at the 1-cell level already gives new information; in fact precisely the information that $a\colon\ DA \to A$ is a D-action. (Clearly this implies 1-cell equality in (5.23) and (5.24); and it is implied by them, as we see by composing (5.23) with $j'A$ and (5.24) with $j'D^2A$.) It is very convenient to separate the 1-cell from the 2-cell information in such equalities, for example in the kind of reasonings used in §4.3 above.

Thus in fine to give a $\tilde{D}$-algebra A is to give the object A of K, with a D-action A and a D'-action a', and to give a 2-cell $\bar{a}'$ as in (5.20) satisfying the four axioms (5.21)-(5.24). We may write the $\tilde{D}$-algebra A as $(A,a,a',\bar{a}')$. (In §1.1 of the Introduction we described such an algebra, in rough terms, as one provided with both a D-action and a D'-action, the D-action being given the structure of a D'-morphism. We have now made this precise.)

Now we consider what it is to give a $\tilde{D}$-morphism $\tilde{F}$: $A \to B$. It is a pair $\tilde{F} = (F',\bar{f})$ where F' is a D'-morphism $A \to B$ and $\bar{f}$ is a D'-2-cell as in

(5.25)
$$\begin{array}{ccc} \tilde{D}A & \xrightarrow{\tilde{a}} & A \\ {\scriptstyle \tilde{D}F'}\downarrow & \overset{\Longrightarrow}{\bar{f}} & \downarrow{\scriptstyle F'} \\ \tilde{D}B & \xrightarrow[\tilde{b}]{} & B \end{array} .$$

If F' is itself the pair $(f,\bar{f}')$: $A \to B$ then as a 2-cell $\bar{f}$ is of the form

(5.26)
$$\begin{array}{ccc} DA & \xrightarrow{a} & A \\ {\scriptstyle Df}\downarrow & \overset{\Longrightarrow}{\bar{f}} & \downarrow{\scriptstyle f} \\ DB & \xrightarrow[b]{} & B \end{array} .$$

The requirement that $\bar{f}$ actually be a D'-2-cell is (cf. [13](3.19))

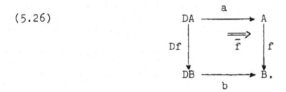

(5.27)

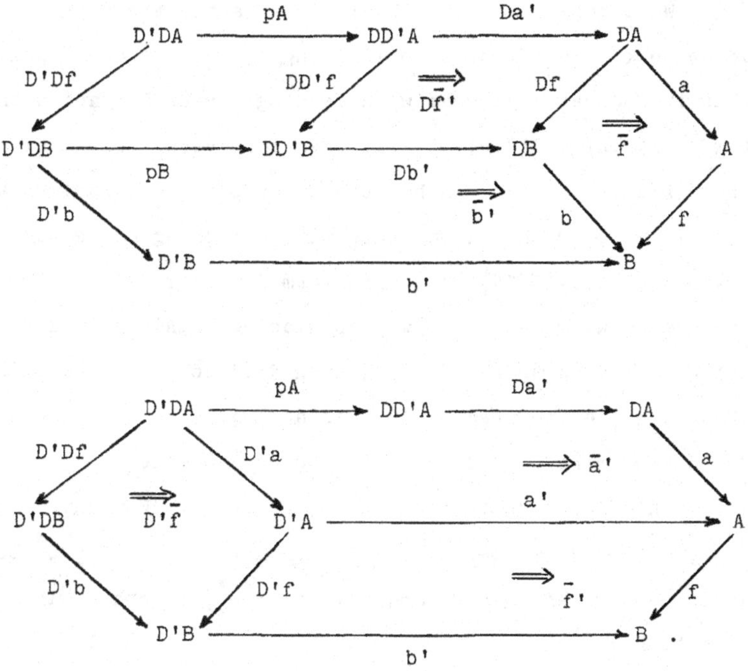

Finally the axioms for $\tilde{F} = (F', \bar{f})$ to be in fact a $\tilde{D}$-morphism, since they assert only an equality at the level of 2-cells, reduce to the axioms for $F = (f, \bar{f})$ to be a D-morphism. Thus in fine to give a $\tilde{D}$-morphism $\tilde{F}$: $(A, a, a', \bar{a}') \to (B, b, b', \bar{b}')$ is to give a triple $\tilde{F} = (f, \bar{f}, \bar{f}')$ such that $F = (f, \bar{f})$: $A \to B$ is a D-morphism, $F' = (f, \bar{f}')$: $A \to B$ is a D'-morphism, and (5.27) is satisfied.

Lastly we consider what a $\tilde{D}$-2-cell γ: $\tilde{F} \Rightarrow \tilde{G}$: $A \to B$ is. It has first of all to be a D'-2-cell γ: $F' \Rightarrow G'$; but then the condition for it to be a $\tilde{D}$-2-cell is, since it is a pure equality of 2-cells, just the condition for it to be a D-2-cell γ: $F \Rightarrow G$. So a $\tilde{D}$-2-cell γ: $(f, \bar{f}, \bar{f}') \Rightarrow (g, \bar{g}, \bar{g}')$ is just a 2-cell γ: $f \Rightarrow g$ that is at once a D-2-cell γ: $F \Rightarrow G$ and a D'-2-cell γ: $F' \Rightarrow G'$.

5.4 We intend to exhibit $\tilde{D}$-algebras as the algebras for a doctrine on K. The first step in doing this is to transform (5.20) so that it becomes a diagram with leading vertex D'D and with terminal vertex the {A,A} of §3.2; it will then involve not the actions $a\colon DA \to A$ and $a'\colon D'A \to A$ but the corresponding doctrine maps $D \to \{A,A\}$ and $D' \to \{A,A\}$, together with something corresponding to $\bar{a}'$. The axioms (5.21)-(5.24) will transform similarly. However we abstract the transform we get in this way, replacing {A,A} by an arbitrary doctrine. If we compare with §4, then §5.3 above is analogons to §§4.2-4.4, while the present §5.4 is analogons to §4.1. Of course we now suppose K to have small homs and to be 2-complete.

This leads us to the following notion. For any doctrine (D^*, m^*, j^*) on K, we define a map K from (D,D',p,π) to D^* to consist of doctrine-maps $k\colon D \to D^*$ and $k'\colon D' \to D^*$, together with a modification $\hat{k}$ as in

(5.28)

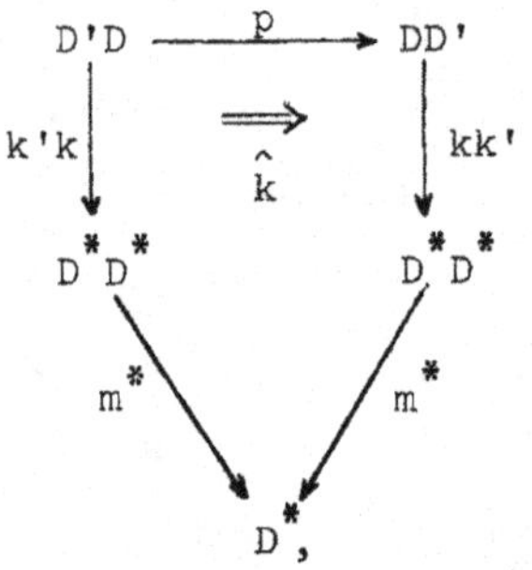

satisfying the following four axioms (which are all automatic at the 1-cell level):

(5.29) 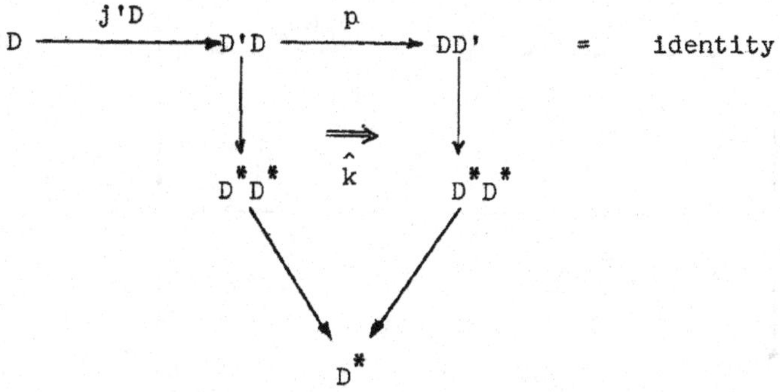

(5.30) 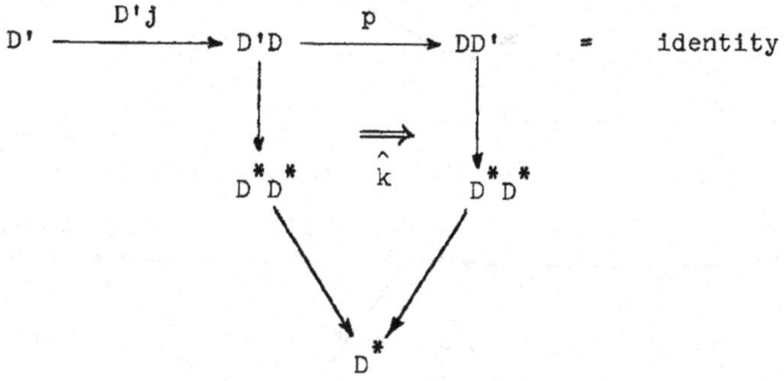

(5.31)

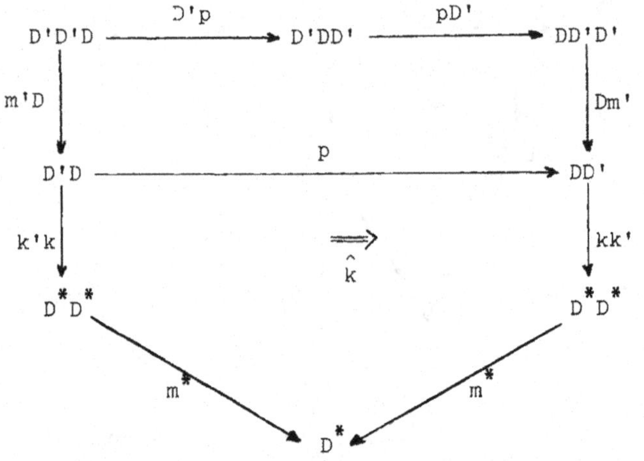

D'D'D
D'p
D'DD'
pD'
DD'D'
m'D
Dm'
D'D
p
DD'
k'k
kk'
D*D*
D*D*
m*
m*
D*

=

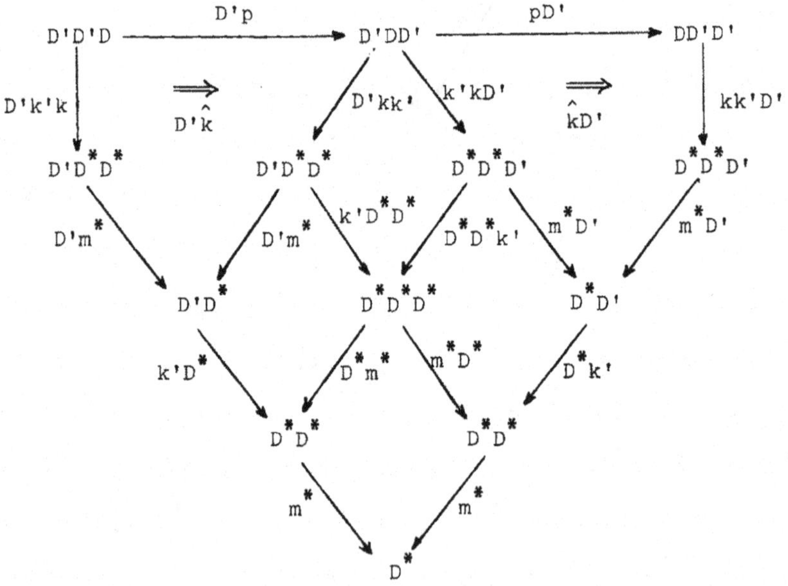

D'D'D
D'p
D'DD'
pD'
DD'D'
D'k'k
D'kk'
k'kD'
kk'D'
D'D*D*
D'D*D*
D*D*D'
D*D*D'
D'm*
D'm*
k'D*D*
D*D*k'
m*D'
m*D'
D'D*
D*D*D*
D*D'
k'D*
D*m*
m*D*
D*k'
D*D*
D*D*
m*
m*
D*

(5.32) 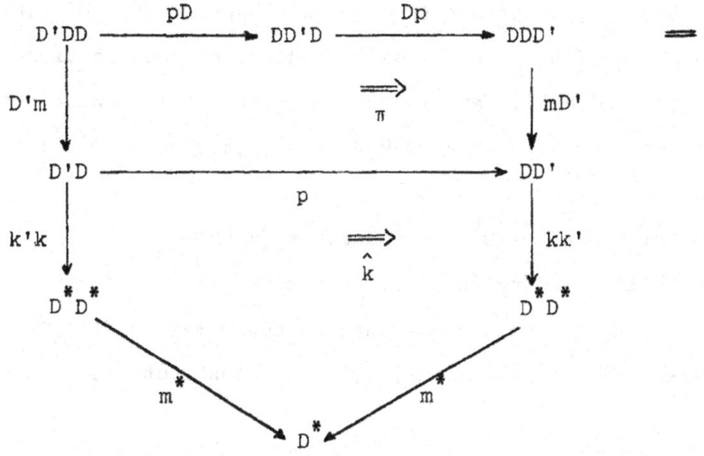
D'DD
pD
DD'D
Dp
DDD'
D'm
π
mD'
D'D
p
DD'
k'k
k̂
kk'
D*D*
m*
m*
D*

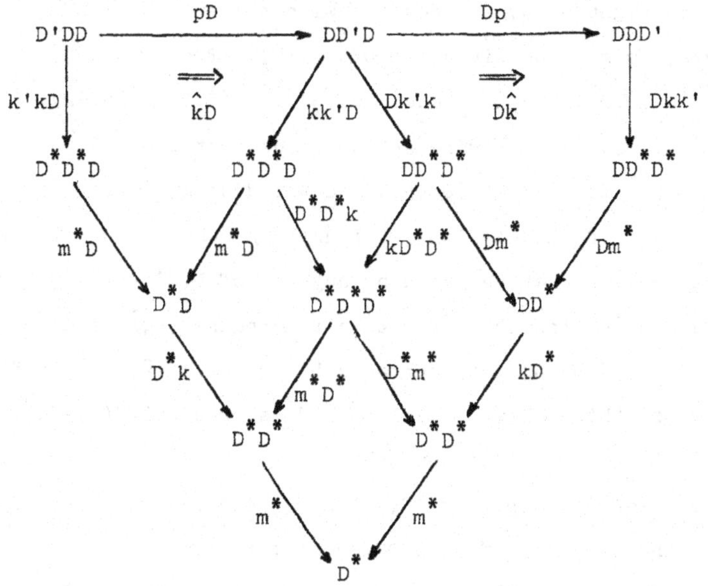
D'DD
pD
DD'D
Dp
DDD'
k'kD
k̂D
kk'D
Dk'k
Dk̂
Dkk'
D*D*D
D*D*D
DD*D*
DD*D*
m*D
m*D
D*D*k
kD*D*
Dm*
Dm*
D*D
D*D*D*
DD*
D*k
m*D*
D*m*
kD*
D*D*
D*D*
m*
m*
D*

If $K = (k,k',\hat{k})$ and $H = (h,h',\hat{h})$ are maps from (D,D',p,π) to D^*, we define a <u>2-cell</u> $H \Rightarrow K$ to consist of doctrine-modifications $\rho\colon\ h \Rightarrow k$ and $\rho'\colon\ h' \Rightarrow k'$, satisfying the evident axioms with respect to $\hat{h}$ and $\hat{k}$. Thus we get a category $Map((D,D',p,\pi),D^*)$, or $M(D^*)$ for short.

A doctrine map $t\colon\ D^* \to E$ induces a functor $M(t)\colon\ M(D^*) \to M(E)$, sending $(k,k',\hat{k})$ to $(tk,tk',t\hat{k})$. A doctrine modification $\tau\colon\ t \Rightarrow t_1$ induces a natural transformation $M(t) \Rightarrow M(t_1)$ whose $(k,k',\hat{k})$-component is $(\tau k,\tau k')$. Thus M constitutes a 2-functor $\mathcal{D}oct \to \mathcal{CAT}$.

Now it is easy to see that a map $K\colon\ (D,D',p,\pi) \to \{A,A\}$ corresponds exactly to a $\tilde{D}$-algebra structure on A as described in (5.20)-(5.24); we have only to replace D^* by $\{A,A\}$ in (5.28), multiply on the right by A, and compose with the evalutation $e\colon\ \{A,A\}A \to A$, to get the situation of (5.20). The axioms (5.29), (5.30), (5.31), (5.32) easily reduce to (5.21), (5.23), (5.22), (5.24) respectively; of course they were set up to do just this.

But now a map from (D,D',p,π) to the doctrine $\langle f,f \rangle$ of §3.3 gives the analogue of (5.20) with f replacing A and the diagram now living in K''. So a,a' get replaced by actions of D,D' on f in K'', that is to say, by enrichments of f to a D-morphism $(f,\bar{f})$ and to a D'-morphism $(f,\bar{f}')$, and $\bar{a}'$ gets replaced by a 2-cell $(\bar{a}',\bar{b}')$ in K''; the condition for this to <u>be</u> a 2-cell in K'' is precisely (5.27). The axioms (5.21)-(5.24) reduce of course to those for $\bar{a}',\bar{b}'$ separately. Hence to give a map $(D,D',p,\pi) \to \langle f,f \rangle$ is to give $\tilde{D}$-algebra structures to A and to B and to enrich f to a $\tilde{D}$-morphism $(f,\bar{f},\bar{f}')\colon\ A \to B$.

Finally, for $\gamma\colon\ f \Rightarrow g\colon\ A \to B$, a map from (D,D',p,π) to the doctrine $[\![\gamma,\gamma]\!]$ of §3.4 is equally easily seen to be just what makes γ a D-2-cell.

Note that a map from (D,D',p,π) to the doctrine $[f,f]$ of §3.3, or equally a map to $\langle f,f\rangle$ that happens to factorize through $[f,f]$, corresponds to a $\tilde{D}$-morphism in which both $\bar{f}$ and $\bar{f}'$ are identities. This is not the same as a strict $\tilde{D}$-morphism, which clearly has $\bar{f} = 1$ but $\bar{f}'$ arbitrary.

5.5 We now pass on to the analogues of §§4.5-4.7. It is clear that the 2-functor M: *Doct* → *CAT* is 2-continuous; it is therefore 2-representable if the appropriate solution-set condition is satisfied. The remarks in §4.5 concerning the conditions under which this is likely to be so apply equally here. It is certainly so in our applications in §6, where D comes from a club in *Cat*/$\underline{\underline{S}}$ and D' from the particular club for symmetric monoidal categories. We henceforth suppose it to be so, and we henceforth take K: $(D,D',p,\pi) \to D^*$ to be the representing map. Thus any map H: $(D,D',p,\pi) \to E$ is tK for a unique doctrine map t: $D^* \to E$.

We want in fact to consider two cases simultaneously. The case we have considered so far may be called the lax case. However, because we have supposed $\tilde{m}A$ to be strong, the doctrine $\tilde{D}$ restricts to one on the sub-2-category D'-*Alg*$_{**}$ of D'-*Alg* where we retain only the strong D'-morphisms. This corresponds to requiring $\bar{a}'$ in (5.20) to be an isomorphism, and hence to requiring $\hat{k}$ in (5.28) to be an isomorphism. We call this the pseudo case. There is no formal difference between the two cases; M is 2-continuous in the latter as in the former; we treat them together with identical notation, although of course it is a different D^* in the latter case. However, in our prime application, D^* will not in the pseudo case come from a club, as it does in the lax case: and I have therefore not the same direct proof of its existence (cf. §1.4 of the Introduction). Of course in the pseudo case we give $\langle f,f\rangle$ its stronger meaning of §3.5, and all our 2-cells are isomorphisms.

In view of §5.4, then, we have an isomorphism between $\tilde{D}$-$\mathcal{A}\ell g$ and D^*-$\mathcal{A}\ell g$. Not, as we saw in the last paragraph of §5.4, between $\tilde{D}$-$\mathcal{A}\ell g_*$ and D^*-$\mathcal{A}\ell g_*$; the strict D^*-morphisms are only <u>some</u> of the strict $\tilde{D}$-morphisms; but we know exactly which ones they are (those with $\bar{f} = 1$ and $\bar{f}' = 1$), and hence we do know D^* implicitly.

In analogy with §4.6, we give the isomorphism explicitly. The D^*-algebra A with action a^*: $D^*A \to A$ corresponds to the $\tilde{D}$-algebra $(A, a, a', \bar{a}')$ where

$$(5.33) \qquad a = a^*.kA, \qquad a' = a^*.k'A, \qquad \bar{a}' = a^*.\hat{k}A.$$

The D^*-morphism $F^* = (f, \bar{f}^*)$: $A \to B$ corresponds to the $\tilde{D}$-morphism $\tilde{F} = (f, \bar{f}, \bar{f}')$ where

$$(5.34) \qquad \bar{f} = \bar{f}^*.kA, \qquad \bar{f}' = \bar{f}^*.k'A;$$

and we have

$$(5.35) \qquad \bar{f}^* = 1 \text{ if and only if } \bar{f} = 1 \text{ and } \bar{f}' = 1.$$

Finally,

$$(5.36) \qquad \gamma\colon\ f \Rightarrow g \quad \text{is a } D^*\text{-2-cell if and only if it is a } \tilde{D}\text{-2-cell.}$$

We can make these statements in the alternative form that every $\tilde{D}$-algebra $(A, a, a', \bar{a}')$ is of the form (5.33) for a unique D^*-action a^*, and so on. The same techniques as in §4.7 allow us to extend this to the more general case where the algebras are not objects A of K but 2-functors A: $L \to K$, which we need to complete our proof; we leave the easy details to the reader, and suppose this done.

<u>5.6</u> In this section (analogous to § 4.8) we proceed to state and prove our second main theorem. The proof itself is not <u>in detail</u> analogous to that of Theorem 4.1; doubtless one could make it so by proceeding as there is a more direct manner, but as we give it we avoid large diagrams by being less direct.

In this section the "underlying objects" of all our algebras are endo-2-functors of K.

<u>Theorem 5.1</u> <u>Let</u> (D,D',p,π) <u>be a pseudo distributive law on the 2-complete 2-category</u> K, <u>and suppose that maps from</u> (D,D',p,π) <u>into doctrines are represented by</u> $K = (k,k',\hat{k})\colon\ (D,D',p,\pi) \to D^*$. <u>Then</u> DD' <u>is a</u> D^*<u>-algebra. Let</u> $s\colon\ D^* \to DD'$ <u>be the unique strict</u> D^*<u>-morphism such that</u>

$$(5.37)\qquad sj^* = jj'.$$

<u>Define</u> $h\colon\ DD' \to D^*$ <u>as the composite</u>

$$(5.38)\qquad DD' \xrightarrow[kk']{} D^*D^* \xrightarrow[m^*]{} D^*.$$

<u>Then</u> $sh = 1$, <u>and there is a modification</u> $\eta\colon\ 1 \Rightarrow hs$ <u>such that</u> $s\eta = 1$ <u>and</u> $\eta h = 1$. <u>This applies to both the lax and the pseudo cases - the latter being that where the third element of a map from</u> (D,D',p,π) <u>to a doctrine is required to be an isomorphism; in the latter case</u> $\hat{k}$ <u>is an isomorphism, and so is</u> η.

<u>Proof</u> As in the proof of Theorem 4.1, the differences in the pseudo case take care of themselves: $\hat{k}$ is an isomorphism, whence the $\bar{a}'$ below is invertible by (5.39) below, whence $\bar{h}$ and $\bar{h}'$ below are invertible; thus $\bar{h}^*$ below is invertible because we use the stronger $\langle h,h \rangle$ of §3.5 in the pseudo case; so finally η is invertible from the details of Theorem 2.6.

For the purpose of naming the various actions on D^* and on DD', we treat these as if their names were A and B respectively.

D^* is a D^*-algebra with action m^*, and is therefore a $\tilde{D}$-algebra $(D^*,a,a',\bar{a}')$ where by (5.33) we have

$$(5.39) \qquad a = m^*.kD^*, \qquad a' = m^*.k'D^*, \qquad \bar{a}' = m^*.\hat{k}D^*.$$

D' is a D'-algebra with action m', so we can form the free $\tilde{D}$-algebra on it, namely $\tilde{D}D'$ with $\tilde{D}$-action $\tilde{m}D'$(also called $\tilde{b} = (b,\bar{b}')$). By (5.12) the underlying object of $\tilde{D}D'$ is DD'. Thus DD' is a $\tilde{D}$-algebra $(DD', b, b', \bar{b}')$; hence it is also a D^*-algebra with action b^*.

Since D^* is the free D^*-algebra on 1, there is as in the statement of the theorem a unique strict D^*-morphism $s\colon\ D^* \to DD'$ satisfying (5.37).

We shall need below the value of the composite

$$(5.40) \qquad D' \xrightarrow{k'} D^* \xrightarrow{s} DD'.$$

As a strict D^*-morphism, s is <u>a fortiori</u> a strict D'-morphism. Moreover the doctrine map k' is a strict D'-morphism; note that the D'-action a' on D^* given by (5.39) is just that arising in the usual way from the doctrine map k'. Thus the composite (5.40) is a strict D'-morphism; whence as in (2.4) and (2.5) it is determined by its composite with $j'\colon\ 1 \to D'$. But $k'j' = j^*$ since k' is a doctrine map, so that using (5.37) we have $sk'j' = sj^* = jj'$, which may be written as

$$1 \xrightarrow{j'} D' \xrightarrow{jD'} DD'.$$

Since jD' too is a strict D'-morphism by §5.2, we conclude that the composite (5.40) is given by

$$(5.41) \qquad sk' = jD'.$$

Since D^* is a $\tilde{D}$-algebra and $\tilde{D}D'$ is a free one, there is by (2.4) and (2.5) a unique strict $\tilde{D}$-morphism $\tilde{H}$: $\tilde{D}D' \to D^*$ satisfying

$$(5.42) \qquad \tilde{H}.\tilde{j}D' = k'\colon \ D' \to D^*$$

and given explicitly as the composite

$$(5.43) \qquad \tilde{D}D' \xrightarrow{\tilde{D}k'} \tilde{D}D^* \xrightarrow{\tilde{a}} D^*.$$

As in §5.3, we write $\tilde{H}$ in the extended form $\tilde{H} = (h,\bar{h},\bar{h}')$. Because it is a <u>strict</u> $\tilde{D}$-morphism, $\bar{h} = 1$ by (5.25). As a D'-morphism, $\tilde{a} = (a,\bar{a}')$ as in §5.3; and $\tilde{D}k'$ is $(Dk',1)$ by (5.14), k' already being strict as a D'-morphism. We conclude that

$$(5.44) \qquad h = a.Dk', \qquad \bar{h} = 1, \qquad \bar{h}' = \bar{a}'.D'Dk'.$$

Substituting in this value for h the value of a from (5.39), we see that h has indeed the value (5.38) in the statement of the theorem.

Being a strict D^*-morphism, s is <u>a fortiori</u> a strict $\tilde{D}$-morphism. Consider the composite strict $\tilde{D}$-morphism

$$\tilde{D}D' \xrightarrow{\tilde{H}} D^* \xrightarrow{s} \tilde{D}D'.$$

As in (2.4) and (2.5), it is determined by its composite with $\tilde{j}D'$. By (5.42) and (5.41), this composite is jD'; which by (5.17) is another name for $\tilde{j}D'$. Hence

$$(5.45) \quad s\tilde{H} = 1.$$

Now let $H^* = (h,\bar{h}^*)$: $DD' \to D^*$ be the D^*-morphism corresponding as in (5.34) to the $\tilde{D}$-morphism $\tilde{H}$. Since this correspondence preserves composition, and since s corresponds to itself by (5.35), we have

$$(5.46) \quad sH^* = 1.$$

So we are now in the situation of §2.3, with D^*,1 replacing D,A respectively. By (5.37) the r: $1 \to DD'$ corresponding to s: $D^* \to DD'$ is jj'. So

$$\begin{aligned} hr &= h.jj' \\ &= m^*.kk'.jj' \quad \text{by (5.38)} \\ &= m^*.j^*j^* \quad \text{since } k,k' \text{ are doctrine maps} \\ &= m^*.D^*j^*.j^* \\ &= j^* \quad \text{since } m^*.D^*j^* = 1 \text{ (doctrine axiom).} \end{aligned}$$

Therefore we are in the special case of §2.3 in which (2.23) commutes. Our desired reflexion result now follows from Theorem 2.6, once we prove that

$$(5.47) \quad \bar{h}^*.D^*jj'.h = 1.$$

The rest of the proof consists in verifying this.

We do this, by hindsight, in several small steps. First, by (5.23),

$$(5.48) \qquad \bar{a}'.D'jD^* = 1.$$

Compose this with $D'k'$: $D'D' \to D'D^*$, and use the naturality of j to replace $D'jD^*.D'k'$ by $D'Dk'.D'jD'$. Using the third equation of (5.44), we now have

$$(5.49) \qquad \bar{h}'.D'jD' = 1.$$

Use the second equation of (5.34) to express $\bar{h}'$ in terms of $\bar{h}^*$ (mentally replacing f: $A \to B$ by h: $DD' \to D^*$); we have $\bar{h}' = \bar{h}^*.k'DD'$, so that (5.49) simplifies to

$$(5.50) \qquad \bar{h}^*.k'jD' = 1.$$

Now one of the axioms satisfied by the D^*-morphism H^* is

$$(5.51) \qquad \begin{array}{ccccccccccc} D^*D^*DD' & \xrightarrow{m^*DD'} & D^*DD' & \xrightarrow{b^*} & DD' & = & D^*D^*DD' & \xrightarrow{D^*b^*} & D^*DD' & \xrightarrow{b^*} & DD' \\ & & \Big\downarrow D^*h & \Rightarrow \bar{h}^* & \Big\downarrow h & & \Big\downarrow D^*D^*h & \Rightarrow D^*\bar{h}^* & \Big\downarrow D^*h & \Rightarrow \bar{h}^* & \Big\downarrow h \\ & & D^*D^* & \xrightarrow[m^*]{} & D^* & & D^*D^*D & \xrightarrow[D^*m^*]{} & D^*D^* & \xrightarrow[m^*]{} & D^*. \end{array}$$

Compose both sides on the front end with $D^*k'jD'$: $D^*D'D' \to D^*D^*DD'$, and use D^* of (5.50) to simplify the right side. We end up with

$$(5.52) \qquad \bar{h}^*.m^*DD'.D^*k'jD' = \bar{h}^*.D^*b^*.D^*k'jD'.$$

(That is, we have used (5.50) to kill the $D^*\bar{h}^*$ on the right of (5.51), and then reverted to linear rather than diagrammatic notation).

Now compose (5.52) with $kD'j'$: $DD' \to D^*D'D'$, which can also be written as

$$DD' \xrightarrow{kD'} D^*D' \xrightarrow{D^*D'j'} D^*D'D'.$$

What we then have on the right side is

$$\bar{h}^*.D^*b^*.D^*k'jD'.D^*D'j'.kD',$$

or $\bar{h}^*.D^*y.kD'$ say, where $y = b^*.k'jD'.D'j'$. By the naturality of k, this is equally $\bar{h}^*.kDD'.Dy$. But by (5.34), $\bar{h}^*.kDD' = \bar{h}$, and by (5.44) this is 1. So after composing (5.52) with $kD'j'$, we get 1 on the right side. Simplifying a little what we get on the left side, we now have

$$(5.53) \qquad \bar{h}^*.m^*DD'.kk'jj' = 1.$$

Write $kk'jj'$ as the composite

$$DD' \xrightarrow{kk'} D^*D^* \xrightarrow{D^*D^*jj'} D^*D^*DD'$$

and the use the naturality of m^* to replace $m^*DD'.D^*D^*jj'$ by $D^*jj'.m^*$. Then (5.53) becomes

$$(5.54) \qquad \bar{h}^*.D^*jj'.m^*.kk' = 1,$$

which by (5.38) is the (5.47) that we seek. □

6. Pseudo-commutative doctrines and clubs

6.1 An endofunctor D of $\mathcal{Cat}$ is made into an endo-2-functor by giving a natural transformation $[A,B] \to [DA,DB]$ (the "strength" of D) subject to two axioms (multiplicative and unitary). From this strength we derive a natural (and in fact 2-natural) transformation

(6.1) $t\colon\ A\times DB \to D(A\times B)$

as the image under adjunction of the composite

(6.2) $A \to [B,A\times B] \to [DB,D(A\times B)]$.

It is well known that giving the strength of D is equivalent to giving t; in fact t and the strength are mates, in the sense of [13] §2.2, under the adjunctions $-\times A \dashv [A,-]$ and $-\times DA \dashv [DA,-]$; so that t is often called the "monoidal strength" of D: of course t must satisfy two axioms corresponding to those for the strength.

If now (D,m,j) is a doctrine on $\mathcal{Cat}$, it is easy to express the monoidal strength of D^2 in terms of t, while that of 1 is the identity. The 2-naturality, as distinct from the naturality, of j and of m can be expressed in terms of the monoidal strengths, in the form of a commutative diagram involving j and t and another involving m and t. So all told, to make a mere monad (D,m,j) on $\mathcal{Cat}$ into a doctrine is to give a natural t as in (6.1) satisfying four axioms.

We use the symbol $^{\#}$ in a general way to denote conjugation under the symmetry $c\colon\ A\times B \to B\times A$ of $\mathcal{Cat}$. So alongside t we also have

(6.3) $t^{\#}\colon\ DA\times B \to D(A\times B)$,

namely the composite

(6.4) $\quad DA\times B \underset{c}{\rightarrow} B\times DA \underset{t}{\rightarrow} D(B\times A) \underset{Dc}{\rightarrow} D(A\times B).$

We can then form the 2-natural

(6.5) $\quad \tilde{d}\colon\ DA\times DB \rightarrow D(A\times B)$

as the composite

(6.6) $\quad DA\times DB \underset{t^{\#}}{\rightarrow} D(A\times DB) \underset{Dt}{\rightarrow} D^2(A\times B) \underset{m}{\rightarrow} D(A\times B),$

as well as its conjugate

(6.7) $\quad \tilde{d}^{\#}\colon\ DA\times DB \rightarrow D(A\times B),$

namely the composite

(6.8) $\quad DA\times DB \underset{t}{\rightarrow} D(DA\times B) \underset{Dt^{\#}}{\rightarrow} D^2(A\times B) \underset{m}{\rightarrow} D(A\times B).$

Moreover, using I for the unit category, we can set

(6.9) $\quad d^{\circ} = jI\colon\ I \rightarrow DI.$

Kock [14] shows that $\tilde{d}$ and d° enrich D to a monoidal 2-functor $(D,\tilde{d},d^{\circ})\colon$ *Cat* → *Cat*. This is not in general symmetric monoidal; the extra condition needed for this is precisely

(6.10) $\quad \tilde{d} = \tilde{d}^{\#}.$

The unit j: 1 → D is always monoidally 2-natural; recall that for natural transformations there is no difference between "monoidally 2-natural" and "symmetric monoidally 2-natural". The multiplication m: $D^2 \to D$ is not in general monoidally 2-natural, but is so <u>if and only if</u> (6.10) is satisfied.

Kock called a doctrine - or more generally a V-monad on a symmetric monoidal closed V - <u>commutative</u> if it satisfied (6.10). Because of the last remark, it has now become rather more common to call them <u>monoidal</u> V-monads; they are the monads on V in the 2-category of monoidal V-categories, and necessarily lie in the 2-category of symmetric monoidal V-categories. For commutative D, under the mildest completeness assumptions on V, the V-category of algebras V^D is symmetric closed, and indeed symmetric monoidal closed if it admits coequalizers; and the forgetful and free V-functors $U\colon\ V^D \to V$, $F\colon\ V \to V^D$ are symmetric monoidal. Conversely, of course, if V^D,U,F are monoidal, so is D = UF. Thus, modulo the always-troublesome matter of the existence of coequalizers in V^D, commutativity of D is the necessary and sufficient condition for V^D,U,F to be symmetric monoidal closed: it generalizes Linton's criterion (cf. [6] p.549) when V = *Sets* and D comes from a finitary theory.

A final remark at this level. Suppose D is a commutative monad on *Sets*. Then, being symmetric monoidal, D takes a commutative monoid A to a commutative monoid DA, and in fact lifts to a monad $\tilde{D}$ on the category of commutative monoids. So if D' is the monad on *Sets* whose algebras are commutative monoids, we get an <u>honest</u> distributive law p: D'D → DD'.

<u>6.2</u> In the doctrine case, commutativity as expressed by (6.10) seems to be rare in natural examples. We call a doctrine <u>pseudo-commutative</u> if there is instead an isomorphism (invertible modification)

$$(6.11) \qquad \gamma\colon\ \tilde{d} \Rrightarrow \tilde{d}^{\#}$$

satisfying suitable axioms (one of which is that γ be involutary; $\gamma^{\#}\gamma\colon\ \tilde{d} \Rrightarrow \tilde{d}^{\#} \Rrightarrow \tilde{d}$ is the identity). Then we get a <u>pseudo-distributive</u>

law (p,π): $D'D \to DD'$ as in §§5.1, 5.2, where now D' is not (as in the last paragraph) the monad for commutative monoids, but rather the doctrine for symmetric monoidal categories.

I shall do no more than sketch this in the above generality, for in all my examples D is $\mathcal{D}\circ-$ for a club $\mathcal{D}$ in $\mathcal{C}at/\underline{\underline{S}}$, and then everything becomes much easier because all the diagrams come down by one dimension. The extra generality would be pleasant, but unless I succeed in finding a more compact treatment of it, my applications do not justify the extra complication.

Let then A be a D'-category, that is, a symmetric monoidal category, and let D be pseudo-commutative with γ as in (6.11) satisfying axioms to be determined. We give a symmetric monoidal structure to DA. Its tensor product and identity object are given by

$$(6.12) \qquad DA \times DA \underset{\tilde{d}}{\to} D(A \times A) \underset{D\otimes}{\to} DA,$$

$$(6.13) \qquad I \underset{d^{\circ}}{\to} DI \underset{DI}{\to} DA;$$

its associativity and right identity isomorphisms are obtained from those for A (called a and r) by composing Da with

$$DA \times DA \times DA \underset{\tilde{d} \times 1}{\to} D(A \times A) \times DA \underset{\tilde{d}}{\to} D(A \times A \times A)$$

and Dr with

$$DA \times I \underset{1 \times d^{\circ}}{\to} DA \times DI \underset{\tilde{d}}{\to} D(A \times I).$$

Its commutativity isomorphism is obtained from that (called c) for A as the composite

(6.14)

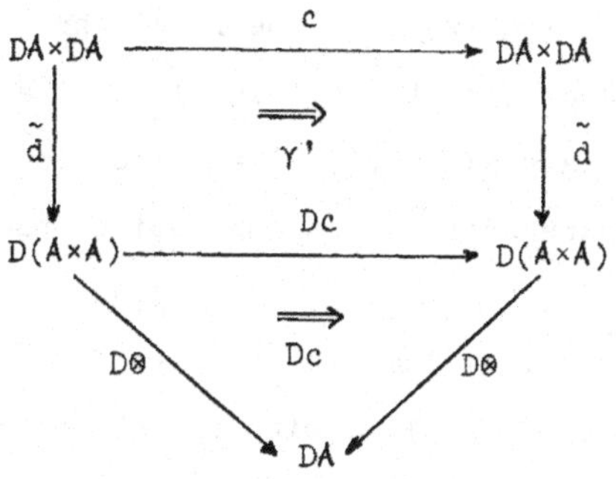

where γ' denotes $Dc.\gamma$. The monoidal-category axioms are immediate because $(D,\tilde{d},d^\circ)$ is monoidal; the symmetric-monoidal-category axioms ($c^2 = 1$ and the hexagonal axiom) require respectively that γ be involutory and that it satisfy a kind of hexagonal axiom (relating $\gamma_{A\times B,C}$ to $\gamma_{A,C}$ and $\gamma_{B,C}$).

The above passage from A to DA respects strict symmetric monoidal functors and the corresponding natural transformations, and therefore gives us a lifting of D as in (5.1), whence also a lifting as in (5.2). The lifting of j at the (5.1) level is automatic, because $(D,\tilde{d},d^\circ)$ is monoidal, except as regards the symmetry: for this it suffices to impose the axiom that the composite of γ with $j\times 1$: $A\times DB \to DA\times DB$ is the identity.

It remains to lift m at the (5.2) level, that is, to enrich m: $D^2A \to DA$ to a strong (but not strict) symmetric monoidal functor $(m,\hat{m},m^\circ)$; here I use $\hat{m}$ in place of the more usual $\tilde{m}$, which unfortunately has another sense in §5, being in fact the name of the triple $(m,\hat{m},m^\circ)$. We can take m° to be the identity. To get an $\hat{m}$, it turns out that we need a modification
μ: $\tilde{d}(m\times m) \Rightarrow m.D\tilde{d}.\tilde{d}$: $D^2A\times D^2B \to D(A\times B)$. The four axioms satisfied by t,m,j allow us to write this as μ: $m.Dt.m.D\tilde{d}.t^\# \Rightarrow m.Dt.m.D\tilde{d}^\#.t^\#$, and we set $\mu = m.Dt.m.D\gamma.t^\#$. Of the three axioms saying that $(m,\hat{m},m^\circ)$ is a symmetric monoidal functor, the one involving m° is satisfied in virtue

of our axiom $\gamma(j\times 1) = 1$, leaving two. Of the final three axioms stating that $(\tilde{D},\tilde{m},\tilde{j})$ is a doctrine, the two involving $\tilde{j}$ are satisfied in virtue of the same axiom $\gamma(j\times 1) = 1$, leaving one. So three more axioms for γ ensure that we get our desired pseudo distributive law.

I suspect that the six axioms we now have for γ reduce to one or two, or at most three, by analogy with the club case that follows: but I have not yet succeeded in so reducing them, and am reluctant to delay this volume longer while trying to do so. So I pass on to the club case without giving a definitive set of axioms for the γ of a "pseudo commutative doctrine". In the club case we establish the pseudo distributive law <u>ab initio</u>.

<u>6.3</u> We now suppose that $D = \mathcal{D}\circ -$ for a club $\mathcal{D}$ in $\mathcal{C}at/\underline{\underline{S}}$, and we use the notation of [9] §10. In particular the augmentation of $\mathcal{D}$ is $\Gamma\colon \mathcal{D} \to \underline{\underline{S}}$; we typically write τ for ΓT, σ for ΓS, and so on; in fact we are running so short of symbols that we shall quite generally use Greek letters for natural numbers, as well as for permutations or functions. It should cause no confusion if we use $m\colon \mathcal{D}\circ\mathcal{D} \to \mathcal{D}$ for the multiplication of $\mathcal{D}$, although to date m has denoted the 2-natural $\mathcal{D}\circ\mathcal{D}\circ - \to \mathcal{D}\circ -$; then m sends $T[S_1 \ldots S_\tau]$ to $T(S_1 \ldots S_\tau)$ and similarly for morphisms. Again as in [9] §10, the image under $j\colon\ \mathcal{J} \to \mathcal{D}$ of the unique object of $\mathcal{J}$ is $\underline{\underline{1}} \in \mathcal{D}$. The associative law for m expresses the equality of $T(S_1(R_{11} \ldots R_{1\sigma_1}) \ldots S_\tau(R_{\tau 1} \ldots R_{\tau\sigma_\tau}))$ with $T(S_1 \ldots S_\tau)(R_{11} \ldots R_{\tau\sigma_\tau})$, and similarly for morphisms; the two unitary laws for m and j express the equalities $T(\underline{\underline{1}} \ldots \underline{\underline{1}}) = T$ and $\underline{\underline{1}}(T) = T$, and similarly for morphisms. The ordered pairs that are the objects and the morphisms of $\mathcal{A}\times\mathcal{B}$ are written $\langle A,B\rangle$ and $\langle f,g\rangle$.

The functor $t\colon \mathcal{A} \times \mathcal{D}\circ\mathcal{B} \to \mathcal{D}\circ(\mathcal{A}\times\mathcal{B})$ of (6.1) is easily seen to be given by

$$t\langle A,\ S[B_1 \ \ldots\ B_\sigma]\rangle = S[\langle A,B_1\rangle \ \ldots\ \langle A,B_\sigma\rangle],$$

$$t\langle f,\ h[g_1 \ \ldots\ g_\sigma]\rangle = h[\langle f,g_1\rangle \ \ldots\ \langle f,g_\sigma\rangle];$$

with analogous formulae for $t^\#$. Hence we calculate $\tilde{d},\tilde{d}^\#$: $\mathcal{D}\circ A \times \mathcal{D}\circ B \to \mathcal{D}\circ(A\times B)$; they send the object $\langle T[A_1 \ \ldots\ A_\tau],\ S[B_1 \ \ldots\ B_\sigma]\rangle$ respectively to

$$(6.15)\qquad T(S \ \ldots\ S)[\langle A_1,B_1\rangle \ \ldots\ \langle A_1,B_\sigma\rangle \ \ldots\ \langle A_\tau,B_1\rangle \ \ldots\ \langle A_\tau,B_\sigma\rangle],$$

$$(6.16)\qquad S(T \ \ldots\ T)[\langle A_1,B_1\rangle \ \ldots\ \langle A_\tau,B_1\rangle \ \ldots\ \langle A_1,B_\sigma\rangle \ \ldots\ \langle A_\tau,B_\sigma\rangle],$$

with similar effects on morphisms. Here S occurs τ times in (6.15) and T occurs σ times in (6.16); the order of the $\langle A_\alpha,B_\beta\rangle$ is in (6.15) the lexicographical order of the $\langle\alpha,\beta\rangle$, and in (6.16) the lexicographical order of the $\langle\beta,\alpha\rangle$.

To give an isomorphism γ as in (6.11) is to give for each $T,S \in \mathcal{D}$ a natural isomorphism

$$(6.17)\qquad z_{TS}\colon\ T(S \ \ldots\ S) \to S(T \ \ldots\ T)$$

whose type $\Gamma z_{TS} = \zeta_{\tau,\sigma}$ is the permutation of $\tau\sigma$ demanded by a comparison of (6.15) and (6.16); then the $\langle T[A_1 \ \ldots\ A_\tau], S[B_1 \ \ldots\ B_\sigma]\rangle$-component of $\gamma = \gamma_{A,B}$ is the morphism $z_{TS}[1,1,\ \ldots,1]$ from (6.15) to (6.16). By the <u>naturality</u> of z_{TS} we mean of course that we have commutativity in

$$(6.18)\qquad \begin{array}{ccc} T(S\ldots S) & \xrightarrow{\;z_{TS}\;} & S(T\ldots T) \\ {\scriptstyle f(g\ldots g)}\downarrow & & \downarrow{\scriptstyle g(f\ldots f)} \\ T'(S'\ldots S') & \xrightarrow[\;z_{T'S'}\;]{} & S'(T'\ldots T'). \end{array}$$

(This diagram is more complicated than it looks: in general $\tau' \neq \tau$, and $\Gamma f = \phi$ is a function from τ to τ'; the α-th g in $f(g\ldots g)$ is a morphism from the α-th S in $T(S\ldots S)$ to the $\phi\alpha$-th S' in $T'(S'\ldots S')$, in accordance with the conventions of [9] §10; however it does make good sense, both legs having the same type.)

Thus does the club case fit into the general situation sketched in §6.2; it remains only to put such axioms on z_{TS} as ensure those we require for γ. However we deal henceforth only with the club case, forgetting γ and referring only to z_{TS}, and carrying out in detail what we barely outlined in §6.2.

6.4 We define then a pseudo-commutative club to be a club $\mathcal{D}$ in $\mathcal{Cat}/\underline{S}$ together with an isomorphism z_{TS} as in (6.17), of the type $\zeta_{\tau,\sigma}$ described above, natural in the sense of (6.18), and satisfying in addition one axiom: namely the commutativity of

$$(6.19)\qquad \begin{array}{ccc} T(S(R_1\ldots R_\sigma)\ldots S(R_1\ldots R_\sigma)) & \xrightarrow{\;z_{T,S(R_1,\ldots R_\sigma)}\;} & S(R_1\ldots R_\sigma)(T\ldots T) \\ \| & & \| \\ T(S\ldots S)(R_1\ldots R_\sigma\ldots R_1\ldots R_\sigma) & & S(R_1(T\ldots T)\ldots R_\sigma(T\ldots T)) \\ {\scriptstyle z_{TS}(1,1,\ldots,1)}\downarrow & & \downarrow{\scriptstyle S(z_{R_1T},\ldots,z_{R_\sigma T})} \\ S(T\ldots T)(R_1\ldots R_1\ldots R_\sigma\ldots R_\sigma) & = & S(T(R_1\ldots R_1)\ldots T(R_\sigma\ldots R_\sigma)). \end{array}$$

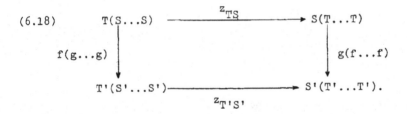

We derive some immediate consequences of (6.19). First, put $S = \underline{\underline{1}}$ (so that $\sigma = 1$) and $R_1 = \underline{\underline{1}}$. The top and the left edges are then both the isomorphism $z_{T\underline{\underline{1}}}$ while the right edge is $z_{\underline{\underline{1}}T}$, whence

$$(6.20) \qquad z_{\underline{\underline{1}}T} = 1.$$

Now in (6.19) put $T = \underline{\underline{1}}$, $S = \underline{\underline{1}}$, $R_1 = R$. Using (6.20) we get

$$(6.21) \qquad z_{R\underline{\underline{1}}} = 1.$$

Finally just put $S = \underline{\underline{1}}$, $R_1 = R$, T arbitrary. Using (6.21) we get

$$(6.22) \qquad z_{RT}z_{TR} = 1.$$

Hence z is involutary and satisfies (6.20).

Before giving examples we make the following remark. Any club $\mathcal{D}$ becomes a strict monoidal category, with $\underline{\underline{1}}$ as identity, if we set $T \otimes S = T(S \ldots S)$. Moreover $\Gamma\colon\ \mathcal{D} \to \underline{\underline{S}}$ is a strict morphism of strict monoidal categories when $\underline{\underline{S}}$ is given the cartesian monoidal structure $\tau \otimes \sigma = \tau \times \sigma = \tau\sigma$. For a pseudo-commutative club, $z_{TS}\colon\ T \otimes S \to S \otimes T$ makes $\mathcal{D}$ symmetric monoidal; the hexagon axiom is got by setting $R_1 = \ldots = R_\sigma = R$ in (6.19), and the other axiom is (6.22). Moreover $\Gamma\colon\ \mathcal{D} \to \underline{\underline{S}}$ is then a strict symmetric monoidal functor, as $\zeta_{\tau,\sigma}$ is the classical symmetry on $\underline{\underline{S}}$. This observation is useful in limiting our search for examples.

<u>Example 6.1</u> The club $\underline{\underline{S}}$ itself. We recall that $\tau(\sigma_1, \ldots, \sigma_\tau)$ is $\sigma_1 + \ldots + \sigma_\tau$ while $\underline{\underline{1}}$ is 1. We take $z_{\tau\sigma}\colon\ \tau(\sigma \ldots \sigma) \to \sigma(\tau \ldots \tau)$ to be the permutation $\zeta_{\tau,\sigma}\colon\ \tau\sigma \to \sigma\tau$ of §6.3. This is easily seen to satisfy the axiom (6.19); in fact if it did not the axiom would not make sense, its two legs being of different types. Note that an $\underline{\underline{S}}$-algebra is just a category with strictly-associative finite coproducts.

<u>Example 6.2</u> The club S whose algebras are categories-with-finite-coproducts. The augmentation Γ: $S \to \underline{\underline{S}}$ is an equivalence, and we define z_{TS} (as we must) to be the unique morphism such that $\Gamma z_{TS} = \zeta_{\tau,\sigma}$.

<u>Example 6.3</u> The sub-club $\underline{\underline{P}}$ of $\underline{\underline{S}}$ with the natural numbers as objects but with permutations as its only morphisms; the augmentation $\underline{\underline{P}} \to \underline{\underline{S}}$ is of course the inclusion. Again $z_{\tau\sigma} = \zeta_{\tau,\sigma}$. The $\underline{\underline{P}}$-algebras are the <u>strict</u> symmetric monoidal categories.

<u>Example 6.4</u> The club P whose algebras are the symmetric monoidal categories. By Mac Lane's original coherence result [23], the augmentation Γ is an equivalence of P with $\Gamma(P) = \underline{\underline{P}} \subset \underline{\underline{S}}$. Again z_{TS} is the unique morphism with $\Gamma z_{TS} = \zeta_{\tau,\sigma}$.

<u>Example 6.5</u> The full sub-club of $\underline{\underline{S}}$ determined by the objects 0 and 1; it is the arrow category $\underline{\underline{2}}$. Its algebras are categories-with-an-initial-object.

<u>Example 6.6</u> The full sub-club of $\underline{\underline{P}}$ determined by the objects 0 and 1; it is the discrete category 2. Its algebras are categories-with-a-distinguished-object.

<u>Example 6.7</u> Our remaining examples are all of the kind where the augmentation Γ: $D \to \underline{\underline{S}}$ is the constant functor at $1 \in \underline{\underline{S}}$. A club of this kind is nothing but a strict monoidal category D, with $T(S) = T\otimes S$ and with $\underline{\underline{1}}$ as the identity for $\otimes$. It is pseudo-commutative if and only if it is <u>symmetric</u> monoidal, in which case we set z_{TS}: $T(S) \to S(T)$ equal to the symmetry c: $T\otimes S \to S\otimes T$. Since $D \circ A$ is just $D\times A$, a D-algebra is a category A with a <u>strictly</u> associative and unitary action $D\times A \to A$ (as for example E acts on K in §3.2 above). Note that the club whose algebras are categories-bearing-a-monad, namely $\underline{\underline{\Delta}}$, is <u>not</u> an example, for the monoidal structure on the simplicial category $\underline{\underline{\Delta}}$ is <u>not</u> symmetric. We pass to some particular cases of this

example.

<u>Example 6.7.1</u> Let C be a symmetric monoidal category and consider the club $\mathcal{D}$ whose algebras are categories A together with a coherently (but not strictly) associative and unitary functor $\otimes$: $C \times A \to A$. For instance, any <u>tensored</u> C-category A is such an algebra (where by C-category I mean "category enriched over C" - a remark necessitated by the fact that I sometimes, for a club $\mathcal{D}$, call a $\mathcal{D}$-algebra a $\mathcal{D}$-category). Since all the operations on A are unary, the augmentation of $\mathcal{D}$ is constant at 1. We have to show that $\mathcal{D}$ is symmetric, and is thus a pseudo-commutative club. But we know just what $\mathcal{D}$ is: it is the canonical strict monoidal category equivalent to C; its objects are n-ads $\langle C_1, \ldots, C_n \rangle$ of objects of C, its tensor product is given on objects by concatenation, and its morphisms $\langle C_1 \ldots, C_n \rangle \to \langle B_1, \ldots, B_m \rangle$ are the morphisms $C_1 \otimes (C_2 \otimes \ldots (C_{n-1} \otimes C_n)) \to B_1 \otimes (B_2 \otimes \ldots (B_{m-1} \otimes B_m))$ in C. It is now clear that $\mathcal{D}$ is symmetric when C is.

<u>Example 6.7.2</u> Let $\mathcal{D}$ be the discrete category $\underline{\underline{N}}$ of natural numbers, which is symmetric monoidal with + as its tensor product. A $\mathcal{D}$-algebra is a category A with an endofunctor E: $A \to A$.

<u>Example 6.7.3</u> Let $\mathcal{D}$ be the discrete category 2 with the symmetric monoidal structure having the usual multiplication of its objects 0 and 1 as tensor product. This is not the same as Example 6.6, for the augmentation is now constant at 1, whereas there it was $\Gamma 0 = 0$, $\Gamma 1 = 1$. An algebra is a category A with an endofunctor E such that $E^2 = E$.

<u>Example 6.7.4</u> Let $\mathcal{D}$ be the arrow category $\underline{\underline{2}}$ with the symmetric monoidal structure given on objects as in the last example. This is again different from Example 6.5, as the augmentation is again constant at 1. An algebra is a category A together with an indempotent comonad. Replacing $\underline{\underline{2}}$ by $\underline{\underline{2}}^{op}$ gives another example, where an algebra is a category bearing an indempotent monad.

<u>6.5</u> Let $\mathcal{D}$ be a pseudo-commutative club as in §6.4, giving the doctrine $D = \mathcal{D}\circ -$; let D' be the doctrine $\mathcal{P}\circ -$ whose algebras are symmetric monoidal (s.m.) categories. We exhibit a pseudo distributive law of D' over D, as described in §§5.1, 5.2. (We could equally produce a pseudo distributive law if we took D' to be the doctrine $\underline{\underline{P}}\circ -$ for <u>strict</u> s.m. categories; this requires only the easy observation that if the s.m. structure on $\mathcal{A}$ is strict, so is that we construct below on $\mathcal{D}\circ\mathcal{A}$; we shall not refer further to this case.)

The first step is to produce the lifting (5.1) of D, and its extension (5.2). Let $(\mathcal{A},\otimes,I,a,r,c)$ be a s.m. category; we give $D\mathcal{A} = \mathcal{D}\circ\mathcal{A}$ a s.m. structure to get $\tilde{D} = (\mathcal{D}\circ\mathcal{A},\bar{\otimes},\bar{I},\bar{a},\bar{r},\bar{c})$. We set

$$(6.23)\quad T[A_1 \ldots A_\tau]\,\bar{\otimes}\,S[B_1 \ldots B_\sigma]$$

$$= T(S\ldots S)[A_1\otimes B_1,\ldots,A_1\otimes B_\sigma,\ldots,A_\tau\otimes B_1,\ldots,A_\tau\otimes B_\sigma],$$

and define $\bar{\otimes}$ similarly on morphisms. We define the rest of the structure by

$$(6.24)\quad \bar{I} = \underline{\underline{1}}[I],$$

$$(6.25)\quad \begin{array}{ccc} (T[A_1\ldots]\,\bar{\otimes}\,S[B_1\ldots])\,\bar{\otimes}\,R[C_1\ldots] & \xrightarrow{\ \bar{a}\ } & T[A_1\ldots]\,\bar{\otimes}\,(S[B_1\ldots]\,\bar{\otimes}\,R[C_1\ldots]) \\ \| & & \| \\ T(S\ldots S)(R\ldots R)[(A_1\otimes B_1)\otimes C_1,\ldots] & \xrightarrow[1[a,a,\ldots,a]]{} & T(S\ldots S)(R\ldots R)[A_1\otimes(B_1\otimes C_1),\ldots], \end{array}$$

$$(6.26)\quad \begin{array}{ccc} T[A_1\ldots]\,\bar{\otimes}\,\bar{I} & \xrightarrow{\ \bar{r}\ } & T[A_1\ldots] \\ \| & & \| \\ T[A_1\ldots]\otimes\underline{\underline{1}}[I] & & \| \\ \| & & \| \\ T[A_1\otimes I,\ldots] & \xrightarrow[1[r,r,\ldots,r]]{} & T[A_1\ldots], \end{array}$$

$$(6.27)\qquad \begin{array}{ccc} T[A_1\ldots]\,\bar{\otimes}\,S[B_1\ldots] & \xrightarrow{\quad\bar{c}\quad} & S[B_1\ldots]\,\bar{\otimes}\,T[A_1\ldots] \\ \Big\| & & \Big\| \\ T(S\ldots S)[A_1\otimes B_1,\ldots] & \xrightarrow[z_{TS}[c,c,\ldots,c]]{} & S(T\ldots T)[B_1\otimes A_1,\ldots]\,. \end{array}$$

The axioms for $\bar{a},\bar{r},\bar{c}$ follow from those for a,r,c together with (6.22) (needed for $\bar{c}^2 = 1$) and the special case of (6.19) where $R_1 = \ldots = R_\sigma = R$ (needed for the hexagonal axiom connecting $\bar{a}$ and $\bar{c}$). (Note that when $\mathcal{A}$ is the unit category $\mathcal{I}$ this is the s.m. structure on $\mathcal{D}\circ\mathcal{I}$, $= \mathcal{D}$ itself, described just before Example 6.1 above.) If $\phi\colon\ \mathcal{A} \to \mathcal{B}$ is a strict s.m. functor so too is $\mathcal{D}\circ\phi\colon\ \mathcal{D}\circ\mathcal{A} \to \mathcal{D}\circ\mathcal{B}$ sending $T[A_1\ldots A_\tau]$ to $T[\phi A_1\ldots\phi A_\tau]$, and similarly for s.m. natural transformations. Thus we have the lifting (5.1) of D.

Its automatic extension (5.2) sends the non-strict s.m. functor $(\phi,\hat{\phi},\phi^\circ)$, where $\hat{\phi}\colon\ \phi A \otimes \phi B \to \phi(A\otimes B)$ and $\phi^\circ\colon\ I_{\mathcal{B}} \to \phi I_{\mathcal{A}}$, to $(\Phi,\hat{\Phi},\Phi^\circ)$ where $\Phi = \mathcal{D}\circ\phi$, where $\hat{\Phi}$ has components $T(S\ldots S)[\hat{\phi},\ldots,\hat{\phi}]$, and where $\Phi^\circ = \underline{\underline{1}}[\phi^\circ]$. (Recall that we are using $\hat{\phi}$ for the more familiar $\tilde{\phi}$ to avoid confusion with the $\tilde{m}$ etc. of §5.) Simiarly $\tilde{D}$ sends the s.m. natural transformation $\alpha\colon (\phi,\hat{\phi},\phi^\circ) \Rightarrow (\psi,\hat{\psi},\psi^\circ)$ to that with $T[A_1\ldots A_\tau]$-component $T[\alpha A_1,\ldots,\alpha A_\tau]$.

The functor $j\circ\mathcal{A}\colon\ \mathcal{A} \to \mathcal{D}\circ\mathcal{A}$ is the functor sending A to $\underline{\underline{1}}[A]$, and is clearly a strict s.m. functor because $z_{\underline{\underline{1}}\,\underline{\underline{1}}} = 1$ by (6.20). This gives us the $\tilde{j}_*$ and the $\tilde{j}$ of §5.1.

The next thing is to enrich $m\circ\mathcal{A}\colon\ \mathcal{D}\circ\mathcal{D}\circ\mathcal{A} \to \mathcal{D}\circ\mathcal{A}$ to a strong s.m. functor $\tilde{m}\mathcal{A} = (m\circ\mathcal{A},\hat{m}\mathcal{A},m^\circ\mathcal{A})$, as in (5.18); we abbreviate $\tilde{m}\mathcal{A}$, $\hat{m}\mathcal{A}$, $m^\circ\mathcal{A}$ to $\tilde{m},\hat{m},m^\circ$ where no confusion is likely. Use $\bar{\bar{\otimes}}$ etc. for the s.m. structure on $\mathcal{D}\circ\mathcal{D}\circ\mathcal{A}$. Since $\bar{\bar{I}} = \underline{\underline{1}}[\bar{I}] = \underline{\underline{1}}[\underline{\underline{1}}[I]] = \underline{\underline{1}}[\underline{\underline{1}}][I]$, we have $(m\circ\mathcal{A})\bar{\bar{I}} = \underline{\underline{1}}[I] = \bar{I}$; so we take $m^\circ = 1$. Now let

$$X = T[P_1[A_{11}\dots A_{1\pi_1}], \dots, P_\tau[A_{\tau 1}\dots A_{\tau\pi_\tau}]]$$

$$Y = S[R_1[B_{11}\dots B_{1\rho_1}], \dots, R_\sigma[B_{\sigma_1}\dots B_{\sigma\rho_\sigma}]]$$

be objects of $\mathcal{D}\circ\mathcal{D}\circ\mathcal{A}$. We need $\hat{m}$: $(m\circ A)X \mathbin{\bar{\otimes}} (m\circ A)Y \to (m\circ A)(X \mathbin{\bar{\bar{\otimes}}} Y)$. Now $(m\circ A)X = T(P_1\dots P_\tau)[A_{11}\dots]$ and

$$(6.28)\quad (m\circ A)X \mathbin{\bar{\otimes}} (m\circ A)Y$$

$$= T(P_1\dots P_\tau)(S(R_1\dots R_\sigma)\dots S(R_1\dots R_\sigma))[A_{11}\otimes B_{11},\dots]$$

$$= T(M_1\dots M_\tau)[A_{11}\otimes B_{11},\dots],$$

where for example

$$(6.29)\quad M_1 = P_1(S\dots S)(R_1\dots R_\sigma\dots R_1\dots R_\sigma).$$

On the other hand,

$$X \mathbin{\bar{\bar{\otimes}}} Y = T(S\dots S)[P_1[A_{11}\dots] \mathbin{\bar{\otimes}} R_1[B_{11}\dots], \dots]$$

$$= T(S\dots S)[P_1(R_1\dots R_1)[A_{11}\otimes B_{11},\dots], \dots],$$

so that

$$(6.30)\quad (m\circ A)(X \mathbin{\bar{\bar{\otimes}}} Y)$$

$$= T(S\dots S)(P_1(R_1\dots R_1)\dots)[A_{11}\otimes B_{11},\dots]$$

$$= T(N_1\dots N_\tau)[A_{11}\otimes B_{11},\dots]$$

where for example

$$(6.31)\quad N_1 = S(P_1\dots P_1)(R_1\dots R_1\dots R_\sigma\dots R_\sigma).$$

We now take the X,Y-component of $\hat{m}$ to be

$$(6.32)\quad T(z_{P_1S}(1,1,\ldots,1),\ \ldots,\ z_{P_\tau S}(1,1,\ldots,1))[1,1,\ldots,1],$$

where for example $z_{P_1S}(1,1,\ldots,1)\colon\ M_1 \to N_1$.

Using the properties (6.18)-(6.22) of z, it is trivial if somewhat tedious to verify that $\tilde{m} = (m\circ A,\hat{m},m^\circ)$ satisfies the axioms for a s.m. functor. It is clearly 2-natural in A by its construction. Finally, using the same properties of z and a large sheet of paper, we easily verify that $(\tilde{D},\tilde{m},\tilde{j})$ satisfies the axioms for a doctrine. Thus we have our distributive law p,π as in §5.2.

We shall not need the explicit values of p and π. One point, however, should be noted. As a 2-natural transformation $\mathcal{P}\circ\mathcal{D}\circ- \to \mathcal{D}\circ\mathcal{P}\circ-$, p comes from a map $\mathcal{P}\circ\mathcal{D} \to \mathcal{D}\circ\mathcal{P}$ in $\mathcal{C}at\int^{*}\underline{\underline{S}}$, that we may as well still call p; see for example Theorem 8.1 of [9]. However, although $\mathcal{P}$ and $\mathcal{D}$ are both clubs in $\mathcal{C}at/\underline{\underline{S}}$, the map p: $\mathcal{P}\circ\mathcal{D} \to \mathcal{D}\circ\mathcal{P}$ does not lie in $\mathcal{C}at/\underline{\underline{S}}$. For instance, it sends $\otimes[T,S]$ to $T(S\ldots S)[\otimes,\otimes,\ldots,\otimes]$; the type of the former is $\tau+\sigma$, while that of the latter is $2\tau\sigma$; so p does not commute with the augmentation Γ. This is in contrast with say §4.10 above, where we did not have to move out of $\mathcal{C}at/\underline{\underline{S}}$; and it is perhaps the first example to show that, even if we are only concerned with clubs in $\mathcal{C}at/\underline{\underline{S}}$, we need the fuller theory of [9]. (This was adumbrated in §10.7 of [9].)

We sum up the result of this section in:

Theorem 6.8 *If $\mathcal{D}$ is a pseudo-commutative club, the construction above defines a pseudo distributive law of D' over D = $\mathcal{D}\circ-$, where D' = $\mathcal{P}\circ-$ is the doctrine for symmetric monoidal categories. Similarly if D' is the doctrine $\underline{\underline{P}}\circ-$ for strict symmetric monoidal categories.*

The examples in §6.4 of pseudo-commutative clubs now give examples of pseudo distributive laws, as promised in §5.1.

6.6 For the special pseudo distributive laws obtained in this way, the $\tilde{D}$-algebras, and hence the D^*-algebras, admit a simpler description than that given in §5.3. We work out from first principles what they are.

To give such an algebra $\mathcal{B}$ is to give a s.m. category $\mathcal{B}$ together with an action $\tilde{b}$: $\tilde{D}\mathcal{B} \to \mathcal{B}$. So $\tilde{b}$ is to be a s.m. functor $\tilde{b} = (b,\hat{b},b^\circ)$: $\mathcal{D}\circ\mathcal{B} \to \mathcal{B}$ that satisfies the axioms for a $\tilde{D}$-action. (we are using $\mathcal{B}$, b rather than $\mathcal{A}$, a to avoid confusion with the associativity a.) The s.m. structure on $\mathcal{D}\circ\mathcal{B}$ is of course that described in §6.4, which we employ without further comment.

At the level of ordinary functors, the action axioms tell us that b: $\mathcal{D}\circ\mathcal{B} \to \mathcal{B}$ is to be a $\mathcal{D}$-action on $\mathcal{B}$ in the ordinary sense. We write the image under b of $T[B_1 \dots B_\tau] \in \mathcal{D}\circ\mathcal{B}$ in the usual way as $T(B_1 \dots B_\tau) \in \mathcal{B}$, and similarly for morphisms. The action axioms now express the equality of $T(S_1(A_{11}\dots) \dots S_\tau(A_{\tau 1}\dots))$ with $T(S_1 \dots S_\tau)(A_{11}\dots A_{\tau 1}\dots)$ and of $\underline{\underline{1}}(A)$ with A, together with the similar equalities for morphisms.

Since $j\circ\mathcal{B}$: $\mathcal{B} \to \mathcal{D}\circ\mathcal{B}$ is the strict s.m. functor sending B to $\underline{\underline{1}}[B]$, the unitary axiom for an action immediately forces b° to be the identity:

$$(6.33) \quad b^\circ = 1: \ I \to \underline{\underline{1}}(I) = I.$$

Since $m^\circ = 1$ too, the associativity axiom for an action is satisfied at the level of $^\circ$-components.

The only extra piece of structure, then, that a D^*-algebra $\mathcal{B}$ has, over and above its symmetric monoidal structure and its $\mathcal{D}$-algebra structure, is the natural transformation

(6.34)
$$\begin{array}{c} T(A_1\dots A_\tau)\otimes S(B_1\dots B_\sigma) \\ \Big\downarrow \hat{b}_{T[A_1\dots A_\tau],S[B_1\dots B_\sigma]} \\ T(S\dots S)(A_1\otimes B_1,\dots,A_1\otimes B_\sigma,\dots,A_\tau\otimes B_1,\dots,A_\tau\otimes B_\sigma), \end{array}$$

which we abbreviate to $\hat{b}_{TS}$ or just to $\hat{b}$.

The axioms for $(b,\hat{b},1)$ to be a s.m. functor reduce to the following three:

(6.35)
$$\begin{array}{ccc} T(A_1\dots A_\tau)\otimes \underline{\underline{1}}(I) & = & T(A_1\dots A_\tau)\otimes I \\ \Big\downarrow \hat{b} & & \\ T(\underline{\underline{1}}\dots\underline{\underline{1}})(A_1\otimes I\dots A_\tau\otimes I) & & \Big\downarrow r \\ \Big\| & & \\ T(A_1\otimes I\dots A_\tau\otimes I) & \xrightarrow[1(r,r,\dots,r)]{} & T(A_1\dots A_\tau). \end{array}$$

(6.36)
$$\begin{array}{ccc} (T(A_1\dots)\otimes S(B_1\dots))\otimes R(C_1\dots) & \xrightarrow{a} & T(A_1\dots)\otimes(S(B_1\dots)\otimes R(C_1\dots)) \\ \Big\downarrow \hat{b}\otimes 1 & & \Big\downarrow 1\otimes\hat{b} \\ T(S\dots S)(A_1\otimes B_1,\dots)\otimes R(C_1\dots) & & T(A_1\dots)\otimes S(R\dots R)(B_1\otimes C_1,\dots) \\ \Big\downarrow \hat{b} & & \Big\downarrow \hat{b} \\ T(S\dots S)(R\dots R)((A_1\otimes B_1)\otimes C_1,\dots) & \xrightarrow[1(a,a,\dots,a)]{} & T(S\dots S)(R\dots R)((A_1\otimes(B_1\otimes C_1),\dots \end{array}$$

(6.37)
$$\begin{array}{ccc} T(A_1\dots)\otimes S(B_1\dots) & \xrightarrow{c} & S(B_1\dots)\otimes T(A_1\dots) \\ \Big\downarrow \hat{b} & & \Big\downarrow \hat{b} \\ T(S\dots S)(A_1\otimes B_1,\dots) & \xrightarrow[z_{TS}(c,c,\dots,c)]{} & S(T\dots T)(B_1\otimes A_1,\dots). \end{array}$$

The $\hat{}$-component of the unitary axiom for an action reduces to

(6.38) $\hat{b}_{\underline{\underline{1}}[A],\underline{\underline{1}}[B]}: \ \underline{\underline{1}}(A)\otimes\underline{\underline{1}}(B) \to \underline{\underline{1}}(\underline{\underline{1}})(A\otimes B)$ is $1: \ A\otimes B \to A\otimes B$.

Finally the $\hat{}$-component of the associativity axiom for an action reduces, in view of (6.32), to

(6.39)

$$\begin{array}{ccc}
T(P_1\ldots P_\tau)(A_{11}\ldots)\otimes S(R_1\ldots R_\sigma)(B_{11}\ldots) & \xrightarrow{\hat{b}_{T(P),S(R)}} & T(P_1\ldots P_\tau)(S(R_1\ldots R_\sigma)\ldots S(R_1\ldots R_\sigma))(A_{11}\otimes B_{11},\ldots) \\
\downarrow \hat{b}_{TS} & & \| \\
T(S\ldots S)(P_1(A_{11}\ldots)\otimes R_1(B_{11}\ldots),\ldots) & & T(P_1(S\ldots S)(R_1\ldots R_\sigma\ldots R_1\ldots R_\sigma),\ldots)(A_{11}\otimes B_{11},\ldots) \\
\downarrow T(S\ldots S)(b_{P_1R_1},\ldots) & & \downarrow T(z_{P_1S}(1,\ldots,1),\ldots)(1,\ldots,1) \\
T(S\ldots S)(P_1(R_1\ldots R_1)(A_{11}\otimes B_{11},\ldots),\ldots) & = & T(S(P_1\ldots P_1)(R_1\ldots R_1\ldots R_\sigma\ldots R_\sigma),\ldots)(A_{11}\otimes B_{11},\ldots)
\end{array}$$

where in the topmost arrow $T(P)$ stands for $T(P_1\ldots P_\tau)$, etc.

Now write

(6.40) $e^B_{T[A_1\ldots A_\tau]}$ for $\hat{b}_{T[A_1\ldots A_\tau],\underline{\underline{1}}[B]}$,

abbreviating it where desirable to e^B_T or e_T or e, so that

(6.41) $e: \ T(A_1\ldots A_\tau)\otimes B \to T(A_1\otimes B,\ldots,A_\tau\otimes B)$.

We show that $\hat{b}$ can be given in terms of e alone, the axioms on $\hat{b}$ reducing to simple axioms on e.

Write

$$(6.42)\quad e^{\#}:\ B\otimes T(A_1\ldots A_\tau) \rightarrow T(B\otimes A_1,\ldots,B\otimes A_\tau)$$

for the conjugate

$$(6.43)\quad e^{\#} = T(c,c,\ldots c).e.c$$

of e under c. Then (6.37) gives

$$(6.44)\quad \hat{b}_{\underline{\underline{1}}T} = e^{\#}.$$

Now in (6.39) set each P_α equal to $\underline{\underline{1}}$ and set $S = \underline{\underline{1}}$. Since $z_{\underline{\underline{1}}\underline{\underline{1}}} = 1$ by (6.20), we find that $\hat{b}_{TR}$ is the composite

$$(6.45)\quad \begin{array}{c}
T(A_1\ldots A_\tau)\otimes R(B_1\ldots B_\rho)\\
\Big\downarrow e\\
T(A_1\otimes R(B_1\ldots B_\rho),\ldots,A_\tau\otimes R(B_1\ldots B_\rho))\\
\Big\downarrow T(e^{\#},\ldots,e^{\#})\\
T(R(A_1\otimes B_1,\ldots,A_1\otimes B_\rho),\ldots,R(A_\tau\otimes B_1,\ldots,A_\tau\otimes B_\rho))\\
\Big\|\\
T(R\ldots R)(A_1\otimes B_1,\ldots,A_1\otimes B_\rho,\ldots,A_\tau\otimes B_1,\ldots,A_\tau\otimes B_\rho).
\end{array}$$

Consider the following five axioms on e:

$$(6.46)\quad \begin{array}{ccc}
T(S_1\ldots S_\tau)(A_{11}\ldots A_{\tau\sigma_\tau})\otimes B & = & T(S_1(A_{11}\ldots)\ldots S_\tau(A_{\tau 1}\ldots))\otimes B\\
 & & \Big\downarrow e_T\\
\Big\downarrow e_{T(S_1\ldots S_\tau)} & & T(S_1(A_{11}\ldots)\otimes B,\ldots S_\tau(A_{\tau 1}\ldots)\otimes B)\\
 & & \Big\downarrow T(e_{S_1},\ldots,e_{S_\tau})\\
T(S_1\ldots S_\tau)(A_{11}\otimes B,\ldots) & = & T(S_1(A_{11}\otimes B,\ldots)\ldots S_\tau(A_{\tau 1}\otimes B,\ldots)).
\end{array}$$

(6.47) $e_{\underline{\underline{1}}}\colon\ \underline{\underline{1}}(A)\otimes B \to \underline{\underline{1}}(A\otimes B)$ is $1\colon\ A\otimes B \to A\otimes B$.

(6.48)
$$\begin{array}{ccc}
T(A_1\ldots A_\tau)\otimes I & & \\
\downarrow{\scriptstyle e_T} & \searrow{\scriptstyle r} & \\
T(A_1\otimes I,\ldots,A_\tau\otimes I) & \xrightarrow[T(r,r,\ldots,r)]{} & T(A_1\ldots A_\tau).
\end{array}$$

(6.49)
$$\begin{array}{ccc}
(T(A_1\ldots A_\tau)\otimes B)\otimes C & \xrightarrow{a} & T(A_1\ldots A_\tau)\otimes(B\otimes C) \\
\downarrow{\scriptstyle e_T\otimes 1} & & \\
T(A_1\otimes B,\ldots)\otimes C & & \downarrow{\scriptstyle e_T} \\
\downarrow{\scriptstyle e_T} & & \\
T((A_1\otimes B)\otimes C,\ldots) & \xrightarrow[T(a,a,\ldots,a)]{} & T(A_1\otimes(B\otimes C),\ldots).
\end{array}$$

(6.50)
$$\begin{array}{ccc}
& T(A_1\ldots A_\tau)\otimes S(B_1\ldots B_\sigma) & \\
\swarrow{\scriptstyle e_T} & & \searrow{\scriptstyle e_S^{\#}} \\
T(A_1\otimes S(B_1\ldots B_\sigma),\ldots) & & S(T(A_1\ldots A_\tau)\otimes B_1,\ldots) \\
{\scriptstyle T(e_S^{\#}\ldots e_S^{\#})}\downarrow & & \downarrow{\scriptstyle S(e_T\ldots e_T)} \\
T(S(A_1\otimes B_1,\ldots)\ldots S(A_\tau\otimes B_1,\ldots)) & & S(T(A_1\otimes B_1,\ldots)\ldots T(A_1\otimes B_\sigma,\ldots)) \\
\| & & \| \\
T(S\ldots S)(A_1\otimes B_1,\ldots) & \xrightarrow[z_{TS}(1,1,\ldots,1)]{} & S(T\ldots T)(A_1\otimes B_1,\ldots).
\end{array}$$

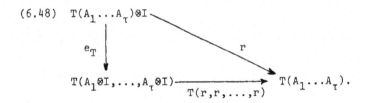

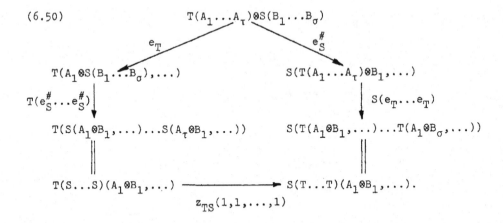

Proposition 6.9 Let $\mathcal{B}$ be a symmetric monoidal category and also a $\mathcal{D}$-algebra for the pseudo-commutative club $\mathcal{D}$. There is a bijection between natural transformations $\hat{b}$ as in (6.34) satisfying (6.35)-(6.39), and natural transformations e as in (6.41) satisfying (6.46)-(6.50) where $e^{\#}$ is defined by (6.43). We get e from $\hat{b}$ by (6.40) and $\hat{b}$ from e by (6.45). A $\mathcal{B}$ with all this structure is precisely an algebra for the doctrine D^{*} of §5.5 (lax case). We pass to the pseudo case by supposing $\hat{b}$, or equivalently e, to be an isomorphism.

Proof We have already shown that if we start with $\hat{b}$ and define e by (6.40) then (6.45) gives us back $\hat{b}$. If on the other hand we start with e and define $\hat{b}$ by (6.45), then (6.40) gives us back e; just put $S = \underline{\underline{1}}$ in (6.50), use (the conjugate of) (6.47), and use (6.21).

The only other point to be verified is the equivalence of the $\hat{b}$-axioms with the e-axioms. Note that each of the e-axioms (6.x) implies its conjugate $(6.x)^{\#}$. First, then, (6.35) is identical with (6.48); (6.37) identical with (6.50); and (6.38) identical with (6.47). Next we get (6.46) from (6.39) by setting $S = R_1 = \underline{\underline{1}}$ in the latter and using (6.21); and we get (6.49) from (6.36) by setting $S = R = \underline{\underline{1}}$ and using (6.38).

It remains, given the e-axioms, to get (6.36) and (6.39). The latter is easy using the definition (6.45) of $\hat{b}$, using (6.46) and $(6.46)^{\#}$, and using (6.50). As for (6.36) we easily get the special case (6.36)' where $T = R = \underline{\underline{1}}$ by using (6.49) and $(6.49)^{\#}$. From this together with (6.46) and (6.49) we next get the special case (6.36)" where $R = \underline{\underline{1}}$. Finally from this, from $(6.46)^{\#}$, and from $(6.49)^{\#}$, we get the desired (6.36). The reader will have no trouble in supplying the details. □

The first four of the e-axioms admit conceptual interpretations, and we are led to the following definitive description of D^{*}-algebras.

Theorem 6.10 Let $\mathcal{D}$ be a pseudo-commutative club and let D^* be the doctrine arising as in §5.5 from the pseudo distributive law of Theorem 6.8. Then a D^*-algebra is a $\mathcal{D}$-algebra $\mathcal{B}$ that is also a symmetric monoidal category, together with for each $B \in \mathcal{B}$ an enrichment of the functor $-\otimes B$: $\mathcal{B} \to \mathcal{B}$ to an op-$\mathcal{D}$-functor $(-\otimes B, e^B)$: $\mathcal{B} \to \mathcal{B}$, satisfying axioms to be given below. We agree to abbreviate the $T[A_1 \ldots A_\tau]$-component

$$(6.51) \quad e^B_{T[A_1 \ldots A_\tau]}: \quad T(A_1 \ldots A_\tau)\otimes B \to T(A_1\otimes B, \ldots, A_\tau \otimes B)$$

of e^B to e_T where convenient; and we observe that conjugation by the symmetry c of $\mathcal{B}$ produces from e^B an enrichment of $B\otimes-$ to an op-$\mathcal{D}$-functor $(B\otimes-, e^{\#B})$. The axioms now are:

(6.52) For f: $B \to C$, the natural transformation $-\otimes f$: $-\otimes B \to -\otimes C$ is op-$\mathcal{D}$-natural.

(6.53) a: $(-\otimes B)\otimes C \to -\otimes(B\otimes C)$ is op-$\mathcal{D}$-natural.

(6.54) r: $-\otimes I \to 1$ is op-$\mathcal{D}$-natural.

(6.55) e and $e^\#$ are related by the commutativity of (6.50), which it suffices to impose for T,S belonging to a set of generators of the discrete club $|\mathcal{D}|$.

The above is for the lax case; for the pseudo case we require the op-$\mathcal{D}$-functor $-\otimes\mathcal{B}$ to be strong, that is, e to be an isomorphism.

Proof By [9] §10.8 to say that $(-\otimes B, e^B)$ is an op-$\mathcal{D}$-functor is to say that the e of (6.51) is natural in T and the A_α, and that it satisfies (6.46) and (6.47) (corresponding to (10.25) and (10.26) of [9]). The axiom (6.52) above just makes e natural in B as well as in T and the A_α. The axioms (6.53) and (6.54) reduce, by (10.27) of [9], to (6.49) and (6.48) respectively. The result now follows from Proposition 6.9, except for the last clause in (6.55).

The proof of that last clause is easy: using (6.46) and (6.19) we see that (6.50) for T,S, where $T = R(P_1 \ldots P_\rho)$, is a consequence of (6.50) for R,S and for P_α,S. □

When $\mathcal{D}$ is presented by a small number of generators and relations (c.f. [9] §10) the above theorem gives a very compact description of D^*-algebras.

Example 6.11 Let $\mathcal{D}$ be itself the club $\mathcal{P}$ for symmetric monoidal categories, which is pseudo-commutative by Example 6.4. Then a D^*-algebra $\mathcal{B}$ has two symmetric monoidal structures: the $\mathcal{D}'$-structure denoted as above by $(\mathcal{B},\otimes,I,a,r,c)$, and the $\mathcal{D}$-structure denoted by $(\mathcal{B},\oplus,N,a',r',c')$. To give the enrichment e^B of $-\otimes B$ to a symmetric op-monoidal functor, we need as usual only to give its components

$$e_\oplus:\quad (A_1 \oplus A_2)\otimes B \to (A_1 \otimes B)\oplus(A_2 \otimes B),$$

$$e_N:\quad N\otimes B \to N,$$

corresponding to the generators for the objects of $\mathcal{D}$, subject to the usual three axioms. The axiom (6.52) merely asks these components to be natural in B as well as the A_α. The condition (6.53) that a be op-s.m.-natural reduces as usual to two diagrams, as does similarly (6.54) for r. Finally (6.55) requires us to impose (6.50) in the three cases $(T,S) = (\oplus,\oplus)$, $(\oplus,N)$, and (N,N) (that for $(N,\oplus)$ being the conjugate of that for $(\oplus,N)$). All told we have ten simple axioms involving $e_{\oplus}$ and e_N; and it is easy to see that these are precisely the independent axioms among the 24 given for this situation by Laplaza [18].

6.7 If we take the description of D^*-algebras given by Theorem 6.10, and modify it by changing the sense of the arrow e in (6.51), at the same time changing its name to $^*e$, so that $^*e$ enriches $-\otimes B$ to a $\mathcal{D}$-functor rather than an op-$\mathcal{D}$-functor, we obtain a description of the algebras for some new doctrine, which we shall call $^*D$.

In the situation of §5 we should have got $^*D$ rather than D^* if from the pseudo distributive law we had constructed a doctrine $\tilde{D}$ not on D'-*Alg* but on the 2-category with D'-algebras as objects and op-D'-morphisms as 1-cells. Since π is invertible this $^*D$ case is exactly like the D^*-case; there is nothing to change except the sense of the 2-cells. In particular Theorem 5.1 in the $^*D$-case gives a coreflexion $s\colon\ ^*D \to DD'$, with $h\colon\ DD' \to {}^*D$, $sh = 1$, $\eta\colon\ hs \Rightarrow 1$, $s\eta = 1$, $\eta h = 1$.

The point of our introducing $^*D$ is that, in the lax case, it is of the form $^*\mathcal{D}\circ -$ for a club $^*\mathcal{D}$ in $\mathcal{C}at/\underline{\underline{S}}$. In the light of the general principles of [9] §10, this is clear from Theorem 6.10; for the type of the natural transformation $^*e$ lies in $\underline{\underline{S}}$.

The $^*D$ for the pseudo case, and the D^* for the lax and the pseudo cases, sometimes arise from clubs in $\mathcal{C}at/\underline{\underline{S}}$, namely when the type of e in (6.51) is in $\underline{\underline{S}}$, which only happens when $\tau = 1$ in (6.51); since this is to be so for all $T \in \mathcal{D}$, it only happens when $\Gamma\colon\ \mathcal{D} \to \underline{\underline{S}}$ is constant at 1.

We can bring Example 6.11 above, the one studied by Laplaza, under the $^*\mathcal{D}$-setting, just by replacing the model $\mathcal{B}$ by $\mathcal{B}^{op}$ (i.e. in effect by passing to the opposite doctrine). This works in this case because $\mathcal{B}^{op}$ is symmetric monoidal when $\mathcal{B}$ is.

<u>Example 6.12</u> Another important example occurs naturally in the ${}^*\mathcal{D}$-setting. Let $\mathcal{D}$ be the club $\mathcal{S}$ of Example 6.2, whose algebras are categories with finite coproducts. Consider any $\mathcal{S}$-algebra $\mathcal{B}$ that is also symmetric monoidal. For $T \in \mathcal{S}$ with $\Gamma T = \tau$, and for $A_\alpha \in \mathcal{B}$, the object $T(A_1 \ldots A_\tau)$ is a coproduct in $\mathcal{B}$ of the A_α, the coprojections being $f_\alpha(1)$ for suitable f_α: $\underline{\underline{1}} \to T$ in $\mathcal{S}$ whose types are the various maps $1 \to \tau$ in $\underline{\underline{S}}$.

It follows that there is a unique ${}^*e$ turning $\mathcal{B}$ into a ${}^*\mathcal{S}$-algebra. For the naturality in T of ${}^*e$ demands commutativity in

$$\begin{array}{ccc} T(A_1 \otimes B, \ldots, A_\tau \otimes B) & \xrightarrow{\;{}^*e_T\;} & T(A_1 \ldots A_\tau) \otimes B \\ {\scriptstyle f_\alpha(1)} \uparrow & & \uparrow {\scriptstyle f_\alpha(1) \otimes 1} \\ \underline{\underline{1}}(A_1 \otimes B) & \xrightarrow[{}^*e_{\underline{\underline{1}}}]{} & \underline{\underline{1}}(A_1) \otimes B \end{array} \quad ;$$

and ${}^*e_{\underline{\underline{1}}}$ is to be 1 by (6.47). So the only possible ${}^*e$ is that whose α-th component is $f_\alpha(1) \otimes 1$. On the other hand it is easily seen that this <u>does</u> satisfy the axioms.

We are led to the conclusion that the doctrine ${}^*\mathcal{S}\circ-$ is the coproduct, in the category of doctrines, of the doctrines $\mathcal{S}\circ-$ and $\mathcal{P}\circ-$; or equally that, in the category of clubs in $\mathcal{C}at/\underline{\underline{S}}$ and club-maps in $\mathcal{C}at/\underline{\underline{S}}$, ${}^*\mathcal{S}$ is the coproduct of $\mathcal{S}$ and $\mathcal{P}$.

<u>6.8</u> We can apply Theorem 5.1 to the present situation, to obtain a reflexion s: $D^* \to \mathcal{D}\circ\mathcal{P}\circ-$, or sA: $D^*A \to \mathcal{D}\circ\mathcal{P}\circ A$, which is an equivalence in the pseudo case: or equally a coreflexion s : ${}^*DA \to \mathcal{D}\circ\mathcal{P}\circ A$ which, in the lax case, taking $A = I$, can be written s: ${}^*\mathcal{D} \to \mathcal{D}\circ\mathcal{P}$.

I shall just indicate briefly how s is calculated. First, however, observe that this special case of Theorem 5.1 can be proved independently of §5, taking Theorem 6.10 as the <u>definition</u> of

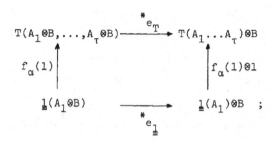

D^*-algebra, and applying Theorem 2.6 directly, using the techniques of the present §6 alone. This gives for instance quite a short proof of Laplaza's result; but it conceals the connexion between "distributivity of $-\otimes B$" in the sense of (6.51) and "(pseudo) distributive law" in the sense of Beck.

$P \circ A$ is a s.m. category, indeed a free one, whence $\mathcal{D} \circ P \circ A$ is also a s.m. category, with the structure given in §6.5. At the same time $\mathcal{D} \circ P \circ A$ is a $\mathcal{D}$-category, in fact a free one. For $T \in \mathcal{D}$ and for $A_1, \ldots, A_\tau, B \in \mathcal{D} \circ P \circ A$ an easy calculation of $T(A_1 \ldots A_\tau) \otimes B$ and of $T(A_1 \otimes B, \ldots, A_\tau \otimes B)$ shows these to be equal. We can therefore set $e = 1$ in (6.51). The axioms (6.46)-(6.49) are trivially satisfied; and (6.50) is in fact satisfied (not trivially because $e^\#$ given by (6.43) is not 1 when $e = 1$; but easily in view of the value of c in $\mathcal{D} \circ P \circ A$ given by (6.27)).

Thus $\mathcal{D} \circ P \circ A$ (and in particular $\mathcal{D} \circ P$) is a D^*-category; even for the pseudo case, hence also for the lax case; and equally a ${}^*D$-category. Since s is a strict D^* (resp. ${}^*D$)-morphism, we can write at once the image $s(\Delta)$ of any diagram Δ, in the free D^*-category D^*A, made out of the elementary data; and in the pseudo case Δ commutes if and only if $s(\Delta)$ does so.

In the lax-${}^*D$-case we have the usual more flexible results expressed in terms of clubs. The objects of ${}^*\mathcal{D}$ are generated by $\otimes$, I, and the objects $T \in \mathcal{D}$; the morphisms by a,r,c, those of $\mathcal{D}$, and ${}^*e$; the functor s: ${}^*\mathcal{D} \to \mathcal{D} \circ P$ respects all this structure, and in particular sends ${}^*e$ to 1. In the other direction h: $\mathcal{D} \circ P \to {}^*\mathcal{D}$ sends $T[V_1 \ldots V_\tau]$ to $T(V_1 \ldots V_\tau)$. Then $sh = 1$ and η: $hs \Rightarrow 1$ for a suitable η with $s\eta = 1$, $\eta h = 1$. To find η directly, and to provide a proof independent of Theorem 5.1, we enrich h to an op-${}^*\mathcal{D}$-functor: we make it strict as a $\mathcal{D}$-functor, and as an op-monoidal-functor $(h, \hat{h}, h^\circ)$ it has $h^\circ = 1$ and $\hat{h}$ formally the same as the $\hat{b}$ of (6.34) (whose sense is now reversed). The η is then formed as in Theorem 2.6

from the $\bar{h}$ of the resulting op-$^*\mathcal{D}$-functor $(h,\bar{h})$.

In the case $\mathcal{D} = \mathcal{P}$ studied by Laplaza, it is easy to see that the composite of s: $^*\mathcal{P} \to \mathcal{P}\circ\mathcal{P}$ with the equivalence $\Gamma\circ\Gamma$: $\mathcal{P}\circ\mathcal{P} \to \underline{\underline{P}}\circ\underline{\underline{P}}$ gives a functor $^*\mathcal{P} \to \underline{\underline{P}}\circ\underline{\underline{P}}$ that is precisely his distortion ([19] §2). So as indicated in §1.4 above, our coreflexion result implies his theorem on page 231 of [19]. (Note that Laplaza is considering diagrams writable in $^*\mathcal{D}$ - that is, involving $^*e$ but not its inverse - but studying their commutativity in a model in which $^*e$ happens to be an isomorphism. He actually says "monomorphism", which since we have dualized would be "epimorphism", but this is not enough as it stands, and is too rare in examples to be worth pursuing.)

The other main result of Laplaza, given in his earlier paper ([18] Proposition 10) follows from the same coreflexion theorem, if we observe that, for a discrete A, two morphisms f,g: $X \to Y$ in $\underline{\underline{P}}\circ\underline{\underline{P}}\circ A$ necessarily commute if $Y = n[m_1[A_{11}\ldots],\ldots,m_n[A_{n1}\ldots]]$ where all the $m_\alpha[A_{\alpha 1}\ldots]$ are different and, for each α, all the $A_{\alpha 1},\ldots,A_{\alpha m_\alpha}$ are different (for the permutations expressing f,g are then determined and equal).

REFERENCES

[1] M. Barr, Coequalizers and free triples, Math. Zeit. 116 (1970), 307-322.

[2] J. Beck, Distributive laws, Lecture Notes in Math. 80 (1969), 119-140.

[3] J. Beck, On coherence isomorphisms, Preprint, Forschungsinst. für Math., E.T.H. Zürich, (1971).

[4] B.J. Day and G.M. Kelly, Enriched functor categories, Lecture Notes in Math. 106 (1969), 178-191.

[5] E. Dubuc, Free monoids, to appear in Journal of Algebra.

[6] S. Eilenberg and G.M. Kelly, Closed categories, Proc. Conf. on Categorical Algebra (La Jolla 1965), Springer-Verlag 1966.

[7] J.R. Isbell, On coherent and strict algebras, Jour. of Algebra 13 (1969), 299-307.

[8] G.M. Kelly, An abstract approach to coherence, Lecture Notes in Math. 281 (1972), 106-147.

[9] G.M. Kelly, On clubs and doctrines, in this volume.

[10] G.M. Kelly, Doctrinal adjunction, in this volume.

[11] G.M. Kelly and S. Mac Lane, Coherence in closed categories, Jour. Pure and Applied Algebra 1 (1971), 97-140.

[12] G.M. Kelly and S. Mac Lane, Closed coherence for a natural transformation, Lecture Notes in Math. 281 (1972), 1-28.

[13] G.M. Kelly and R. Street, Review of the elements of 2-categories, in this volume.

[14] A. Kock, Monads on symmetric monoidal closed categories, Arch. Math. 21 (1970), 1-10.

[15] J. Lambek, Deductive systems and categories I. Syntactic calculus and residuated categories, Math. Systems Theory 2 (1968), 287-318.

[16] J. Lambek, Deductive systems and categories II. Standard constructions and closed categories, Lecture Notes in Math. 86 (1969), 76-122.

[17] M.L. Laplaza, Coherence for associativity not an isomorphism, Jour. Pure and Applied Algebra 2 (1972), 107-120.

[18] M.L. Laplaza, Coherence for distributivity, Lecture Notes in Math. 281 (1972), 29-65.

[19] M.L. Laplaza, A new result of coherence for distributivity, Lecture Notes in Math. 281 (1972), 214-235.

[20] F.W. Lawvere, Ordinal sums and equational doctrines, Lecture Notes in Math. 80 (1969), 141-155.

[21] G. Lewis, Coherence for a closed functor, Lecture Notes in Math. 281 (1972), 148-195.

[22] G. Lewis, Coherence for a closed functor, Ph.D. Thesis, University of New South Wales, 1974.

[23] S. Mac Lane, Natural associativity and commutativity, Rice University Studies 49 (1963), 28-46.

[24] R. Street, Fibrations and Yoneda's lemma in a 2-category, in this volume.

[25] M.E. Szabo, Proof-theoretical investigations in categorical algebra, Ph.D. Thesis, McGill Univ., 1970.

[26] M.E. Szabo, A categorical equivalence of proofs, to appear.

[27] R. Voreadu, A coherence theorem for closed categories, Ph.D. Thesis, Univ. of Chicago, 1972.

[28] R. Voreadu, Some remarks on the subject of coherence, to appear in Cahiers de Topologie.

[29] R. Voreadu, Non-commutative diagrams in closed categories, to appear.

Vol. 247: Lectures on Operator Algebras. Tulane University Ring and Operator Theory Year, 1970–1971. Volume II. XI, 786 pages. 1972. DM 40,-

Vol. 248: Lectures on the Applications of Sheaves to Ring Theory. Tulane University Ring and Operator Theory Year, 1970–1971. Volume III. VIII, 315 pages. 1971. DM 26,-

Vol. 249: Symposium on Algebraic Topology. Edited by P. J. Hilton. VII, 111 pages. 1971. DM 16,-

Vol. 250: B. Jónsson, Topics in Universal Algebra. VI, 220 pages. 1972. DM 20,-

Vol. 251: The Theory of Arithmetic Functions. Edited by A. A. Gioia and D. L. Goldsmith VI, 287 pages. 1972. DM 24,-

Vol. 252: D. A. Stone, Stratified Polyhedra. IX, 193 pages. 1972. DM 18,-

Vol. 253: V. Komkov, Optimal Control Theory for the Damping of Vibrations of Simple Elastic Systems. V, 240 pages. 1972. DM 20,-

Vol. 254: C. U. Jensen, Les Foncteurs Dérivés de $\varprojlim$ et leurs Applications en Théorie des Modules. V, 103 pages. 1972. DM 16,-

Vol. 255: Conference in Mathematical Logic – London '70. Edited by W. Hodges. VIII, 351 pages. 1972. DM 26,-

Vol. 256: C. A. Berenstein and M. A. Dostal, Analytically Uniform Spaces and their Applications to Convolution Equations. VII, 130 pages. 1972. DM 16,-

Vol. 257: R. B. Holmes, A Course on Optimization and Best Approximation. VIII, 233 pages. 1972. DM 20,-

Vol. 258: Séminaire de Probabilités VI. Edited by P. A. Meyer. VI, 253 pages. 1972. DM 22,-

Vol. 259: N. Moulis, Structures de Fredholm sur les Variétés Hilbertiennes. V, 123 pages. 1972. DM 16,-

Vol. 260: R. Godement and H. Jacquet, Zeta Functions of Simple Algebras. IX, 188 pages. 1972. DM 18,-

Vol. 261: A. Guichardet, Symmetric Hilbert Spaces and Related Topics. V, 197 pages. 1972. DM 18,-

Vol. 262: H. G. Zimmer, Computational Problems, Methods, and Results in Algebraic Number Theory. V, 103 pages. 1972. DM 16,-

Vol. 263: T. Parthasarathy, Selection Theorems and their Applications. VII, 101 pages. 1972. DM 16,-

Vol. 264: W. Messing, The Crystals Associated to Barsotti-Tate Groups: With Applications to Abelian Schemes. III, 190 pages. 1972. DM 18,-

Vol. 265: N. Saavedra Rivano, Catégories Tannakiennes. II, 418 pages. 1972. DM 26,-

Vol. 266: Conference on Harmonic Analysis. Edited by D. Gulick and R. L. Lipsman. VI, 323 pages. 1972. DM 24,-

Vol. 267: Numerische Lösung nichtlinearer partieller Differential- und Integro-Differentialgleichungen. Herausgegeben von R. Ansorge und W. Törnig. VI, 339 Seiten. 1972. DM 26,-

Vol. 268: C. G. Simader, On Dirichlet's Boundary Value Problem. IV, 238 pages. 1972. DM 20,-

Vol. 269: Théorie des Topos et Cohomologie Etale des Schémas. (SGA 4). Dirigé par M. Artin, A. Grothendieck et J. L. Verdier. XIX, 525 pages. 1972. DM 50,-

Vol. 270: Théorie des Topos et Cohomologie Etale des Schémas. Tome 2. (SGA 4). Dirigé par M. Artin, A. Grothendieck et J. L. Verdier. V, 418 pages. 1972. DM 50,-

Vol. 271: J. P. May, The Geometry of Iterated Loop Spaces. IX, 175 pages. 1972. DM 18,-

Vol. 272: K. R. Parthasarathy and K. Schmidt, Positive Definite Kernels, Continuous Tensor Products, and Central Limit Theorems of Probability Theory. VI, 107 pages. 1972. DM 16,-

Vol. 273: U. Seip, Kompakt erzeugte Vektorräume und Analysis. IX, 119 Seiten. 1972. DM 16,-

Vol. 274: Toposes, Algebraic Geometry and Logic. Edited by. F. W. Lawvere. VI, 189 pages. 1972. DM 18,-

Vol. 275: Séminaire Pierre Lelong (Analyse) Année 1970–1971. VI, 181 pages. 1972. DM 18,-

Vol. 276: A. Borel, Représentations de Groupes Localement Compacts. V, 98 pages. 1972. DM 16,-

Vol. 277: Séminaire Banach. Edité par C. Houzel. VII, 229 pages. 1972. DM 20,-

Vol. 278: H. Jacquet, Automorphic Forms on GL(2). Part II. XIII, 142 pages. 1972. DM 16,-

Vol. 279: R. Bott, S. Gitler and I. M. James, Lectures on Algebraic and Differential Topology. V, 174 pages. 1972. DM 18,-

Vol. 280: Conference on the Theory of Ordinary and Partial Differential Equations. Edited by W. N. Everitt and B. D. Sleeman. XV, 367 pages. 1972. DM 26,-

Vol. 281: Coherence in Categories. Edited by S. Mac Lane. VII, 235 pages. 1972. DM 20,-

Vol. 282: W. Klingenberg und P. Flaschel, Riemannsche Hilbertmannigfaltigkeiten. Periodische Geodätische. VII, 211 Seiten. 1972. DM 20,-

Vol. 283: L. Illusie, Complexe Cotangent et Déformations II. VII, 304 pages. 1972. DM 24,-

Vol. 284: P. A. Meyer, Martingales and Stochastic Integrals I. VI, 89 pages. 1972. DM 16,-

Vol. 285: P. de la Harpe, Classical Banach-Lie Algebras and Banach-Lie Groups of Operators in Hilbert Space. III, 160 pages. 1972. DM 16,-

Vol. 286: S. Murakami, On Automorphisms of Siegel Domains. V, 95 pages. 1972. DM 16,-

Vol. 287: Hyperfunctions and Pseudo-Differential Equations. Edited by H. Komatsu. VII, 529 pages. 1973. DM 36,-

Vol. 288: Groupes de Monodromie en Géométrie Algébrique. (SGA 7 I). Dirigé par A. Grothendieck. IX, 523 pages. 1972. DM 50,-

Vol. 289: B. Fuglede, Finely Harmonic Functions. III, 188. 1972. DM 18,-

Vol. 290: D. B. Zagier, Equivariant Pontrjagin Classes and Applications to Orbit Spaces. IX, 130 pages. 1972. DM 16,-

Vol. 291: P. Orlik, Seifert Manifolds. VIII, 155 pages. 1972. DM 16,-

Vol. 292: W. D. Wallis, A. P. Street and J. S. Wallis, Combinatorics: Room Squares, Sum-Free Sets, Hadamard Matrices. V, 508 pages. 1972. DM 50,-

Vol. 293: R. A. DeVore, The Approximation of Continuous Functions by Positive Linear Operators. VIII, 289 pages. 1972. DM 24,-

Vol. 294: Stability of Stochastic Dynamical Systems. Edited by R. F. Curtain. IX, 332 pages. 1972. DM 26,-

Vol. 295: C. Dellacherie, Ensembles Analytiques, Capacités, Mesures de Hausdorff. XII, 123 pages. 1972. DM 16,-

Vol. 296: Probability and Information Theory II. Edited by M. Behara, K. Krickeberg and J. Wolfowitz. V, 223 pages. 1973. DM 20,-

Vol. 297: J. Garnett, Analytic Capacity and Measure. IV, 138 pages. 1972. DM 16,-

Vol. 298: Proceedings of the Second Conference on Compact Transformation Groups. Part 1. XIII, 453 pages. 1972. DM 32,-

Vol. 299: Proceedings of the Second Conference on Compact Transformation Groups. Part 2. XIV, 327 pages. 1972. DM 26,-

Vol. 300: P. Eymard, Moyennes Invariantes et Représentations Unitaires. II, 113 pages. 1972. DM 16,-

Vol. 301: F. Pittnauer, Vorlesungen über asymptotische Reihen. VI, 186 Seiten. 1972. DM 18,-

Vol. 302: M. Demazure, Lectures on p-Divisible Groups. V, 98 pages. 1972. DM 16,-

Vol. 303: Graph Theory and Applications. Edited by Y. Alavi, D. R. Lick and A. T. White. IX, 329 pages. 1972. DM 26,-

Vol. 304: A. K. Bousfield and D. M. Kan, Homotopy Limits, Completions and Localizations. V, 348 pages. 1972. DM 26,-

Vol. 305: Théorie des Topos et Cohomologie Etale des Schémas. Tome 3. (SGA 4). Dirigé par M. Artin, A. Grothendieck et J. L. Verdier. VI, 640 pages. 1973. DM 50,-

Vol. 306: H. Luckhardt, Extensional Gödel Functional Interpretation. VI, 161 pages. 1973. DM 18,-

Vol. 307: J. L. Bretagnolle, S. D. Chatterji et P.-A. Meyer, Ecole d'été de Probabilités: Processus Stochastiques. VI, 198 pages. 1973. DM 20,-

Vol. 308: D. Knutson, λ-Rings and the Representation Theory of the Symmetric Group. IV, 203 pages. 1973. DM 20,-

Vol. 309: D. H. Sattinger, Topics in Stability and Bifurcation Theory. VI, 190 pages. 1973. DM 18,-

Vol. 310: B. Iversen, Generic Local Structure of the Morphisms in Commutative Algebra. IV, 108 pages. 1973. DM 16,-

Vol. 311: Conference on Commutative Algebra. Edited by J. W. Brewer and E. A. Rutter. VII, 251 pages. 1973. DM 22,-

Vol. 312: Symposium on Ordinary Differential Equations. Edited by W. A. Harris, Jr. and Y. Sibuya. VIII, 204 pages. 1973. DM 22,-

Vol. 313: K. Jörgens and J. Weidmann, Spectral Properties of Hamiltonian Operators. III, 140 pages. 1973. DM 16,-

Vol. 314: M. Deuring, Lectures on the Theory of Algebraic Functions of One Variable. VI, 151 pages. 1973. DM 16,-

Vol. 315: K. Bichteler, Integration Theory (with Special Attention to Vector Measures). VI, 357 pages. 1973. DM 26,-

Vol. 316: Symposium on Non-Well-Posed Problems and Logarithmic Convexity. Edited by R. J. Knops. V, 176 pages. 1973. DM 18,-

Vol. 317: Séminaire Bourbaki – vol. 1971/72. Exposés 400–417. IV, 361 pages. 1973. DM 26,-

Vol. 318: Recent Advances in Topological Dynamics. Edited by A. Beck. VIII, 285 pages. 1973. DM 24,-

Vol. 319: Conference on Group Theory. Edited by R. W. Gatterdam and K. W. Weston. V, 188 pages. 1973. DM 18,-

Vol. 320: Modular Functions of One Variable I. Edited by W. Kuyk. V, 195 pages. 1973. DM 18,-

Vol. 321: Séminaire de Probabilités VII. Edité par P. A. Meyer. VI, 322 pages. 1973. DM 26,-

Vol. 322: Nonlinear Problems in the Physical Sciences and Biology. Edited by I. Stakgold, D. D. Joseph and D. H. Sattinger. VIII, 357 pages. 1973. DM 26,-

Vol. 323: J. L. Lions, Perturbations Singulières dans les Problèmes aux Limites et en Contrôle Optimal. XII, 645 pages. 1973. DM 42,-

Vol. 324: K. Kreith, Oscillation Theory. VI, 109 pages. 1973. DM 16,-

Vol. 325: Ch.-Ch. Chou, La Transformation de Fourier Complexe et L'Equation de Convolution. IX, 137 pages. 1973. DM 16,-

Vol. 326: A. Robert, Elliptic Curves. VIII, 264 pages. 1973. DM 22,-

Vol. 327: E. Matlis, 1-Dimensional Cohen-Macaulay Rings. XII, 157 pages. 1973. DM 18,-

Vol. 328: J. R. Büchi and D. Siefkes, The Monadic Second Order Theory of All Countable Ordinals. VI, 217 pages. 1973. DM 20,-

Vol. 329: W. Trebels, Multipliers for (C, α)-Bounded Fourier Expansions in Banach Spaces and Approximation Theory. VII, 103 pages. 1973. DM 16,-

Vol. 330: Proceedings of the Second Japan-USSR Symposium on Probability Theory. Edited by G. Maruyama and Yu. V. Prokhorov. VI, 550 pages. 1973. DM 36,-

Vol. 331: Summer School on Topological Vector Spaces. Edited by L. Waelbroeck. VI, 226 pages. 1973. DM 20,-

Vol. 332: Séminaire Pierre Lelong (Analyse) Année 1971-1972. V, 131 pages. 1973. DM 16,-

Vol. 333: Numerische, insbesondere approximationstheoretische Behandlung von Funktionalgleichungen. Herausgegeben von R. Ansorge und W. Törnig. VI, 296 Seiten. 1973. DM 24,-

Vol. 334: F. Schweiger, The Metrical Theory of Jacobi-Perron Algorithm. V, 111 pages. 1973. DM 16,-

Vol. 335: H. Huck, R. Roitzsch, U. Simon, W. Vortisch, R. Walden, B. Wegner und W. Wendland, Beweismethoden der Differentialgeometrie im Großen. IX, 159 Seiten. 1973. DM 18,-

Vol. 336: L'Analyse Harmonique dans le Domaine Complexe. Edité par E. J. Akutowicz. VIII, 169 pages. 1973. DM 18,-

Vol. 337: Cambridge Summer School in Mathematical Logic. Edited by A. R. D. Mathias and H. Rogers. IX, 660 pages. 1973. DM 42,-

Vol. 338: J. Lindenstrauss and L. Tzafriri, Classical Banach Spaces. IX, 243 pages. 1973. DM 22,-

Vol. 339: G. Kempf, F. Knudsen, D. Mumford and B. Saint-Donat, Toroidal Embeddings I. VIII, 209 pages. 1973. DM 20,-

Vol. 340: Groupes de Monodromie en Géométrie Algébrique. (SGA 7 II). Par P. Deligne et N. Katz. X, 438 pages. 1973. DM 40,-

Vol. 341: Algebraic K-Theory I, Higher K-Theories. Edited by H. Bass. XV, 335 pages. 1973. DM 26,-

Vol. 342: Algebraic K-Theory II, "Classical" Algebraic K-Theory, and Connections with Arithmetic. Edited by H. Bass. XV, 527 pages. 1973. DM 36,-

Vol. 343: Algebraic K-Theory III, Hermitian K-Theory and Geometric Applications. Edited by H. Bass. XV, 572 pages. 1973. DM 38,-

Vol. 344: A. S. Troelstra (Editor), Metamathematical Investigation of Intuitionistic Arithmetic and Analysis. XVII, 485 pages. 1973. DM 34,-

Vol. 345: Proceedings of a Conference on Operator Theory. Edited by P. A. Fillmore. VI, 228 pages. 1973. DM 20,-

Vol. 346: Fučík et al., Spectral Analysis of Nonlinear Operators. II, 287 pages. 1973. DM 26,-

Vol. 347: J. M. Boardman and R. M. Vogt, Homotopy Invariant Algebraic Structures on Topological Spaces. X, 257 pages. 1973. DM 22,-

Vol. 348: A. M. Mathai and R. K. Saxena, Generalized Hypergeometric Functions with Applications in Statistics and Physical Sciences. VII, 314 pages. 1973. DM 26,-

Vol. 349: Modular Functions of One Variable II. Edited by W. Kuyk and P. Deligne. V, 598 pages. 1973. DM 38,-

Vol. 350: Modular Functions of One Variable III. Edited by W. Kuyk and J.-P. Serre. V, 350 pages. 1973. DM 26,-

Vol. 351: H. Tachikawa, Quasi-Frobenius Rings and Generalizations. XI, 172 pages. 1973. DM 18,-

Vol. 352: J. D. Fay, Theta Functions on Riemann Surfaces. V, 137 pages. 1973. DM 16,-

Vol. 353: Proceedings of the Conference on Orders, Group Rings and Related Topics. Organized by J. S. Hsia, M. L. Madan and T. G. Ralley. X, 224 pages. 1973. DM 20.-

Vol. 354: K. J. Devlin, Aspects of Constructibility. XII, 240 pages. 1973. DM 22,-

Vol. 355: M. Sion, A Theory of Semigroup Valued Measures. V, 140 pages. 1973. DM 16,-

Vol. 356: W. L. J. van der Kallen, Infinitesimally Central-Extensions of Chevalley Groups. VII, 147 pages. 1973. DM 16,-

Vol. 357: W. Borho, P. Gabriel und R. Rentschler, Primideale in Einhüllenden auflösbarer Lie-Algebren. V, 182 Seiten. 1973. DM 18,-

Vol. 358: F. L. Williams, Tensor Products of Principal Series Representations. VI, 132 pages. 1973. DM 16,-

Vol. 359: U. Stammbach, Homology in Group Theory. VIII, 183 pages. 1973. DM 18,-

Vol. 360: W. J. Padgett and R. L. Taylor, Laws of Large Numbers for Normed Linear Spaces and Certain Fréchet Spaces. VI, 111 pages. 1973. DM 16,-

Vol. 361: J. W. Schutz, Foundations of Special Relativity: Kinematic Axioms for Minkowski Space Time. XX, 314 pages. 1973. DM 26,-

Vol. 362: Proceedings of the Conference on Numerical Solution of Ordinary Differential Equations. Edited by D. Bettis. VIII, 490 pages. 1974. DM 34,-

Vol. 363: Conference on the Numerical Solution of Differential Equations. Edited by G. A. Watson. IX, 221 pages. 1974. DM 20,-

Vol. 364: Proceedings on Infinite Dimensional Holomorphy. Edited by T. L. Hayden and T. J. Suffridge. VII, 212 pages. 1974. DM 20,-

Vol. 365: R. P. Gilbert, Constructive Methods for Elliptic Equations. VII, 397 pages. 1974. DM 26,-

Vol. 366: R. Steinberg, Conjugacy Classes in Algebraic Groups (Notes by V. V. Deodhar). VI, 159 pages. 1974. DM 18,-

Vol. 367: K. Langmann und W. Lütkebohmert, Cousinverteilungen und Fortsetzungssätze. VI, 151 Seiten. 1974. DM 16,-

Vol. 368: R. J. Milgram, Unstable Homotopy from the Stable Point of View. V, 109 pages. 1974. DM 16,-

Vol. 369: Victoria Symposium on Nonstandard Analysis. Edited by A. Hurd and P. Loeb. XVIII, 339 pages. 1974. DM 26,-

Vol. 370: B. Mazur and W. Messing, Universal Extensions and One Dimensional Crystalline Cohomology. VII, 134 pages. 1974. DM 16,-